ADVANCES IN

Applied
Microbiology

VOLUME 20

CONTRIBUTORS TO THIS VOLUME

Bernard J. Abbott

George R. Anderson

Rudolph J. Allgeier

Jack Cameron

Hubert A. Conner

Yukio Fujisawa

J. Fuska

Toshihiko Kanzaki

Charles R. Manclark

Stephen I. Morse

Charlotte Parker

B. Proksa

M. Sternberg

ADVANCES IN

Applied Microbiology

Edited by D. PERLMAN

School of Pharmacy
The University of Wisconsin
Madison, Wisconsin

VOLUME 20

 1976

ACADEMIC PRESS, New York   San Francisco   London

A Subsidiary of Harcourt Brace Jovanovich, Publishers

ACADEMIC PRESS, INC.
111 Fifth Avenue, New York, New York 10003

United Kingdom Edition published by
ACADEMIC PRESS, INC. (LONDON) LTD.
24/28 Oval Road, London NW1

LIBRARY OF CONGRESS CATALOG CARD NUMBER: 59–13823

ISBN 0–12–002620–1

PRINTED IN THE UNITED STATES OF AMERICA

CONTENTS

LIST OF CONTRIBUTORS ix

The Current Status of Pertussis Vaccine: An Overview

CHARLES R. MANCLARK

Text 1

Biologically Active Components and Properties of *Bordetella pertussis*

STEPHEN I. MORSE

I. Introduction 9
II. Sensitization to Histamine 10
III. β-Adrenergic Blockade 12
IV. Lymphocytosis-Promoting Factor 14
V. Adjuvanticity 17
VI. Lipopolysaccharide Endotoxin 19
VII. Heat-Labile Toxin 19
VIII. Hemagglutinin 20
IX. Mitogenicity for Lymphocytes 20
X. Interrelationships between Biological Activities 21
XI. Pathogenesis of Pertussis 23
References 24

Role of the Genetics and Physiology of *Bordetella pertussis* in the Production of Vaccine and the Study of Host–Parasite Relationships in Pertussis

CHARLOTTE PARKER

I. Introduction 27
II. Genetics 30
III. Physiology 31
References 41

Problems Associated with the Development and Clinical Testing of an Improved Pertussis Vaccine

GEORGE R. ANDERSON

I. Do We Need an Improved Pertussis Vaccine? 44
II. Why Don't We Have an Improved or Fully Acceptable Pertussis Vaccine Now? 48

v

III. Where Do We Go From Here with the Development of an Improved
      Pertussis Vaccine? 51
IV.   Can We Really Expect to Get New Pertussis Vaccines Approved,
      Marketed, and in General Distribution? 54
      References 55

Problems Associated with the Control Testing
of Pertussis Vaccine

JACK CAMERON

I.   Introduction 57
II.   Colonial Variation in *Bordetella pertussis* 58
III. Opacity Standards 65
IV.   Potency Assay 66
V.   Role of Adjuvant 69
VI.   Agglutinins 72
VII. Toxicity 73
VIII. Summary 77
      References 79

Vinegar: Its History and Development

HUBERT A. CONNER AND RUDOLPH J. ALLGEIER

I.   Introduction 82
II.   In the Beginning 83
III. Nonfood Uses of Vinegar through the Centuries 87
IV.   Vinegar in Food Processing 89
V.   Production Methods 90
VI.   Major Types of Vinegar 100
VII. The Industry 105
VIII. Microbiology of Vinegar Production 110
IX.   Mechanism of the Biological Conversion of Ethanol to Acetic Acid 120
X.   Composition of Vinegar 124
XI.   Summary 126
      References 127

Microbial Rennets

M. STERNBERG

I.   Introduction 135
II.   Microbial Rennets from *Endothia parasitica, Mucor miehei,* and *Mucor
      pusillus* var. *Lindt* 136
III. Screening for New Microbial Rennets 146
IV.   Conclusions 151
      References 153

Biosynthesis of Cephalosporins

TOSHIHIKO KANZAKI AND YUKIO FUJISAWA

I.	Introduction	159
II.	Occurrence of Cephalosporins and Related Metabolites	160
III.	Genetic Studies and Strain Improvement	169
IV.	Fermentation	172
V.	Biosynthesis	176
VI.	Enzymes Related to Cephalosporin Metabolism	196
VII.	Conclusions	196
	References	198

Preparation of Pharmaceutical Compounds by Immobilized Enzymes and Cells

BERNARD J. ABBOTT

I.	Introduction	203
II.	Penicillins	206
III.	Cephalosporins	217
IV.	Amino Acids	225
V.	Other Products	240
VI.	Concluding Remarks	249
	References	252

Cytotoxic and Antitumor Antibiotics Produced by Microorganisms

J. FUSKA AND B. PROKSA

I.	Introduction	259
II.	Antibiotics and Their Synonyms	260
III.	List of Tumors	274
	References	335

SUBJECT INDEX	371
CONTENTS OF PREVIOUS VOLUMES	374

LIST OF CONTRIBUTORS

Numbers in parentheses indicate the pages on which the authors' contributions begin.

BERNARD J. ABBOTT, *The Lilly Research Laboratories, Eli Lilly and Company, Indianapolis, Indiana* (203)

GEORGE R. ANDERSON, *Division of Biologic Products, Bureau of Disease Control and Laboratory Services, Michigan Department of Public Health, Lansing, Michigan* (43)

RUDOLPH J. ALLGEIER, *Wheaton Place, Catonsville, Maryland* (81)

JACK CAMERON, *Connaught Laboratories Ltd., Willowdale, Toronto, Canada* (57)

HUBERT A. CONNER, *Department of Physical Sciences, Northern Kentucky University, Highland Heights, Kentucky* (81)

YUKIO FUJISAWA, *Microbiological Research Laboratories, Central Research Division, Takeda Chemical Industries, Ltd., Jusohonmachi, Yodogawa-ku, Osaka, Japan* (159)

J. FUSKA, *Department of Microbiology and Biochemistry, Faculty of Chemistry, Slovak Polytechnical University, Bratislava Czechoslovakia* (259)

TOSHIHIKO KANZAKI,* *Microbiological Research Laboratories, Central Research Division, Takeda Chemical Industries, Ltd., Jusohonmachi, Yodogawa-ku, Osaka, Japan* (159)

CHARLES R. MANCLARK, *Bureau of Biologics, Food and Drug Administration, Bethesda, Maryland* (1)

STEPHEN I. MORSE, *Department of Microbiology and Immunology, State University of New York, Downstate Medical Center, Brooklyn, New York* (9)

CHARLOTTE PARKER, *Microbiology Department, University of Texas, Austin, Texas* (27)

B. PROKSA, *Institute of Experimental Pharmacology, Slovak Academy of Sciences, Bratislava, Czechoslovakia* (259)

M. STERNBERG, *Marschall Division Research, Miles Laboratories, Inc., Elkhart, Indiana* (135)

---

* Present address: Corporate Planning Division, Takeda Chemical Industries, Ltd., Nihonbashi, Chuo-ku, Tokyo, Japan.

The Current Status of Pertussis Vaccine: An Overview[1]

CHARLES R. MANCLARK

Bureau of Biologics, Food and Drug Administration, Bethesda, Maryland

Whooping cough has been a major cause of infant morbidity and mortality in the United States. With the introduction of an improved and standardized pertussis vaccine in the 1940s, there followed a remarkable decline in pertussis in the United States, most of the Western world, and Australia, New Zealand, and Japan. Clinicians and public health authorities agree that the use of pertussis vaccine has been and remains the most effective method of controlling whooping cough. Antimicrobials are of limited value during the prodromal stages of the disease; they may be useful in the control of bacterial infections which are secondary to *Bordetella pertussis* infection, but they have had a negligible effect on the control of whooping cough. In addition to vaccines, an observed decrease in the amount and severity of pertussis and the use of modern medical supportive procedures have contributed to the apparent control of pertussis.

The early days of pertussis research were occupied with determining the etiologic agent of the disease and methods for its culture and maintenance. Eventually production culture methods evolved and a vaccine was developed, standardized, and shown to be effective in the British clinical trials conducted during the 1940s and 1950s. Much of pertussis vaccine research and development occurred prior to the advent of modern biochemistry, physiology and microbial genetics. Because the vaccine appears to be effective, few attempts have been made to change the early formulations. With the exception of one extracted vaccine, all pertussis vaccines produced in the United States are whole cell products. As much as these vaccines resemble each other, the variability in the manufacture of vaccines from manufacturer to manufacturer and the variation from lot to lot from the same manufacturer is usually not appreciated.

Pertussis vaccine is one of the more troublesome products to produce and assay. As an example of this, pertussis vaccine has one of the highest failure rates of all products submitted to the Bureau of Biologics for testing and release. Approximately 15–20% of all lots which pass the manufacturers' tests fail to pass the Bureau's tests. Many of the problems associated with the production of pertussis vaccine result from the fact that little is known about the physiology and genetics of *B. pertussis*. The usual pure culture techniques are not employed when transferring

[1] This introductory statement and the four articles that follow are from a seminar entitled "Pertussis Vaccine: Current Status" held during the 75th Annual Meeting of the American Society for Microbiology, New York City, April 28, 1975.

or propagating cultures, nor are there any simple markers of production cultures to insure that they are typical of the prototroph or that cultures of high protective potency and low reactogenicity are selected for use in the production of vaccine. Recent studies in our laboratory have shown that *B. pertussis* cultures and vaccines contain high levels of adenylate cyclase and that this enzyme is related to a number of biological activities. Assays for adenylate cyclase are easily done and may provide a useful marker for vaccine development and for definitive genetic and physiologic studies.

Since a genetic study has not been done and prototrophic *B. pertussis* cannot be properly characterized, it follows that a definitive study of the physiology of *B. pertussis* has not been done. Consequently, there is evidence that current culture methods select against prototrophic cells in favor of nutritionally and antigenically deficient mutants.

Even though whooping cough has been controlled in those countries which have employed pertussis vaccine prophylaxis, many unanswered questions concerning the host–parasite relationship remain. A panel of experts has been convened at the Bureau of Biologics to study the efficacy and safety of all biologics, including those with a pertussis vaccine component. Much current interest by the panel and other public health researchers centers around the following topics.

Does pertussis vaccine protect against whooping cough and/or prevent *B. pertussis* infection? Very little is known about the pathogenesis of *B. pertussis,* and even less is known about the mechanism and/or duration of vaccine- or disease-induced immunity. Immunity is probably not mediated by serum antibody, but secretory antibody *may* play a role. In the absence of evidence to the contrary, by default, immunity is usually considered to be cellular. What is the reservoir or infection between epidemics? What is the prevalence of asymptomatic carriage?

Pertussis disease and infection are underreported. One of the difficulties in evaluating the efficacy or failure of pertussis vaccine is caused by problems in the laboratory and in the clinical diagnosis of pertussis. In the absence of the whoop pertussis may be undiagnosed. Some viral infections may mimic whooping cough. A definitive diagnosis requires the isolation and identification of the etiologic agent. Unfortunately, diagnosis is complicated by deficiencies in existing laboratory procedures. Also, most laboratories are not properly equipped or trained to use the methods that do exist. The result is that pertussis is either not diagnosed or is misdiagnosed.

We know that pertussis vaccines passing the required toxicity and safety tests can cause adverse reactions in children. Local reactions are relatively common and include edema, erythema, induration, pain, and sometimes ulceration at the injection site. Systemic reactions are less common, but include fever, collapse, seizures, persistent screaming, and, rarely, paralysis and death. Adverse reaction rates are not accurately reported, but more adverse reactions are probably experienced with the use of pertussis vaccine than with other biologicals.

With control of the disease, we may be approaching a time in which more vaccine-related problems than those due to the disease will be experienced. The higher rate of vaccine reactions than disease has been responsible for decisions in several countries making pertussis immunization an optional procedure. It is the opinion of public health authorities in the United States that the benefits of vaccination outweigh the risks of disease. Unfortunately, the reduced use of pertussis vaccine in those countries where it has been made an option may provide the ultimate proof of the efficacy of pertussis vaccine.

Much work with improved pertussis vaccines is based on the results of the aforementioned British clinical vaccine trials done in the 1940s and 1950s. These trials showed, among other things, that the clinical efficacy of pertussis vaccine correlated with the mouse potency test. The premise that evolved was that the protective antigen measured in the mouse potency test was *the* antigen that conferred immunity in infants. Since then other biological entities in pertussis vaccines have been identified. A partial list would include: protective antigen (or mouse protective antigen), histamine sensitizing factor (HSF), lymphocytosis promoting factor (LPF), hemagglutinin, hemolysin, endotoxin, dermonecrotic toxin (heat labile toxin), late-appearing toxin (late weight loss factor), adjuvant (especially for IgE-like antibody), active and passive anaphylaxis-promoting activity factor, shock-promoting activity factor, and lethal toxin.

As a result of studies done since the British trials, it is known that although there is not an absolute correlation between the mouse protective antigen and HSF or LPF, vaccines can be ranked with regard to these components. For example, vaccines containing high levels of mouse protective antigen generally would contain high levels of HSF, LPF, and so on. It is likely, therefore, that the vaccines used in the British trials that tested high for mouse protective antigen probably would have contained high levels of HSF, LPF, and so on, if those tests had been done.

Pertussis is a localized *infection* without invasiveness. Septicemia is not produced. Only the ciliated epithelium of the respiratory tract is involved. There is evidence that pathologic changes, probably due to soluble substances released by the organism or to diffuse physiologic changes initiated by the disease, occur at sites distant from the site of infection. One approach to understanding the host–parasite relationship would be to determine the mechanisms responsible for the organism's predilection for and attachment in the upper respiratory tract and to study those substances (toxins, sensitogens, metabolites, and so on) that mediate the symptoms of disease production at sites distant from the site of infection.

Some investigators consider the mouse protective antigen to be a distinct entity that can be separated from the other biological activities of the cell. What little is known about mouse protective antigen would indicate that it is a rather innocuous substance and it would be difficult to understand its role in the *disease* process. If one subscribes to the thought that protective antigen is the same as HSF, LPF, or other substances in the cell, then it is apparent that the roles of these metabolites would have to be determined in human disease and/or immunity to disease.

The various factors and cellular components of B. *pertussis* have not been directly related to human disease or reactions to vaccine. These relations require that the various components be isolated and characterized.

One area that has received little direct consideration is the mechanism by which *infection* is established and a determination of how it may be prevented. One approach to such a problem is to determine the characteristics of B. *pertussis* cell surfaces that confer specificity and are related to attachment and localization in the target tissues of the respiratory tract.

Although pili have not been satisfactorily demonstrated for B. *pertussis*, it may be productive to consider the possibility that pili may exist and that they play a role in the attachment process. Piliated gram-negative bacteria have been associated with a number of infectious processes of similar surfaces. In many instances the presence of pili has been associated with a number of infectious processes of similar surfaces. In many instances the presence of pili has been associated with hemagglutinin. *Bordetella pertussis* has a hemagglutinin. Extracted mouse protective antigens have been shown to have hemagglutinating activity and electron micrographs demonstrate a fine fiberlike structure. If it can be shown that B. *pertussis* is piliated, then the role of such structures in pertussis infection or immunity should be determined.

The ultimate goal of any investigations of B. *pertussis* and the host–parasite relationship in pertussis should be the development of an improved vaccine. Because of the ethical and logistic problems in carrying out field trials on an improved pertussis vaccine, every attempt should be made to include all possible approaches to vaccine immunity in a single trial. One of the several dilemmas facing the manufacturer or developer of a new vaccine is justifying the development and use of a new vaccine if present vaccines are judged to be safe and effective. In addition, pertussis vaccine is a relatively low-profit item and the cost of developing and field testing it would be very high.

The clinical testing of a new product poses the most difficult problems. It may be possible to field test a vaccine in one of the underdeveloped countries where whooping cough is more common than in the United States, but would such trials be ethical? Would the testing of a vaccine in an underdeveloped country be a proper evaluation of a product that is to be used in the United States? Would an underdeveloped country have the laboratory facilities and technical competence necessary to evaluate the efficacy of the vaccine? If a meaningful clinical trial could be conducted in the United States or elsewhere, since the vaccine recipient is an infant, what are the problems associated with informed consent? Most important, are present laboratory procedures sufficient to assure that a new vaccine is safe?

The costs of developing and testing an improved vaccine could be borne by the manufacturer and passed on to the consumer. The government could pay for the development of a vaccine through the mechanism of grants or contracts, and the costs could be spread to the taxpayer. There are advantages and disadvantages of each approach, but if the government supported the research, the newly developed process probably would become public property.

It may be that the problems associated with the proper and ethical clinical testing of an improved pertussis vaccine are so great as to be considered insurmountable, and some other approach to make pertussis vaccine more effective and less reactogenic may be necessary. Some may conclude that any clinical trial involving infants is immoral and unethical. However, it is possible that it may turn out to be more immoral and unethical to abstain from such clinical studies if it is demonstrated that the probability of risk is low and of success, high.

On April 16, 1973, the Department of Health, Education, and Welfare announced the formation of a Panel on Review of Bacterial Vaccines and Toxoids with Standards of Potency. The panel was composed of physicians and scientists who were among the leaders in the fields of microbiology, immunology, preventive medicine, and public health, with a wide range of experience in clinical medicine and research.[2] These experts have

[2] Panel on Review of Bacterial Vaccines and Toxoids with Standards of Potency:

Gene H. Stollerman, M.D.
Professor and Chairman
Department of Medicine
University of Tennessee
College of Medicine

Theodore C. Eickhoff, M.D.
Acting Chairman
Department of Medicine
University of Colorado Medical Center

Geoffrey Edsall, M.D.
Professor of Microbiology and Head
Department of Microbiology
London School of Hygiene and
    Tropical Medicine
London, England

John C. Feeley, Ph.D.
Chief, Bacterial Immunology Section

Bacteriology Branch
Center for Disease Control

Edward A. Mortimer, Jr., M.D.
Professor and Chairman of the Department of Community Health, Professor
of Pediatrics
Case Western Reserve University

Hjordis M. Foy, M.D., Ph.D.
Associate Professor
Department of Epidemiology and International Health
School of Public Health and Community
    Medicine
University of Washington

Jay P. Sanford, M.D.
Dean
Uniformed Services University

been charged with the responsibility of reviewing, among other things, the safety and efficacy of vaccines containing a pertussis component. After many months of study and deliberation, a Provisional Generic Statement on Pertussis Vaccine was released on November 21, 1975. The following recommendations have been abstracted from the panel's statement. These recommendations represent a concensus of present thought and provide reasonable goals and guidelines for research to improve pertussis vaccine, as well as for the clarification of our understanding of the host–parasite relationship in pertussis.

A.   The panel strongly recommended that adequate public support be provided for studies of the pathogenesis of pertussis and the biology of the organism, particularly as related to the immunology of pertussis, the complications of the disease, and the untoward reactions to immunization. Without such basic studies a more effective and safer pertussis vaccine cannot be developed.

B.   Surveillance of pertussis in well-defined populations should be undertaken. Such surveillance would have three purposes: first, to determine the incidence in the United States, including distribution by age and vaccine status; second, to evaluate the possibility that a change in serotypes of B. pertussis in a community causes outbreaks of pertussis in individuals previously immunized with serotypes formerly present; and third, to determine whether the current infrequency of the disease in the United States may ultimately result in a population of older children and adults whose immunity has waned because of a lack of repeated exposure to the organism.

Further, the panel is convinced that currently employed surveillance systems to identify adverse reactions to pertussis are inadequate and recommends that definitive steps be taken by the appropriate subdivisions of the Public Health Service to improve them. Several alternatives are available. Perhaps the same channels as those proposed for reporting of adverse drug reactions can be utilized. Special field stations with sufficient populations under surveillance may have to be established and funded.

C.   Specific recommendations of the panel regarding the production, use, and evaluation of pertussis vaccines include the following:

The weight-gain test in mice used to determine toxicity of pertussis vaccine needs revision to include specifications regarding mouse strain(s) to be used. Studies should be undertaken to develop other assays predictive of human reactivity. Obviously, better definition of the organism's biological characteristics would facilitate prediction and prevention of reactogenicity in man.

The agglutination test used to determine vaccine response in humans should be standardized. It is recommended that a reference serum be used for comparison. A reference laboratory should be available at the Bureau of Biologics. The interval between immunization and obtaining serum for testing of the serologic response must be specified. An acceptable titer obtained by a standardized method should be defined; fold titer rises or geometric mean titers are not adequate indicators of induced immunity.

The regulation for manufacturers regarding the human dose should be updated to reflect current recommendations and practices. To achieve this, a requirement that pertussis vaccine have a potency of "4 units per single human dose" could be substituted.

The vaccine label should warn that if shock, encephalopathic symptoms, convulsions, or thrombocytopenia follow a vaccine injection, no additional injections with pertussis antigens should be given (immunizations can be continued with DT). The label should also include a cautionary statement about fever, excessive screaming, and somnolence.

Any fractionated vaccine that differs from the original whole cell vaccine should be field tested if possible until better laboratory methods for evaluating immunogenicity in man are developed. Field testing should include agglutination testing and, if possible, evaluation of clinical efficacy in man.

D.   Pertussis vaccine is one of the immunizing agents for which it is strongly urged that legislation be enacted to provide reasonable federal compensation to the few individuals injured and disabled by meritorious public health programs. Such legislation would protect manufacturers and physicians against liability in situations in which the injury was not a consequence of defective or inappropriate manufacture or administration of the vaccine.

The scientific community and public health authorities are in general agreement that pertussis vaccine is the most effective measure employed in the control of pertussis disease. Recent concern has been expressed that vaccines currently in use have not benefited from modern genetic, physiologic, and immunologic knowledge and are less immunogenic and more reactogenic than is desirable or necessary. These and other concerns are reflected in the recommendations of the panel and are discussed in greater detail by Doctors Morse, Parker, Anderson, and Cameron in the following presentations.

Biologically Active Components and Properties of *Bordetella pertussis*[1]

STEPHEN I. MORSE

Department of Microbiology and Immunology, State University of New York, Downstate Medical Center, Brooklyn, New York

I.	Introduction	9
II.	Sensitization to Histamine	10
III.	β-Adrenergic Blockade	12
IV.	Lymphocytosis-Promoting Factor	14
V.	Adjuvanticity	17
VI.	Lipopolysaccharide Endotoxin	19
VII.	Heat-Labile Toxin	19
VIII.	Hemagglutinin	20
IX.	Mitogenicity for Lymphocytes	20
X.	Interrelationships between Biological Activities	21
XI.	Pathogenesis of Pertussis	23
	References	24

I. Introduction

A variety of biological effects may be observed in experimental animals following injection of phase I *Bordetella pertussis* cells or products. A partial list includes: sensitization to the lethal effects of the pharmacological agents histamine and serotonin as well as to nonspecific stresses such as cold and peptone shock; unresponsiveness to the induction of hyperglycemia by epinephrine, which has been attributed to the causation of a β-adrenergic blockade; leukocytosis with predominating lymphocytosis; adjuvanticity with respect to both antibody production—including reaginic antibody (IgE)—and cell-mediated immunity; and direct, acute toxic effects which are in general related to either a distinct heat-labile toxin or to heat-stable lipopolysaccharide endotoxin.

In addition, components of *B. pertussis* can be shown to induce striking effects in *in vitro* systems. Some of these *in vitro* effects are clearly relevant to *in vivo* events, e.g., adjuvant activity in spleen cell cultures and depression of cyclic adenosine 3′,5′-monophosphate (cAMP) formation. Others, such as hemagglutination, hemolysis, and mitogenic stimulation of murine, as well as human lymphocytes, have yet to be shown to be related to *in vivo* occurrences.

Paradoxically, while clinical pertussis has become of decreasing importance where effective vaccines are routinely used, there has been an increased interest in the biological activities of *B. pertussis*. This interest is

[1] Supported in part by research grant AI 09683 from the National Institutes of Health.

especially manifest in the numerous studies designed to isolate and characterize the responsible bacterial factors and to determine the underlying
mechanisms involved in the observed reactions. Although there is still a
long way to go in both cases, a sense is emerging that what apparently
appear to be quite disparate biological activities in some instances may be
caused by a single component inducing a single basic specific biochemical
effect which, in turn, is manifested by superficially unrelated events (see
Section X). Heretofore, each event has been attributed to the activity of
a distinct bacterial component with a distinct mechanism of action.

Interest has also been furthered by the interrelated facts that we com-
prehend neither the pathogenesis of whooping cough nor the compo-
nent(s) of pertussis vaccine which engenders protection. It is likely that
identification of the bacterial determinant of pathogenicity and virulence
and the corollary definition of the protective antigen will ultimately lead
to the development of an effective, well-defined immunizing preparation.
Moreover, it is possible that such a preparation would be freer of adverse
reactions than current products. Admittedly, although the true incidences
of serious reactions to pertussis vaccine in infants is not known with
precision, they are generally conceded to be extremely rare. However,
it is now realized that protection is not lifelong and that previously
immunized older children and adults are as susceptible as those not
previously immunized. The disease is often milder, but it may be serious;
morbidity is apparent in terms of lost time from work, school, and so on;
and the patient is a source of infection to others. It is in this population
that local and systemic complications of vaccination are more frequent
and severe (Linnemann *et al.*, 1975).

The purpose of this article is to review some of the information on
the biologically active components of phase I *B. pertussis* and their
effects. A certain amount of selectivity of topics and references is necessary because of the vast number of reports in the field, and some of the
selectivity may reflect the writer's interests, and perhaps his biases.

For background information, the reader is referred to excellent review
articles by Kind (1958), Munoz and Bergman (1968), Pittman (1970),
and Munoz (1971).

II.   Sensitization to Histamine

Parfentjev and his co-workers (1947a,b,c) found that in mice previously
injected with pertussis vaccine, the subsequent injection of pertussis
vaccine or pertussis extracts was followed by lethal shock. Subsequently, Parfentjev and Goodline (1948) demonstrated that pertussis-
vaccinated mice became highly sensitive to the lethal effects of histamine.
The trivial term for the factor or factors of *B. pertussis* which causes
sensitization is histamine-sensitizing factor or HSF.

Histamine sensitization is readily induced in mice, which are normally resistant to the lethal effects of this amine, while rats, which are also normally resistant, are slightly less susceptible to sensitization. In contrast, in guinea pigs and rabbits which are normally sensitive to the action of histamine, it is difficult to heighten sensitivity with pertussis.

There is a marked variation in HSF activity in different strains of mice and it was proposed that responsiveness is under the control of a single, dominant autosomal gene (Wardlaw, 1970). However, more recent evidence suggests a far more complex mode of inheritance, possibly involving polygenic transmission (Ovary and Caiazza, 1975).

The mechanism of action of HSF is uncertain, but it appears to be unrelated to aberrations in histamine metabolism such as depressed levels of histaminease (Munoz, 1971) or increased levels of histidine decarboxylase (Szentivanyi et al., 1968). While the precise mode of action of histamine at the cellular level under normal circumstances is uncertain, it is conceivable that it is at this site that HSF acts, either by potentiating histamine action or by inhibiting the activity of a regulatory substance. Support for the latter hypothesis derives from the ability of exogenous epinephrine to protect HSF-treated mice from lethal histamine challenge (Bergman and Munoz, 1966, 1971). Conversely, β-adrenergic antagonists such as propranolol can induce histamine sensitization (Fishel and Szentivanyi, 1963; Fishel et al., 1962, 1968). This subject will be considered further in Section III.

Purification of HSF to homogeneity has been difficult to achieve. A major cause of the difficulty has been insolubility which becomes more pronounced as purification proceeds. In addition, inactivation, unrelated to aggregation and insolubility, appears to occur on storage. Some of these problems can be overcome by using hyperosmolar buffers, usually at an alkaline pH, or buffers containing 4 M urea at pH 6–7 (Munoz and Hestekin, 1962; Morse and Bray, 1969; Homma et al., 1969; Parker and Morse, 1973; Lehrer et al., 1974, 1975; Morse and Morse, 1976).

The best evidence available suggests that HSF owes its activity to a protein moiety since it is susceptible to heat (Kind, 1956) and to proteolytic enzymes (Pieroni et al., 1965). Highly purified preparations isolated from intact cells by Lehrer et al. (1974) contained considerable amounts of lipid whereas HSF isolated in our laboratory from culture supernatant fluids have been lipid-free (Morse and Morse, 1976). Whether this discrepancy is the result of differences in homogeneity or whether excreted or secreted HSF has a different chemical composition than extracted, cell-bound HSF is unclear. Carbohydrate is not found to any appreciable extent and studies utilizing sodium periodate indicate that carbohydrate is not required for biological activity (Wardlaw and Jakus, 1966).

When soluble material is employed the intravenous route of administration is the most effective in inducing sensitization, while intraperitoneal injection and the subcutaneous route are less effective (Munoz and Bergman, 1966). Following intravenous inoculation, sensitization to histamine is apparent within 90 minutes; peak sensitization is found 3–5 days after injection and sensitization can persist for a number of weeks (Munoz and Bergman, 1966).

Utilizing female mice of the same age and our most highly purified preparations, we have found the dose which will sensitize 50% of the animals (SD_{50}) to the lethal effects of 1 mg of histamine, as the free base, to range from 0.01 to 0.33 μg, depending upon the strain of mouse.

III. β-Adrenergic Blockade

The thesis that *B. pertussis* induces β-adrenergic blockade derives from studies performed by Parfentjev and Schleyer (1949) and Stronk and Pitman (1955) which showed that pertussis-treated mice, in addition to becoming sensitive to histamine, also did not undergo the hyperglycemic effect usually following administration of histamine. Szentivanyi and his collaborators (1963) then showed that histamine-induced hyperglycemia is essentially an epinephrine-mediated response; the same was true for hyperglycemia caused by 5-hydroxytryptamine (serotonin). Moreover, they found that not only were histamine and serotonin unable to cause hyperglycemia in pertussis-sensitized mice but epinephrine also failed to elevate the blood glucose of the pertussis-treated mice. As was the case with histamine sensitization, epinephrine unresponsiveness also occurred in animals treated with β-adrenergic antagonists.

A central role for cAMP in these reactions has been postulated. Intracellular levels of cAMP are regulated both by the activity of adenylate cyclase, which causes the synthesis of cAMP from ATP, and by the activity of a nucleotide phosphodiesterase which breaks down cAMP to 5' AMP. Many of the effects of β-adrenergic agonists are mediated by cAMP which is increased in amount due to the activation of adenylate cyclase by these agents.

In this regard Ortez *et al.* (1975) have found the cAMP content of the spleens of pertussis-treated mice to be significantly lower than that of normal mice and that splenic adenylate cyclase stimulation by epinephrine, norepinephrine, isoproterenol, and fluoride is markedly diminished.

One simple hypothesis would place the action of the active component of *B. pertussis* at the receptor site for β-adrenergic agonists, thereby attenuating the ability of the agonist to bind to the cell and to activate adenylate cyclase. Thus, the expected increment in intracellular cAMP and the metabolic consequences thereof would not occur.

However, it is abundantly clear that the actual events which take place are far more complex. Parker and Morse (1973) employed human lymphocytes in an *in vitro* system in order to determine the effects of a highly purified preparation of the *B. pertussis* lymphocytosis-promoting factor (LPF), which also had HSF activity, on cAMP metabolism. It was found that pretreatment with LPF abrogated the increase in cAMP levels which is normally stimulated by the β-adrenergic agonist isoproterenol. Similarly LPF-treated lymphocytes did not accumulate cAMP when prostaglandin E1 was added, whereas normal lymphocytes did. These compounds have different receptors, suggesting that the activity is not due to specific blockade of β-adrenergic receptors. The mechanism of action therefore may be to inhibit adenylate cyclase directly or, less likely, to stimulate cAMP phosphodiesterase. It was of interest that the basal level of cAMP, at least after short-term incubation, was unchanged. Studies on the effect of pertussis on guanylate cyclase and cyclic guanosine $3',5'$-monophosphate (cGMP) are in progress and are of importance in view of the apparent reciprocal relationship between cAMP and cGMP.

The precise mechanism responsible for the failure of epinephrine to raise the blood sugar in pertussis-treated mice is also not clear. Presumably activation of liver phosphorylase and subsequent glycogenolysis and glucose formation is dependent upon α-adrenergic activity, while glycogenolysis in muscle is dependent upon β-adrenergic receptors, the end product of metabolism being lactic acid. In pertussis-treated mice, and in mice given β-adrenergic blocking agents, epinephrine does stimulate hepatic glycogenolysis but the expected lactic acidemia and increment in unesterified fatty acids does not occur (Szentivanyi *et al.*, 1963; Keller and Fishel, 1967). The former effect is likely to be caused by the α-adrenergic effects of epinephrine since α-adrenergic antagonists block the effect on hepatic glycogenolysis. Thus, it appears that although muscle glycogenolysis is blocked, hepatic glycogenolysis is, if anything, greater than normal, suggesting that increased peripheral utilization of glucose occurs. In this regard it is noteworthy that Gulbenkian *et al.* (1968) found an increase in serum levels of biologically active insulin in rats and that Ortez *et al.* (1975) found increased splenic uptake and utilization of glucose in pertussis-treated mice.

If the effect of pertussis sensitization is to diminish the β-adrenergic stimulation of cAMP one might conceivably expect the administration of cAMP to abrogate the effect. Cronholm and Fishel (1968) observed that at least in the CFW mouse, exogenous cAMP causes hyperglycemia in normal but not in pertussis-sensitized mice. However, cAMP is poorly taken up by normal cells and obviously the question of uptake in cells of pertussis-treated mice must be answered. In addition, the hyperglycemia effect of cAMP in normal mice may not be a primary one but may be mediated by hormones whose release is caused by the cyclic nucleotide. It is this effect which is in turn abrogated in pertussis-treated mice.

IV. Lymphocytosis-Promoting Factor

One of the cardinal clinical signs of pertussis is relative and absolute lymphocytosis, first reported by Fröhlich in 1897. Lymphocytosis occurs in the majority of patients over 6 months of age, but is less frequent in younger patients (Lagergren, 1963). The cells are for the most part normal-appearing mature small lymphocytes. Although polymorphonuclear leukocytosis may occur to a certain degree, significant elevation of the neutrophil count suggests a secondary bacterial infection.

Both experimental infection and injection of B. pertussis and its products have been shown to induce leucocytosis; in some instances a predominantly granulocytic response was observed (Bradford et al., 1956) while in other cases small lymphocytes were in the majority (Tuta, 1937; Morse, 1965). In the past several years studies in our laboratory have been concerned with both the mechanism of pertussis-induced leukocytosis and lymphocytosis and the bacterial product responsible for the response. Initial studies utilized intact phase I B. pertussis cells or crude supernatant fluids of liquid cultures.

When either whole cells (Morse, 1965) or crude supernatant fluids (Morse and Bray, 1969) are injected intravenously into mice, a leukocytosis develops which reaches maximum values 3–5 days later and then declines to base-line levels in 2–3 weeks. The peak leukocytosis may exceed 200,000 cells/mm^3. The predominant cells in the circulation are mature-appearing small lymphocytes. However, polymorphonuclear leukocytosis also occurs and indeed the relative increase in polymorphonuclear leukocytes (PMNs) is greater than that of the lymphocytes. In studies using these crude materials, an early 6–12 hours, marked but short-lived increase in PMNs was found to occur, but this early phase reaction has been shown to be due to another component, probably endotoxin, since it does not take place after administration of highly purified lymphocytosis-promoting factor.

A variety of techniques have been employed to determine the mechanism of the lymphocytosis. Repeated doses of tritiated thymidine were given during the early stages of the reaction and at the time of the peak response autoradiographic analysis of the peripheral leukocytes was performed (Morse and Riester, 1967a). The exent of nuclear labeling was the same in both normal and pertussis-treated mice. Thus, it appeared that the lymphocytosis was not due to an increment of newly derived cells, but rather was the consequence of a redistribution of cells. (In contrast, the PMN response was due in part to both redistribution and proliferation.)

Gowans and his co-workers (Gowans, 1959; Gowans and Knight, 1964; Gesner and Gowans, 1962) had elegantly defined the recirculation of lymphocytes in rats and mice and had shown that the circulating lymphocytes traffic from blood to lymphatic tissue and then, in the case of lymph nodes, reenter the blood after passing through the thoracic duct. Evidence that the normal recirculation of lymphocytes was impaired in pertussis-treated mice was obtained in studies employing mice with indwelling thoracic duct cannulae (Morse and Riester, 1967b). In animals undergoing pertussis-induced lymphocytosis, the content of thoracic duct lymphocytes was far less than in normal mice. Thus, either the lymphocytes could not leave the blood in the normal fashion or the normal rate or they could not leave the lymph nodes. Alternatively, the route of recirculation could have been altered so as to bypass the lymphatics, the so-called "direct entry" which pertains to normal recirculation of lymphocytes in the spleen where lymphatics play little or no role. The thoracic duct experiments, however, were not performed in splenectomized animals which do undergo lymphocytosis, so this question remains moot.

Several lines of evidence have indicated that a marked decrease in the ability of circulating lymphocytes to "home" from the blood into lymphoid tissue occurs in pertussis-treated mice. There is a marked decrease in the cellularity of the lymph nodes and white pulp of the spleen, although the red pulp of the latter is engorged (Morse, 1965). On histological examination of the lymph nodes of pertussis-treated mice, few lymphocytes are found associated with the postcapillary venules through which the cells traverse to reenter lymphoid tissue (Athanassiades and Morse, 1973). Moreover, when circulating lymphocytes are removed from pertussis-treated mice, labeled *in vitro* with uridine and then injected into normal syngeneic mice, the labeled cells tend to remain in the circulation to a greater extent than is observed when normal cells are injected (Morse and Barron, 1970). Finally, thoracic duct or lymph node lymphocytes incubated *in vitro* with B. *pertussis* culture supernatants or LPF also do not home normally when injected into syngeneic animals (Iwasa *et al.*, 1970; Taub *et al.*, 1972).

Taken together these observations indicate that in pertussis-treated animals the lymphocytes do not emigrate from the blood normally. Because LPF has been shown to reversibly combine with the surface of a number of different cell types, in addition to lymphocytes (Morse and Adler, 1973), it is not clear whether the primary site of action is on the lymphocyte or on the postcapillary venules.

In unpublished studies performed in collaboration with Drs. C. Bianco and V. Nuzzenzweig, both B and T cells were found to be increased in the circulation. The T cells remain the predominant cell type, but the relative increase in B cells is greater; there is a reciprocal change in the cell population in the spleen. Mice congenitally lacking a thymus and T cells, "nude" mice, also respond with a lymphocytosis to *B. pertussis* and purified LPF (Wortis, 1971; Morse and Morse, 1976).

Studies on the properties of LPF have revealed it to be sensitive to proteolytic enzymes, albeit it at a slower rate than one sees with most proteins at the enzyme–substrate ratios used (Morse and Bray, 1969). Periodate has no effect on activity, suggesting the absence of a carbohydrate group essential for biological activity.

The isolation and purification of LPF has the same attendant difficulties as those associated with HSF. The material has been found to be insoluble in H_2O and isotonic buffers and requires an alkaline pH, with the addition of 0.5 or 1 M NaCl, or 4 M urea buffer. Even so, aggregation and apparent loss of activity due to aggregation and/or spontaneous degradation tend to occur in time. Using preparations obtained from either intact organisms or culture supernatant fluids, Sato and Arai (1972) as well as Lehrer *et al.* (1974), have suggested that lipid, in as high an amount as 50%, is a component of LPF. However, we have isolated LPF from culture supernatant fluids of phase I *B. pertussis* and found no lipid present. Briefly, our procedure for obtaining LPF consists of: (1) addition of solid $(NH_4)_2SO_4$ to culture supernatant fluids to achieve a saturation of 90% at 4°C and pH 6.4; (2) harvesting the $(NH_4)_2SO_4$ precipitate and discarding any material soluble in water and 0.15 M NaCl; (3) dissolving the precipitate in 0.05 M or 0.1 M Tris-0.5 M NaCl (Tris-NaCl), pH 10.0, and discarding any insolubles; (4) layering the Tris-NaCl soluble fraction on a discontinuous CsCl gradient of densities 1.5, 1.3, 1.25, and 1.2 and centrifuging at 50,000 g for 3.5 hours; (5) removing the load volume; and (6) passing the load volume through Sephadex G150 using Tris-NaCl as the eluting buffer.

The first included peak contains the LPF. On double-diffusion reactions in gel using heterospecific antipertussis antiserum, one immunologically reactive component is usually found. Occasionally, a second, minor precipitin band is found but this immunoreactive contaminant, unlike LPF, is soluble in water and in isotonic buffers and the LPF can be separated from it by precipitation under conditions in which the osmotic strength and pH of the solvent are decreased. The purified LPF gives a single precipitin band in immunodiffusion and immunoelectrophoresis and is 70–80% homogeneous by polyacrylamide gel analysis and isoelectric focusing. Some of the impurities have been proven to be degradative products.

As indicated above, LPF isolated in this fashion is lipid-free and in addition contains less than 1% carbohydrate. The percentage of nitrogen is 14.5, again suggesting that LPF is in fact a protein.

When LPF is dissociated in sodium dodecylsulfate (SDS) and 2-mercaptoethanol, and then subjected to polyacrylamide gel electrophoresis, four polypeptide bands are found with molecular weights of 13,400, 17,400, 19,300, and 23,500. The molecular weight of LPF as determined by this technique is 73,600. However, there is preliminary evidence that the polypeptides are not present in equimolar proportions and that there may be 2 moles of the smallest subunit for each one of the others, which would put the molecular weight at approximately 87,000. It is of interest that a four- or five-membered structure is also suggested by high-resolution electron microscope examination of negatively stained preparations.

Antisera against purified LPF has been prepared and yields only a single band. This monospecific antiserum is effective in neutralizing the effect of LPF whether used to block the activity of preparations incubated with the antiserum *in vitro* prior to inoculation, or when administered passively to animals before injecting LPF.

V. Adjuvanticity

Bordetella pertussis cells and/or products clearly modify a variety of immunological processes. Most interest has centered on those situations in which there is a heightened immunological response when compared to normal; i.e., adjuvanticity.

Adjuvanticity of *B. pertussis* may be seen for production of IgG and IgM antibody, reaginic antibody, and cell-mediated immunity in such systems as experimental allergic encephalomyelitis.

Dresser (1972) has adduced evidence indicating that at low or moderate doses of antigen, the effect of *B. pertussis* on IgG antibody production by bone marrow-derived lymphocytes (B cells) is mediated through thymus-dependent lymphocytes (T cells). However, at high doses of antigen, the increase in IgM antibody production appears to be due to a direct effect on B cells. The general effect of *B. pertussis* given under appropriate conditions is to increase the immunoglobulin response to a thymus-dependent antigen, such as sheep erythrocytes, and this can be shown *in vitro* as well as *in vivo* (Maillard and Bloom, 1972; Murgo and Athanassiades, 1975).

However, the situation with thymus-independent antigens appears to be somewhat different. Highly purified Type III pneumococcal polysaccharide (SIII), a thymus-independent antigen, was used as an immunogen in mice and *B. pertussis* cells were employed as adjuvant. In normal mice, the SIII response consisted almost entirely of 19 S antibody.

When an optimal immunizing dose of SIII was given with *B. pertussis* cells significant amounts of 7 S SIII antibody appeared in the circulation, and there was a net fall in the amount of 19 S immunoglobulin as compared to that found in normal mice (Kong and Morse, 1976). The overall result was a decrease in the total amount of anti-SIII antibody as measured by a sensitive radioimmunoassay technique. In contrast, when high doses of SIII were given which produced "treadmill paralysis," the administration of *B. pertussis* cells caused the appearance of SIII antibody, primarily of the 7 S type, and a net increase in total antibody.

Killed *B. pertussis* cells were shown by Mota (1958, 1964) and Mota and Peixoto (1966) to exert an adjuvant effect on the production in rats and mice of homocytotropic antibodies, similar in characteristics to human IgE-type reaginic antibodies. As in the case of other tests for adjuvanticity, the experimental systems involve the administration of an antigen, e.g., ovalbumin, dinitrophenylated *Ascaris serum* extract, and the inoculation of *B. pertussis* cells or products. The homocytotropic antibody response is measured by injecting serial dilutions of sera into the skin of animals of the same species. In the case of mouse homocytotropic antibodies, rats can be used for testing. After a 48–72 hour latent period, an appropriate amount of the antigen is injected intravenously together with a dye, such as Evans blue dye, and the extent of dye leakage around the skin injected with serum is measured. To date, studies have been confined to determinating the level of homocytotropic antibody directed against the nonpertussis antigen, and it is not known whether homocytotropic antibody levels against *B. pertussis* components themselves are increased.

Bordetella pertussis enhances the susceptibility of mice to fatal active or passive anaphylaxis, and it is likely that homocytotropic antibody is involved, although it has been postulated that immune complexes are also responsible.

Bordetella pertussis is quite a useful adjuvant for the induction of experimental allergic encephalomyelitis (Levine *et al.*, 1966), a disease which in great measure is due to the direct activity of sensitized lymphocytes rather than serum antibody and is therefore a product of a cell-mediated immune reaction. *Bordetella pertussis* has also been reported to decrease the growth of tumors implanted into experimental animals (Likhite, 1974), and this effect may also be due to enhanced cell-mediated immunity, although critical evidence on this point has not been obtained. There are also other situations in which *B. pertussis* apparently enhances the growth rate of tumors (Floersheim, 1967; Likhite, 1974).

The factor or factors of *B. pertussis* involved in these immunologic reactions has not been defined, and there is no compelling reason to believe that a single product is responsible for all of these events. Candi-

dates for adjuvants include *B. pertussis* endotoxin, since lipopolysaccharide endotoxins from other gram-negative organisms are known adjuvants. However, a heat-labile component is also believed to be important
in pertussis-induced adjuvanticity, and it has been suggested that this
component plays a role in the cell-mediated reactions (Levine *et al.*,
1966; Pieroni and Levine, 1966).

A rather strong case has been made for the activity of LPF in en-
hancing the production of homocytotropic antibody (Section X).

VI. Lipopolysaccharide Endotoxin

A wide range of biological activities can be produced by lipopolysaccharide endotoxins from gram-negative organisms. These include fever,
shock, leukopenia followed by leukocytosis, nonspecific protection against
infection, and preparation and elicitation of the Shwartzman phenomenon. The basic structure of the macromolecule consists of a glycolipid
(Lipid A) connected by ketodeoxyoctonate (KDO) to a core polysaccharide to which specific sugar residues are attached (O antigens).

Comparatively little attention has been given to the structure and function of *B. pertussis* endotoxins as compared to those of the enterobacteria.
Although hexosamine, heptose, hexose, in addition to KDO and lipid,
have been found to be present (MacLennan, 1960; Nakase *et al.*, 1970;
Aprile and Wardlaw, 1973a), the precise nature and organization of the
monosaccharide units is not known. The arrangement and structure are
obviously important because of the number of serologically specific reactions which can be obtained with the isolated LPS (Aprile and Wardlaw,
1973b).

Purified *B. pertussis* O antigen is antigenic, toxic, and pyrogenic, and,
in addition, has an adjuvant effect on the production of diphtheria antitoxin (Nakase *et al.*, 1970).

VII. Heat-Labile Toxin

There is a product of *B. pertussis* which produces acute death in
experimental animals and dermonecrosis. Heating at 56°C destroys these
biological activites and thus distinguishes it from LPS, which is heat
stable. The toxin has been designated heat-labile toxin or HLT. HLT has
not been isolated in homogeneous form and consequently the nature of
the component and its mechanism of action are unknown, although there
is reasonably good evidence that HLT is a protein (Banerjee and Munoz,
1962; Munoz, 1971).

One of the striking characteristics of HLT, in contrast to many other
bacterial protein toxins, is that it is not a true exotoxin. The bulk of

toxic activity residues intracellularly (Munoz, 1971). Moreover, HLT may be present within the bacterial cell as a precursor or zymogen which must be activated before toxicity occurs.

VIII. Hemagglutinin

Most freshly isolated strains of *B. pertussis* are hemagglutinating. In the early stages of growth in liquid medium the hemagglutinin (HA) is cell-associated, but in the later stages of growth the greatest amount is found in the extracellular medium (Masry, 1952). Erythrocytes from a wide variety of species, including man, mouse, rat, guinea pig, rabbit, and chicken, are agglutinated by either intact *B. pertussis* cells or culture supernatant fluids. However, there is no evidence of *in vivo* red blood cell agglutination in either natural or experimental infection.

The nature of the hemagglutinin has been unclear but relevant information has been obtained in recent studies on the structure of *B. pertussis*. Electron microscope examination of the surface of phase I *B. pertussis* cells has revealed the presence of fine filaments commensurate with pili (J. H. Morse and Morse, 1970). The pili measure 20–25 Å in diameter and 60–100 nm in length and are found attached to the cells in early stages of culture in liquid medium or on solid medium. After 24 hours of growth, the pili are sparse on the cell surface, and they are found in increasingly higher concentration in the culture medium. There is abundant evidence that the increase in medium content of both pili and HA is not due to cell lysis but rather to natural loss and, perhaps, at some stage of the growth cycle, to unbalanced synthesis, an event that appears to affect the cell membrane of *B. pertussis* (J. H. Morse and Morse, 1970).

Pili in other bacteria are in some instances hemagglutinating and there is little doubt that *B. pertussis* pili and the HA are identical; e.g., Sato *et al.* (1974) have shown that pili attach to erythrocytes. Little is known of the mechanism of attachment and the receptor site for the pertussis HA, but studies in our laboratory have shown that pretreatment of erythrocytes with crude *V. cholerae* neuraminidase does not prevent *B. pertussis* culture supernatant fluid whereas, as expected, such treatment does prevent agglutination by myxoviruses (Morse and Morse, 1976).

IX. Mitogenicity for Lymphocytes

Recent work in our laboratory under the direction of A. S. Kong (Kong and Morse, 1975, 1976) has revealed that *B. pertussis* culture supernatant fluids contain a potent mitogen for murine lymphocytes. The active material appears to be identical to LPF. The mitogen is somewhat more active than the phytohemagglutinin from *Phaseus vulgaris* (PHA) and slightly less stimulatory than Concanavalin A (Con A). Spleen and lymph node cells respond fully. However, like PHA and unlike Con A, the pertussis mitogen does not stimulate normal thymocytes. As in the case of PHA, cortisone-resistant thymocytes do proliferate in the presence of the mitogen. However, other data indicate that different subpopulations of lymphocytes are affected by PHA and the pertussis mitogen.

Pretreatment of the mitogen with antiserum renders it capable of producing a proliferative response, and addition of antiserum to the cell cultures up to a period of 8–12 hours blunts the response; after that there is no effect of antiserum.

A variety of experimental procedures has been employed to delineate the proliferating cell. Thus, it has been found that spleen and lymph node cells from congenitally athymic "nude" mice which lack T cells are not stimulated. When anti-Thy-1 antiserum (anti-θ antiserum) and complement are added to normal spleen cells to lyse away T cells, the residual cells do not respond. Similarly, the proliferating cells are lysed by Thy-1 antiserum. Thus there is clear evidence that it is the T cells that respond, although a second cell type, probably the macrophage, is required to initiate the stimulatory events, perhaps by processing the mitogen. Recently, studies by J. H. Morse, A. S. Kong, and S. I. Morse have shown that human lymphocytes also proliferate in response to the pertussis mitogen. This response was not due to sensitized lymphocytes, since cord blood cells were also stimulated. The type of human lymphocytes involved, B or T cell, has not as yet been established.

X. Interrelationships between Biological Activities

There is increasing evidence that a simplication and codification of the various "factors" and biologically active products of *B. pertussis* will soon be forthcoming. As isolation and characterization of specific components in homogeneous form is achieved, it will not be surprising to find that a single component produces several biological effects as the result of a single, fundamental, underlying mechanism of action.

Certain associations have been made on the basis of less compelling evidence, but they have thus far stood up. Thus, it seems clear, even without the basis of complete purification, that HSF is also responsible for the induction in mice of sensitivity to 5-hydroxytryptamine (serotonin) and nonspecific stress. In the case of LPS, its heat stability makes it an unlikely candidate for many of the reactions that are dependent upon heat-labile components, but there is little doubt that LPS can serve as an adjuvant though it would appear not to be the only adjuvant principle of *B. pertussis*.

Of most interest to our group is the interrelationship of LPF with other biologically active products. It has previously been shown that a highly purified HSF also had LPF activity and evoked the enhanced production of homocytotropic but not conventional antibody (Lehrer et al., 1974, 1975). Tada et al. (1972) also found that highly purified LPF served as an adjuvant for reaginic antibody production. However, it is not an adjuvant for other antibody classes for cell-mediated immunity and indeed LPF may suppress these immunological responses (Asakawa, 1969; Ochiai et al., 1972).

Our preparations of LPF are greater than 70–80% homogeneous and induce histamine sensitization, hypoglycemia and unresponsiveness to the hyperglycemic effect of epinephrine, and in vitro proliferation of lymphocytes. That these reactions are in fact produced by a single substance is supported by studies in which a single band on polyacrylamide electrophoresis was eluted and found to cause all of the effects. Moreover, a monospecific antiserum neutralized all of the reactions. (We have not thus far looked at enhanced reagenic activity.)

If histamine sensitization, leukocytosis, lymphocytosis, epinephrine refactoriness, mitogenesis of lymphocytes, and enhanced reaginic antibody formation are all caused by a single substance, elucidation of the nature of the underlying mechanism of action will be of great interest. It is known that LPF affects cAMP metabolism, that cyclic nucleotides affect in vitro lymphocyte proliferation, and that there is a relationship between both HSF activity and epinephrine refractoriness to β-adrenergic effects which are in turn cAMP mediated. Taken together these relationships suggest that cyclic nucleotide metabolism is of importance, although a relationship between cyclic nucleotides and control of lymphocyte recirculation has thus far not been made. With respect to reaginic antibody formation, Tada and his co-workers (Tada et al., 1971; Okumura and Tada, 1971; Taniguchi and Tada, 1971) have shown that a suppressor population of lymphocytes is probably involved in the regulation of this type of antibody response. Thus the redistribution of lymphocytes with a loss of suppressor control may be the primary effect rather than stimulation of IgE-producing cells.

The B. pertussis hemagglutinin can now be identified as pili. Although Sato and his co-workers (1973) have suggested that HA and LPF are the same, we have clearly separated the two and there is no biological or immunological relationship between them (Morse and Morse, 1974, 1976). Sato et al. (1974) have also stated that protection against intracerebral challenge in mice, the standard method of assaying vaccine potency, can be induced by their HA–LPF fraction. Preliminary studies in our laboratory using crude HA, free of LPF, have shown protection to be induced by HA. As a historical note, Keogh and North (1948) suggested that HA was a protective antigen but neither Pillemer (1950) nor Masry (1952) could confirm this finding.

HLT remains an unknown with respect to its nature and mode of action. It is true that some bacterial toxins, e.g., the α toxin of *Staphylococcus aureus*, are hemolytic, as well as lethal, and one can speculate that the hemolysin of *B. pertussis* and HLT might be related.

An unanswered question is whether after one excludes the adjuvant effect of pertussis endotoxin and the enhanced production of reaginic antibody by LPF, there is an additional bacterial component with adjuvanticity.

XI.   Pathogenesis of Pertussis

Purposefully eliminated thus far from this discussion is the most important biological property of *B. pertussis* in the context of human health, the ability to induce disease with a high morbidity and significant mortality in infants. A corollary property is the capacity of appropriately prepared whole cell vaccines or soluble extracts to induce protective immunity.

Pertussis is a disease in which the organisms do not invade tissue, whether the infection is of the tracheobronchial tree as in the natural disease in man, or of the cerebral ventricles as in experimental infection in mice. Both are prevented by active immunization, yet the "protective antigen" is not known.

It is obvious that certain of the manifestations of whooping cough are due to biological components described here. Lymphocytosis and hypoglycemia occur and it is possible that the characteristic paroxysmal cough, which may persist long after the acute phase, is related to altered function of the β-adrenergic system. There is also some evidence that HLT may cause lesions near the site of bacterial multiplication but the natural disease is not the sort of toxicosis that is seen when *B. pertussis* is inoculated intraperitoneally in animals. Moreover, antitoxin is not protective against intracerebral infections.

It is difficult to conceive that factors such as LPF, HSF, or HLT are responsible for the primary initiation of disease. The pathogenesis of whooping cough requires that the organisms become fixed, particularly to ciliated epithelium, and multiply in an essentially unrestricted fashion during the acute phase. One might therefore expect that effective immunization is predicated upon restricted attachment, or multiplication of the organisms, or both. Clearly, the organisms do not undergo unrestricted multiplication in immunized mice challenged intracerebrally; there is a modest increase in the number of organisms for a few days and then the bacterial population decreases.

Protection might be induced by antibody acting alone to inhibit attachment by coating the bacterial surface in a nonspecific fashion or by specifically binding to pili which are known to attach the organisms to cell surfaces. Antibodies might also induce protection by causing complement-mediated bactericidal effects, but there is no correlative evidence for this thesis (Dolby and Dolby, 1969).

The role of cellular defense mechanisms has not been defined and it is possible that opsonic antibodies and phagocytic cells are important in immune protection. Very little is known of the susceptibility of B. pertussis to phagocytosis and killing either in vitro or in vivo. It is known that the capsule of phase I B. pertussis does not serve the organism in the same way that the type-specific polysaccharides of pneumococci are responsible for virulence, for concentrated B. pertussis capsular material does not induce protection (Evans and Adams, 1952).

Surely there is gradual but steady clarity emerging from recent studies on the manifold and striking effects of B. pertussis on animal tissues and cells. Delineation of the nature and mechanism of the pathogenesis of pertussis and the basis of vaccine protection appears to be lagging behind . . . but perhaps not too far behind.

References

Aprile, M. A., and Wardlaw, A. C. (1973a). Can. J. Microbiol. 19, 231–239.
Aprile, M. A., and Wardlaw, A. C. (1973b). Can. J. Microbiol. 19, 536–538.
Asakawa, S. (1969). Jpn. J. Med. Sci. Biol. 22, 23–42.
Athanassiades, T. J., and Morse, S. I. (1973). Blood 42, 611–621.
Banerjee, A., and Munoz, J. (1962). J. Bacteriol. 84, 269–274.
Bergman, R. K., and Munoz, J. (1966). Proc. Soc. Exp. Biol. Med. 122, 428–433.
Bergman, R. K., and Munoz, J. (1971). Life Sci. 10, 561–568.
Bradford, W. L., Scherp, H. W., and Tinker, M. R. (1956). Pediatrics 18, 64–70.
Cronholm, L. S., and Fishel, C. W. (1968). J. Bacteriol. 95, 1993–1999.
Dolby, J. M., and Dolby, D. E. (1969). Immunology 16, 737–747.
Dresser, D. W. (1972). Eur. J. Immunol. 2, 50–57.
Evans, D. G., and Adams, M. O. (1952). J. Gen. Microbiol. 7, 169–174.
Fishel, F. W., and Szentivanyi, A. (1963). J. Allergy 34, 439–454.
Fishel, C. W., Szentivanyi, A., and Talmage, D. W. (1962). J. Immunol. 89, 8–18.
Fishel, C. W., Keller, K. F., and O'Bryan, B. S. (1968). J. Immunol. 101, 679–686.
Floersheim, G. L. (1967). Nature (London) 216, 1235–1236.
Fröhlich, J. (1897). J. Kinderkrankh. 44, 53–58.
Gesner, B. M., and Gowans, J. L. (1962). Br. J. Exp. Pathol. 43, 424–430.
Gowans, J. L. (1959). J. Physiol. (London) 146, 54–69.
Gowans, J. L., and Knight, E. J. (1964). Proc. R. Soc. London, Ser. B 159, 257–282.
Gulbenkian, L. S., Schobert, L., Nixon, C., and Tabachnick, I. I. A. (1968). Endocrinology 83, 885–892.
Homma, R., Shimazaki, Y., Funasaka, I., and Kuratsuka, K. (1969). Jpn. J. Med. Sci. Biol. 22, 123–126.
Iwasa, S., Yoshikawa, T., Fukumura, K., and Kurokawa, M. (1970). Jpn. J. Med. Sci. Biol. 23, 47–60.

Keller, K. F., and Fishel, C. W. (1967). *J. Bacteriol.* 94, 804–811.
Keogh, E. V., and North, E. A. (1948). *Aust. J. Exp. Biol. Med. Sci.* 26, 315–322.
Kind, L. S. (1956). *J. Immunol.* 77, 115–118.
Kind, L. S. (1958). *Bacteriol. Rev.* 22, 173–182.
Kong, A. S., and Morse, S. I. (1975). *Fed. Proc., Fed. Am. Soc. Exp. Biol.* 34, 951.
Kong, A. S., and Morse, S. I. (1976). *J. Immunol.* 116, 989–993.
Lagergren, J. (1963). *Acta Paediat. Scand.* 52, 405–409.
Lehrer, S. B., Tan, E. M., and Vaughan, J. H. (1974). *J. Immunol.* 113, 18–26.
Lehrer, S. B., Vaughn, J. H., and Tan, E. M. (1975). *J. Immunol.* 114, 34–39.
Levine, S., Wenk, E. J., Devlin, H. B., Pieroni, R. E., and Levine, L. (1966). *J. Immunol.* 97, 363–368.
Likhite, V. (1974). *Cancer Res.* 34, 1027–1030.
Linnemann, C. C., Jr., Perlstein, P. H., Ramundo, N., Minton, S. D., Englender, G. S., McCormick, J. B., and Hayes, P. S. (1975). *Lancet* 2, 540–543.
MacLennan, A. P. (1960). *Biochem. J.* 74, 398–409.
Maillard, J., and Bloom, B. R. (1972). *J. Exp. Med.* 136, 185–190.
Masry, F. L. G. (1952). *J. Gen. Microbiol.* 7, 201–210.
Morse, J. H., and Morse, S. I. (1970). *J. Exp. Med.* 131, 1342–1357.
Morse, S. I. (1965). *J. Exp. Med.* 121, 49–68.
Morse, S. I., and Adler, A. (1973). *Infect. Immun.* 7, 461–467.
Morse, S. I., and Barron, B. A. (1970). *J. Exp. Med.* 132, 663–672.
Morse, S. I., and Bray, K. K. (1969). *J. Exp. Med.* 129, 523–558.
Morse, S. I., and Morse, J. H. (1974). *Fed. Proc., Fed. Am. Soc. Exp. Biol.* 33, 763.
Morse, S. I., and Morse, J. H. (1976). *J. Exp. Med.* 143, 1483.
Morse, S. I., and Riester, S. K. (1967a). *J. Exp. Med.* 125, 401–408.
Morse, S. I., and Riester, S. K. (1967b). *J. Exp. Med.* 125, 619–628.
Mota, I. (1958). *Nature (London)* 182, 1021–1022.
Mota, I. (1964). *Immunology* 7, 681–699.
Mota, I., and Peixoto, J. M. (1966). *Life Sci.* 5, 1723–1728.
Munoz, J. (1971). *Microb. Toxins* 2A, 271–300.
Munoz, J., and Bergman, R. K. (1966). *J. Immunol.* 97, 120–125.
Munoz, J., and Bergman, R. K. (1968). *Bacteriol. Rev.* 32, 103–126.
Munoz, J., and Hestekin, B. M. (1962). *Nature (London)* 196, 1192–1193.
Murgo, A. J., and Athanassiades, T. J. (1975). *J. Immunol.* 115, 928–931.
Nakase, Y., Tateisi, M., Sekiya, K., and Kasuga, T. (1970). *Jpn. J. Microbiol.* 14, 1–8.
Ochai, T., Okumura, K., Tada, T., and Iwasa, S. (1972). *Int. Arch. Allergy Appl. Immunol.* 43, 196–206.
Okumura K., and Tada, T. (1971). *J. Immunol.* 106, 1019–1025.
Ortez, R. A., Seshachalam, D., and Szentivanyi, A. (1975). *Biochem. Pharmacol.* 24, 1297–1302.
Ovary, Z., and Caiazza, S. S. (1975). *Int. Arch. Allergy Appl. Immunol.* 48, 11–15.
Parfentjev, I. A., and Goodline, M. A. (1948). *J. Pharmacol. Exp. Ther.* 92, 411–413.
Parfentjev, I. A., and Schleyer, W. L. (1949). *Arch. Biochem.* 20, 341–346.
Parfentjev, I. A., Goodline, M. A., and Virion, M. E. (1947a). *J. Bacteriol.* 53, 597, 601.
Parfentjev, I. A., Goodline, M. A., and Virion, M. E. (1947b). *J. Bacteriol* 53, 603–611.
Parfentjev, I. A., Goodline, M. A., and Virion, M. E. (1947c). *J. Bacteriol.* 53, 613–619.
Parker, C. W., and Morse, S. I. (1973). *J. Exp. Med.* 137, 1078–1090.
Pieroni, R. E., and Levine, L. (1966). *Nature (London)* 211, 1419–1420.

Pieroni, R. E., Broderick, E. J., and Levine, L. (1965). *J. Immunol.* **95**, 643–650.

Pillemer, L. (1950). *Proc. Soc. Exp. Biol. Med.* **75**, 704–705.

Pittman, M. (1970). *In* "Infectious Agents and Host Reactions" (S. Mudd, ed.), pp. 239–270. Saunders, Philadelphia, Pennsylvania.

Sata, Y., and Arai, H. (1972). *Infect. Immun.* **6**, 899–904.

Sata, Y., Arai, H., and Suzuki, K. (1973). *Infect. Immun.* **7**, 992–999.

Sato, Y., Arai, H., and Suzuki, K. (1974). *Infect. Immun.* **9**, 801–810.

Stronk, M. G., and Pittman, M. (1955). *J. Infect. Dis.* **96**, 152–161.

Szentivanyi, A., Fishel, C. W., and Talmage, D. W. (1963). *J. Infect. Dis.* **113**, 86–98.

Szentivanyi, A., Katsh, S., and McGarry, B. (1968). *Fed. Proc., Fed. Am. Soc. Exp. Biol.* **27**, 294.

Tada, T., Taniguichi, M., and Okumura, K. (1971). *J. Immunol.* **106**, 1012–1018.

Tada, T., Okumura, K., Ochiai, T., and Iwasa, S. (1972). *Int. Arch. Allergy Appl. Immunol.* **43**, 207–216.

Taniguchi, M., and Tada, T. (1971). *J. Immunol.* **107**, 579–585.

Taub, A. N., Rosett, W., Adler, A., and Morse, S. I. (1972). *J. Exp. Med.* **136**, 1581–1593.

Tuta, J. A. (1937). *Folia Haematol.* **57**, 122–128.

Wardlaw, A. C. (1970). *Int. Arch. Allergy Appl. Immunol.* **38**, 573–589.

Wardlaw, A. C., and Jakus, C. M. (1966). *Can. J. Microbiol.* **12**, 1105–1114.

Wortis, H. (1971). *Clin. Exp. Immunol.* **8**, 305–317.

Role of the Genetics and Physiology of *Bordetella pertussis* in the Production of Vaccine and the Study of Host–Parasite Relationships in Pertussis[1]

CHARLOTTE PARKER

University of Texas at Austin, Austin, Texas

CharLOTTE Parker

University of Texas at Austin, Austin, Texas

I. Introduction 27
   A. Fastidious Nature of *Bordetella pertussis* 27
   B. Phenotypic Changes of *Bordetella pertussis in Vitro*.. 28
   C. Vaccine Production Problems due to *in Vitro* Changes of *Bordetella pertussis* 29
II. Genetics 30
   A. Genetic Hypothesis of Degraded Strains 30
   B. Bacteriocins 31
   C. Transformation 31
III. Physiology 31
   A. Growth Requirements 31
   B. Metabolism 34
   C. Growth Inhibitors 35
   D. Defined Medium 36
   References 41

I. Introduction

A. Fastidious Nature of *Bordetella pertussis*

Bordetella pertussis is a finicky, fastidious, slow-growing bacterium that is difficult to isolate in primary culture. Bordet and Gengou (1906) first isolated the organism on a glycerolated potato-extract agar medium containing 50% blood. Since that time numerous efforts have been made to devise a different medium for primary isolation, but a medium similar or identical to Bordet–Gengou (B–G) is still the most effective. If the blood content is reduced below 30%, or if peptones are incorporated into the medium, or if serum is used instead of blood, freshly isolated strains of *B. pertussis* may show no growth or inferior growth (Dawson *et al.*, 1951). The addition of penicillin to B–G (to reduce the numbers of contaminating gram-positive organisms) is the only major modification of the primary isolation medium in the past 70 years.

Bordetella pertussis will grow on media unsuitable for primary isolation as a result of several different mechanisms. One method by which growth can be obtained on suboptimal medium is by the use of a large inoculum. Media that will not support the growth of small numbers of organisms are often suitable for vaccine production, provided a large

[1] Funded by the University Research Institute, University of Texas, Austin, Texas.

inoculum is used. The broth described by Cohen and Wheeler (1946) is a widely used medium requiring an inoculum of 10^6 colony-forming units (cfu) per ml for growth of pertussis. This broth is free of blood or serum, but contains soluble starch, casein hydrolyzate, salts, and yeast dialyzate. Certain media containing charcoal are also suitable for growth of small numbers of organisms (see Rowatt, 1957b), while others are not.

B. PHENOTYPIC CHANGES OF *Bordetella pertussis* in *Vitro*

A second reason why *B. pertussis* grows well on media other than B–G agar is the rapid adaptation of the organism to *in vitro* culture conditions. *Bordetella pertussis* has been known to show rapid change in several phenotypic properties, including ability to grow on media other than B–G, since 1910 (Bordet and Sleeswyck, 1910). One can select for the ability to grow on blood agar (BA) or nutrient agar (NA) by serial passage. One of the major reasons for use of B–G in strain maintenance is that it tends to produce organisms of the original smooth, fastidious type. Passage on BA tends to select for antigenic and cultural variants. Even on B–G, fresh isolates may show rapid changes in properties. Stand-fast (1951) showed that agglutinability in specific antisera, ability to grow on BA, hemagglutination, virulence for mice, and other properties varied rapidly and independently. In these experiments Standfast passaged strains on B–G containing peptone. This medium probably accentuated the speed of variation, since peptone is inhibitory to fresh isolates.

The antigenic changes that occur with repeated passage were characterized by Leslie and Gardner (1931). They designated phase I as the very homogeneous fresh isolates, and phase IV as completely degraded, antigenically changed, avirulent strains. Phases II and III were intermediate. Since their terminology depended completely on antigenic properties, although other properties also were changed, and since a complete series of reference strains is not available, this usage is unsatisfactory. We suggest that the terms *fresh isolate, intermediate strain,* and *degraded strain* be used. Aprile (1972) found that only five of eight "phase IV" strains she obtained were *B. pertussis,* indicating that caution must be used when attempting to study degraded strains. (Strain 190 from the American Type Culture Collection, Rockville, Maryland, was shown to resemble a *Lophomonas* culture, while strains 8631 and 8632 from the National Type Culture Collection, Colindale, London, were identified as *Brucella melitensis.*)

Although Standfast (1951) showed conclusively that many properties vary independently in the dissociation process, a progression can be seen in which fresh isolates tend to be unable to grow on media other than B–G, are virulent for man, have specific agglutinating antigens, and are able to induce protective immunity. Degraded strains, on the other hand, are able to grow on ordinary medium, are usually avirulent, are unable to induce protective immunity, and have usually lost the specific agglutinating antigens of fresh strains. Intermediate strains vary in their properties.

The genetic basis of the many changes induced in B. pertussis by in vitro culture has not been studied. Variation in surface antigens has been noted by numerous workers, and surface antigens can be affected by several cultural conditions. Even if strains are not degraded, variations in composition of the medium can cause antigenic changes. Lacey (1960) showed that the salt composition of the medium and the incubation temperature exerted profound effects on surface antigens. He termed this reversible variability modulation. Pusztai and Joó (1967) showed that varying the level of nicotinic acid in the culture medium affected the antigenic properties of B. pertussis.

C. Vaccine Production Problems due to in Vitro Changes of Bordetella pertussis

Such extreme variability of surface antigens, both from reversible modulation and genetic changes leading to degradation of strains, has been a troublesome problem for vaccine manufacturers. Although strains can be tested for specific agglutinating antigens, there is no easy way to detect the presence or absence of the protective antigen(s) which is lost independently. Vaccines vary in their effectiveness, and this variation reflects differences in content of protective antigen. Since degraded strains grow better in artificial medium than do fresh isolates, a process of selection continually occurs. This selection tends to produce variants which are of no value for vaccines.

When a suitable strain of B. pertussis is chosen on the basis of potency tests, it is unlikely to retain its properties if subcultured extensively. Lyophilization and freezing at −70°C are satisfactory methods of strain preservation. Nevertheless, most vaccine strains have been maintained in vitro for many years and show differences from fresh isolates in their cultural characteristics. Even though they maintain the property of producing protective antigen, they should be considered intermediate strains. We consider that "good" vaccine strains are merely those that tend to retain protective antigen(s) through many subcultures. These strains are not suitable for defining properties of pertussis, however, as they usually differ from fresh isolates in a variety of ways.

Potency and safety testing of pertussis vaccines is difficult and expensive. The models of choice are the intracerebral mouse challenge for potency testing (Additional Standards: Pertussis Vaccine, 1968) and the mouse weight-gain test for safety testing (Pittman, 1952). Because of the variety of active substances present in pertussis vaccine, production strains must contain large amounts of the protective antigen. Otherwise, vaccines that cannot meet the safety standards may result.

For potency testing a standard challenge strain (18323) is used. Problems of variability and maintenance for this strain are similar to those for vaccine strains. This strain is clearly an intermediate strain, since it shows changes in its virulence for mice by both the intranasal and intracerebral routes, and agglutinates poorly in antiserum against fresh isolates (Pittman, 1970). Methods for cultivation and detoxification of pertussis vaccine have been developed empirically. We still lack the knowledge of the organism and its products that would allow rational experiments to produce a more protective and less reactive vaccine.

II. Genetics

A. Genetic Hypothesis of Degraded Strains

A satisfactory hypothesis for understanding the rapid changes in *B. pertussis* during serial culture is to envision a series of mutations which are independently selected for by inhibitors present in artificial medium. Changes in specific heat-labile agglutinogens occur with a high frequency, including both gain and loss mutations, and may occur independently of inhibitory substances (Stanbridge and Preston, 1974). Other mutations seem to be of the loss type; loss of protective antigen, loss of the hemagglutinin, loss of virulence, etc. The changes observed by Leslie and Gardner (1931) are analogous to a series of loss mutations; encapsulated to smooth to rough. It has been shown that the requirement for enriched medium for the isolation of *B. pertussis* is due to the need for substances that neutralize inhibitors present in the medium (see Section III,A). Earlier, fresh isolates were thought to have complex nutrient requirements, while degraded strains did not. We know now that all *B. pertussis* strains have rather simple nutritional needs, but that fresh isolates, intermediate strains, and degraded strains differ in their sensitivity to inhibitors (Rowatt, 1957b). Blood, albumen, starch, charcoal, or anion-exchange resins which have been used in culture media serve to inactivate or to sequester one or more inhibitors.

Since artificial media contain several different inhibitors (see Section III,C) there is selection in the population of *B. pertussis* cells for resistance to inhibitors. A resistant mutant, when it occurs, tends to overgrow the susceptible parental cells. Then the mutant becomes predominant and probably becomes resistant to yet another inhibitor. This hypothesis suggests that intermediate strains may show any of a variety of changes, but that changes occur in succession, leading finally to a degraded strain. Degraded strains presumably show irreversible changes because they have multiple loss mutations. The studies of Standfast (1951) and Leslie and Gardner (1931) are consistent with this hypothesis.

B. Bacteriocins

Bacteriocin activity was demonstrated in B. pertussis by Litkenhaus and Liu in 1967. They showed that an inhibitory substance produced by two rough (i.e., degraded) strains of B. pertussis showed activity against all smooth (fresh isolates and/or intermediate strains) strains tested. This substance was inactive when tested against two other rough strains, or B. parapertussis and B. bronchiseptica. It is possible that the ability to produce an inhibitory substance is gained during the process of degradation. This would suggest that some or all degraded strains have an additional advantage in artificial medium. Not only are they less sensitive to inhibition by medium components, but they also produce inhibitors which affect the growth of fresh isolates or intermediate strains. Several workers have commented on the possible production of inhibitors by B. pertussis during culture (Pollack, 1949; Sutherland and Wilkinson, 1961). It is not clear whether the bacteriocin activity is related to this phenomenon.

C. Transformation

Almost no formal genetics have been done in B. pertussis, and no genetic investigation of variation has been possible. Transformation of streptomycin resistance was demonstrated by Branefors in 1964 using a single strain. Theoretically, a study of inhibitor-resistant mutants could be performed to test the validity of the explanation we have given for the rapid variation occurring in vitro, but this study has not yet been undertaken.

Fresh isolates of B. pertussis have been demonstrated to be piliated (Morse and Morse, 1970). In addition to the importance of such cell-surface antigens in variation of antigenic characters, piliation may be important in transformation. In Neisseria gonorrhea, loss of piliation is accompanied by both lowered virulence and decreased ability to be transformed (Sparling, 1966; Catlin, 1974).

III. Physiology

A. Growth Requirements

We have some rudimentary understanding of the metabolism and growth of B. pertussis, but a detailed knowledge has yet to be established.

In early experiments, it was assumed that pertussis had exacting growth requirements. Later studies suggested (Hornibrook, 1939) and then demonstrated (Jebb and Tomlinson, 1955) that *B. pertussis* can grow in liquid medium containing only a few amino acids, nicotinamide, and salts, from a small inoculum.

In discussing any experiments concerned with growth requirements of *B. pertussis*, certain problems of interpretation occur. First, it must be determined whether the required substance serves as a nutrient, or merely to neutralize an inhibitory substance; second, small inocula must be used; third, fresh isolates (not merely laboratory strains) must be tested. The numerous discrepancies one finds in the literature concerning pertussis nutrition can usually be explained by one or more of these factors. A fourth problem is the unique inhibition by agar itself. As discussed in Section III,C, agar shows strong inhibitory properties. Therefore work done in broth cannot ordinarily be repeated on agar-containing medium (Verwey *et al.*, 1949). This means that all the physiological studies to date have been performed on population of cells. Since changes occur during growth *in vitro*, the use of solid media has obvious advantages, as individual clones can be studied, but such studies have not yet been done.

Using the simplest defined medium developed so far, Stainer and Scholte (1971) have shown that only two amino acids are required by *B. pertussis*. One must be either cystine or cysteine, in agreement with the results of Jebb and Tomlinson (1957). Glutamic acid or proline (or presumably glutamine) can then be added to obtain excellent growth. This process also agrees with earlier work (Fukumi *et al.*, 1953; Vajdic *et al.*, 1966). Thus pertussis shows an absolute requirement for cystine or cysteine, but can use any one of several other amino acids as the primary source of energy, carbon, and nitrogen. (The need for a sulfhydryl animo acid presumably satisfies the sulfur requirement, but may in fact be related to inhibitors or to the reducing agents required by *B. pertussis*.)

Bordetella pertussis can metabolize several additional amino acids, but it is not known whether any of these can serve as a replacement for glutamic acid. Serine, alanine, aspartic acid, asparagine, and possibly histidine and glycine are used appreciably by growing cells. The sulfur-containing tripeptide glutathione is also used up during growth. (Lane, 1970; see also Jebb and Tomlinson, 1955; Meyer and Cameron, 1957; Goldner *et al.*, 1966). Other amino acids appear not to be attacked.

Nicotinamide has been known to be a required vitamin for *B. pertussis* since early work by Hornibrook (1940). Other vitamin or vitaminlike compounds needed for growth include ascorbic acid and glutathione. Since Lane (1970) showed that glutathione is used up, one supposes it is not acting catalytically. But it should be considered with the vitamins because it is present in small amounts. In Stainer and Scholte's (1971) liquid medium, both ascorbic acid and glutathione are required for maximum cell yield. We found that when agarose solidified minimal medium was used (see Section III,D), ascorbic acid was not required for growth of small inocula from fresh isolates. Glutathione was required for growth of small inocula, but not for gowth when a heavy inoculum was used. Since both ascorbic acid and glutathione are reducing agents, they may be needed for the provision of reducing power, rather than as vitamins, even if required.

The very high cell yields obtained by Stainer and Scholte (1971) in their liquid medium indicate that their salts mixture is adequate. However, they use large inocula in their experiments. Studies of the effects of salts on the growth of *B. pertussis* are not common. Cohen and Wheeler (1946) systematically examined salts concentrations in the formulation of their medium. Most other workers have adapted Cohen and Wheeler's salts mixture. The salts and buffer composition of three media—that of Cohen and Wheeler (1946), of Wilson (1963), and of Stainer and Scholte (1971)—is given in Table I. Notice that Wilson's salt composition is adapted from that of Cohen and Wheeler. The major change made by Wilson was the addition of a Tris [tris(hydroxymethyl)aminomethane] buffer. Stainer and Scholte then used Wilson's formulation, but omitted the copper sulfate.

Added iron has been used in liquid culture media since Cohen and Wheeler (1946) reported that it enhanced growth. Whether iron is a required nutrient or merely serves to detoxify peroxides is not known (Rowatt, 1957b). Differences in growth response can be noted when iron is filter sterilized and added separately to the medium, rather than auto-

TABLE I

SALTS AND BUFFER CONTENT OF VARIOUS MEDIA

Component (mg/liter)	Cohen and Wheeler (1946)	Wilson (1963)	Stainer and Scholte (1971)
NaCl	2500	2500	2500
KH_2PO_4	500	500	500
KCl	—	200	200
$MgCl_2 \cdot 6H_2O$	400	100	100
$CaCl_2$	10	20	20
$FeSO_4 \cdot 7H_2O$	10	10	10
$CuSO_4 \cdot 5H_2O$	0.5	5	—
Tris(hydroxymethyl) aminomethone	—	6075	6075[a]

[a] Tris has now been reduced to 1519 mg/liter (D. W. Stainer, personal communication).

claved. Magnesium sulfate was used to replace sodium chloride in some
instances (Lacey, 1960). This change induced extreme antigenic variation
in cultures of *B. pertussis,* but was reversed by culture on the usual
sodium chloride medium. The exact nature of these changes is unclear,
but it is evident that salts exert a profound effect on the growth of
pertussis.

Manganese is also of some importance. Cameron and Dunning (J.
Cameron, personal communication) showed that brands of agar vary in
their manganese content. Cohen and Wheeler medium, solidified with
agar, will support growth from an inoculum of 10 organisms if low
manganese (Difco) agar is used. However, 10^6 or more cells are required
to initiate growth on the same medium solidified with high manganese
(Oxoid) agar. The manganese effect can be overcome by adding red
blood cells, or heme, but not by adding ferric citrate.

Potassium chloride, calcium chloride, magnesium chloride, and sodium
chloride seem to be important in keeping cells dispersed in liquid culture
(Wilson, 1963) and their concentrations have been modified in various
liquid media. It is important to note that casein hydrolyzate contains an
appreciable quantity of sodium chloride.

Buffers are not necessary for growth, but they play a large role in the
final cell yield of *B. pertussis.* During growth, pertussis produces ammonia
from amino acids, and the pH of the medium rises. Thus, although a pH
of 7.1–7.3 is optimum, cultures that are not buffered often reach a final
pH of 8.0 (Rowatt, 1957b). Some workers (Rowatt, 1957b) used a starting pH as low as 6.9 to insure maximum yield in unbuffered medium.
Since high pH can inhibit further growth, it is possible that pH effects
have been involved in some instances in determining growth requirements.

Reducing agents, such as ascorbic acid, glutathione, and ferrous ions
may play an important role for growth in defined medium. Although it
has not been determined why these substances are stimulatory, perhaps
their role is due to an absolute requirement, or to their reducing potential.

Oxygen is required by *B. pertussis,* but whether carbon dioxide is required is unknown. Oxygen is apparently used in the oxidation of glutamic acid, the preferred energy source. Glutamic acid seems to be the
preferred carbon, nitrogen, and energy source (Jebb and Tomlinson,
1951).

B. METABOLISM

The metabolism of *B. pertussis* is based on the oxidation of amino acids
to provide energy and raw materials. Carbohydrates appear not to be
attacked (Jebb and Tomlinson, 1951). The first step in oxidation of glutamic acid is deamination, yielding α-ketoglutarate. α-Ketoglutarate accu-

mulates in the presence of arsenious oxide. When α-ketoglutarate was tested as a substrate, it was also oxidized (Jebb and Tomlinson, 1951). Succinate and lactate appeared to show a rapid initial rate of oxidation, but pyruvate, oxaloacetate, fumarate, malate, and acetate did not greatly alter the rate of oxygen uptake of cells (Jebb and Tomlinson, 1951; Proom, 1955). Although glutamic acid (or proline or glutamine) appears to be the preferred substrate for growth, several other amino acids are also oxidized. Serine, alanine, and glycine are reported to be rapidly used (Lane, 1970). These amino acids may be capable of serving as the energy source for growth, although glutamate is always preferentially used if present. Cystine or cysteine is required, but its function in pertussis metabolism is unknown.

Since it was shown that ammonia and carbon dioxide are produced from glutamate, and ammonia is produced from other amino acids, we know that the first step in the breakdown of amino acids is oxidative deamination (Rowatt, 1955). How the resulting keto acids are used, whether there is an intact citric acid cycle, how carbohydrates for nucleic acid and for cell wall biosynthesis are made—these questions, and others, remain unanswered. We know that some sort of control mechanisms occur, as the oxidation of other amino acids is slow until glutamate is used (Vajdic et al., 1966).

One can speculate that oxygenase enzymes may play a role in the amino acid metabolism of B. pertussis. This class of enzymes fixes molecular oxygen, rather than removing hydrogen, and often shows a requirement for ascorbic acid and ferrous ions. The ferrous ion–ascorbic acid dependent oxygenases are also dependent on α-ketoglutarate (Abbot and Udenfriend, 1974). The dual requirements of B. pertussis for oxygen and for reduced compounds such as ferrous sulfate, ascorbic acid, and glutathione suggest that oxygenases may play a role in metabolism. Further research on metabolism is needed to determine whether this is true.

Only a few attempts have been made to manipulate the growth of pertussis vaccine strains for vaccine production. The minimal medium described by Stainer and Scholte (1971) is a notably successful effort. Lane (1970) supplemented Cohen and Wheeler broth with monosodium glutamate, and was successful in increasing the cell yield for vaccine production.

C. Growth Inhibitors

Pollack (1947) showed that an intermediate strain of B. pertussis which grew well on BA was sensitive to inhibition by fatty acids. Albumen in blood binds fatty acids, thus neutralizing the inhibitory effect. Agar, peptones, meat extracts, cotton plugs, and other media constituents may contain fatty acids. The use of clean glassware with metal caps will solve the problem with the volatile fatty acids from cotton plugs, but the fatty acids in other substances are not easy to remove. Extraction with organic solvents is not always successful, perhaps because other inhibitors remain. Starch, charcoal, and anion-exchange resins have been used to counter the effects of inhibitors (Rowatt, 1957b; Sutherland and Wilkinson, 1961). Presumably all of these agents, as well as serum proteins, are able to neutralize fatty acids.

Autoclaved cysteine, presumably containing colloidal sulfur compounds, is inhibitory to pertussis. Autoclaved cysteine inhibition can be neutralized by albumen, filter-sterilized cysteine, activated charcoal, or by large numbers of red blood cells (Proom, 1955; Rowatt, 1957a). Agar contains large amounts of sulfur. Therefore it is possible that the notorious inhibitory effect of agar on B. pertussis is due to colloidal sulfur compounds formed during autoclaving. Casein hydrolyzate or other nutrient sources may also form inhibitory sulfur compounds when autoclaved.

Another type of inhibition, called "peroxide" inhibition, was first suggested by Mazloum and Rowley (1955). Rowatt (1957a) suggested that this inhibition was due to organic peroxides formed in media during autoclaving, rather than to hydrogen peroxide. Rowatt also showed that small amounts of lysed red blood cells, hemin, or filtered (but not autoclaved) ferrous sulfate could neutralize this inhibitor. It could be differentiated from sulfur-containing inhibitors because of the much smaller amount of red blood cells required to neutralize the peroxide inhibitor.

The inhibition by manganese in agar media, described by Cameron and Dunning (see Section III,B), is presumably different from Rowatt's peroxide inhibition. Rowatt used only liquid medium, prepared each batch from the same ingredients, and found differences in inhibition from batch to batch. However, the possibility that manganese is the third inhibitor described by Rowatt in her review (Rowatt, 1957a) cannot be excluded.

D. Defined Medium

Stainer and Scholte liquid medium is completely defined, free of complex materials such as starch or agar, and supports the growth of B. pertussis vaccine strains through serial culture. Large inocula are used. This medium contains only salts, amino acids, vitamins, glutathione, and Tris buffer. Its usefulness has been confirmed in other laboratories (Novotny and Brooks, 1975; C. Parker, unpublished results). Although the medium maintains viability for numerous subcultures, the level of protective antigen falls in cells passaged more than eight times in this medium. Thus it is not the perfect medium for vaccine production and strain maintenance. It is, however, suitable for many physiologic investigations, and must be regarded as a major step forward in the study of B. pertussis.

In our laboratory, we have developed a solid medium, based on the Stainer and Scholte liquid medium. A defined solid medium has numerous advantages for genetic studies. Such a medium makes it possible to isolate mutants resistant to inhibitors, to isolate mutants deficient in amino acid metabolism, or to select for recombinants of various types after transformation. Therefore, we have attempted to solidify the defined liquid medium with various grades of agar, i.e., chloroform-extracted agar, and the like. These media would not support the growth of small inocula of B. pertussis. In addition, we have used agarose to solidify the defined medium. Agarose is a nonsulfur-containing polysaccharide obtained from agar. [Agaropectin, the other main component of agar, contains up to 9% sulfate (Araki, 1959). Agarose is used for gels because of its low electro-osmosis, since it has few or no charged components.] The agarose-containing medium, designated minimal solid medium (MSM), successfully supported the growth of B. pertussis from small inocula. The composition of the medium is given in Table II.

The components of the medium are identical to those used by Stainer and Scholte for their liquid medium, with the following additions:

1. We used both proline and glutamic acid because preliminary experiments showed that this combination gave the heaviest growth. No controlled experiments on the effects of proline–glutamine mixtures have been performed.

2. We added sodium acetate because Bundealy and Rao (1965) found it stimulatory for pertussis strains when used in small amounts.

3. We added Dowex-1 chloride, 8% cross-linked, in an attempt to remove any inhibitors remaining in the medium, or any produced during growth of the organisms (Sutherland and Wilkinson, 1961).

4. We solidified the medium with 1% agarose (Bio-Rad).

This medium was then tested in a number of ways for its ability to support growth of colonies when a small inoculum was used. The inoculum was grown on B–G slants, diluted into defined liquid medium, and plated to give isolated colonies on B–G plates. Such experiments showed the following results:

1. MSM supported the growth of both vaccine strains (intermediate strains) and fresh isolates. Some of the fresh isolates tested were obtained from Dr. Larry Baroff, USC Medical Center, Los Angeles, California. We obtained the others during a local outbreak of whooping cough. In one case, we were able to culture a positive nasopharyngeal swab directly on MSM, and isolate B. pertussis.

2. Nicotinamide and glutathione were required for formation of colonies from small inocula.

TABLE II
COMPOSITION OF MINIMAL SOLID MEDIUM[a]

Component	Molarity	Mg/liter
Salts		
NaCl	4.3×10^{-2}	2500
KH$_2$PO$_4$	3.7×10^{-3}	500
KCl	2.6×10^{-3}	200
MgCl$_2$·6H$_2$O	4.9×10^{-4}	100
CaCl$_2$	1.8×10^{-4}	20
FeSo$_4$·7H$_2$O	3.6×10^{-5}	10
Amino acids		
L-Proline	2.0×10^{-2}	2400
L-Glutamic acid	4.6×10^{-3}	670
Glutathione (reduced)	3.3×10^{-4}	100
L-Cystine	1.7×10^{-4}	40
Vitamins		
Ascorbic acid	1.1×10^{-4}	20
Nicotinamide	3.3×10^{-5}	4
Buffer		
Tris(hydroxymethyl)aminomethone	5.0×10^{-2}	6075
Other		
Sodium acetate	2.4×10^{-3}	200
Dowex-1 chloride, 8% cross-linked	—	1000
Agarose	—	10,000

[a] Ferrous sulfate, ascorbic acid, cystine, nicotinamide, and sodium acetate were filter sterilized and added separately to the molten medium. Dowex-1 chloride was prepared by washing with 1 M NaOH, washing extensively with water, and autoclaving in water. Reduced glutatione was dissolved in water, filter sterilized, and added to the molten medium immediately prior to pouring plates. Powdered Bio-Rad agarose was dissolved by gentle heating at less than 95°C in sterile distilled water and mixed with the other ingredients without further treatment.

3. Neither ferrous sulfate, ascorbic acid, nor Dowex was required for the formation of colonies from small inocula.

4. The total number of cfu on B–G and MSM was comparable. Furthermore, the growth rate was similar, and colonies reached the same approximate diameter after 7 days incubation (see Fig. 1).

5. Sometimes colonies showed areas resembling central autolysis (see Fig. 1).

6. No growth was obtained when cystine was omitted from the medium, but cysteine at one-fourth the molar ratio could replace cystine.

These results suggest that this medium is suitable for extensive work on the genetics and physiology of *B. pertussis*. The most serious problem

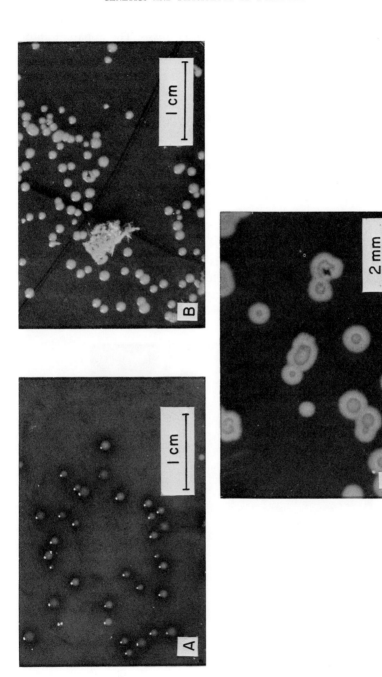

Fig. 1. *Bordetella pertussis* colonies. (A) Strain 10536 on B–G agar, 7 days incubation at 37°C. (B) Strain 10536 on MSM agar, 7 days incubation at 37°C. Note that colony size is similar to that on B–G. (C) Strain USC 70 on MSM agar without Dowex, 7 days incubation at 37°C. Note the central areas which resemble zones of autolysis.

with this medium is the cost of the agarose. We hope to continue to improve the medium for use in studying *B. pertussis*. The apparent autolysis seen on old plates supports our feelings that this medium can still be improved.

We have also done experiments of another type. By spreading plates with cells suspended in liquid medium, we obtained lawns of growth. We have tested a few compounds by placing them in wells cut in the MSM. Figure 2 demonstrates the inhibition of *B. pertussis* by lauric acid. We plan to use this technique to study various substances for their inhibitory, stimulatory, or inhibitor-neutralizing properties. This technique might also be adapted for diffusion-type antibiotic susceptibility tests. Unfortunately, blank paper disks are inhibitory by themselves. Therefore, unless we can find an absorbent, but noninhibitory material, we must use wells in the agar or cups on the agar.

We believe that MSM is relatively free from inhibitors. We plan, therefore, to attempt to isolate mutants resistant to fatty acids, etc., on this medium. We also plan to characterize fresh isolates, intermediate strains, and degraded strains by their response to inhibitors.

We have used MSM (solidified with 0.5% agarose) as a holding medium

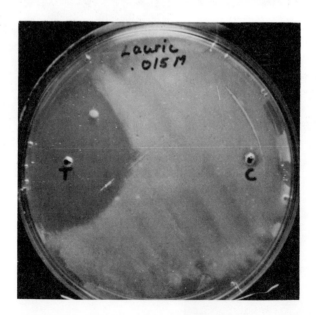

FIG. 2. Inhibition of *Bordetella pertussis* by lauric acid. A plate of MSM was uniformly inoculated with *B. pertussis*, strain 10536, using cells grown on B–G. A calcium alginate swab was used to spread the inoculum. Small wells were cut in the MSM. To the test well (T) 0.01 ml of 0.15 *M* lauric acid in ethanol was added. To the control well (C) 0.01 ml of ethanol was added.

for nasopharyngeal swabs from whooping cough cases. It appears to be somewhat better than casein hydrolyzate as a holding medium. We are continuing with these experiments. Development of a more effective holding medium would aid in the clinical study of pertussis, as specimens could be sent to a central laboratory for culturing.

We hope that this medium will be suitable for genetic studies, utilizing the transformation system of pertussis. If the MSM medium is to prove useful, it must be utilized primarily in research. The cost of agarose prohibits its use for vaccine production. Nevertheless, genetic and metabolic studies can be performed in liquid medium or on MSM, and the information obtained may be vital for vaccine improvement. It may also allow careful studies on host–parasite interactions. Does passage *in vivo* tend to reverse the mutations that occur in serial *in vitro* culture? We suspect that it does. Perhaps defined solid medium will allow us to investigate this and other questions related to the pathogenesis of whooping cough.

ACKNOWLEDGMENTS

We thank Dr. Charles Manclark for his suggestions and encouragement in both the research and the writing of this article. Persons who worked in our laboratory on this project include two undergraduate students, Mrs. Donnamae Pazirandeh and Mr. Richard Dalton. Much of the original work described was done by Mrs. Leanne Field. Dr. Larry Baroff (USC Medical Center, Los Angeles), Mrs. Bertie Pittman (CDC, Atlanta, Georgia), and Dr. Charles Manclark (FDA, Bethesda, Maryland) generously provided cultures.

REFERENCES

Abbott, M. T., and Udenfriend, S. (1974). *In* "Molecular Mechanisms of Oxygen Activation" (O. Hayashi, ed.), p. 167. Academic Press, New York.

Additional Standards: Pertussis Vaccine. (1968). *Fed. Regist.* 33, 8818.

Aprile, M. A. (1972). *Can. J. Microbiol.* 18, 1793.

Araki, C. (1959). *Proc. Int. Congr. Biochem., 4th, 1958,* V. 1, p. 15.

Bordet, J., and Gengou, O. (1906). *Ann. Inst. Pasteur, Paris* 20, 731.

Bordet, J., and Sleeswyk (1910). *Ann. Inst. Pasteur, Paris* 24, 476.

Branefors, P. (1964). *Acta Pathol. Microbiol. Scand.* 62, 249.

Bundealy, A. E., and Rao, S. (1965). *Indian J. Exp. Biol.* 4, 23.

Catlin, B. W. (1974). *J. Bacteriol.* 120, 203.

Cohen, S. M., and Wheeler, M. W. (1946). *Am. J. Public Health* 36, 371.

Dawson, B., Farnworth, E. H., McLeod, J. W., and Nicolson, D. E. (1951). *J. Gen. Microbiol.* 5, 408.

Fukumi, H., Sayama, E., Tomizawa, J., and Uchida, T. (1953). *Jpn. J. Med. Sci. Biol.* 6, 587.

Goldner, M., Jakus, C. M., Rhodes, H. K., and Wilson, R. J. (1966). *J. Gen. Microbiol.* 44, 439.

Hornibrook, J. W. (1939). *Public Health Rep.* 54, 1847.

Hornibrook, J. W. (1940). *Proc. Soc. Exp. Biol.* 45, 598.

Jebb, W. H., and Tomlinson, A. H. (1951). *J. Gen. Microbiol.* 5, 951.

Jebb, W. H., and Tomlinson, A. H. (1955). *J. Gen. Microbiol.* 13, 1.

Jebb, W. H., and Tomlinson, A. H. (1957). *J. Gen. Microbiol.* **17**, 59.

Lacey, B. W. (1960). *J. Hyg.* **58**, 57.

Lane, A. G. (1970). *Appl. Microbiol.* **19**, 512.

Leslie, P. H., and Gardner, A. D. (1931). *J. Hyg.* **31**, 423.

Litkenhaus, C., and Liu, P. V. (1967). *J. Bacteriol.* **93**, 1484.

Mazloum, H. A., and Rowley, D. (1955). *J. Pathol. Bacteriol.* **70**, 439.

Meyer, M. E., and Cameron, H. S. (1957). *J. Bacteriol.* **13**, 158.

Morse, J. H., and Morse, S. I. (1970). *J. Exp. Med.* **131**, 1342.

Novotny, P., and Brooks, J. E. (1975). *J. Biol. Stand.* **3**, 11.

Pittman, M. (1952). *J. Immunol.* **69**, 201.

Pittman, M. (1970). *In* "Infectious Agents and Host Reactions" (S. Mudd, ed.), p. 239. Saunders, Philadelphia, Pennsylvania.

Pollack, M. R. (1947). *Br. J. Pathol.* **28**, 295.

Pollack, M. R. (1949). *Symp. Soc. Exp. Biol.* **3**, 193.

Proom, H. (1955). *J. Gen. Microbiol.* **12**, 63.

Pusztai, Z., and Joó, I. (1967). *Ann. Immunol. Hung.* **10**, 63.

Rowatt, E. (1955). *J. Gen. Microbiol.* **13**, 552.

Rowatt, E. (1957a). *J. Gen. Microbiol.* **17**, 279.

Rowatt, E. (1957b). *J. Gen. Microbiol.* **17**, 297.

Sparling, P. F. (1966). *J. Bacteriol.* **92**, 1364.

Stainer, D. W., and Scholte, M. J. (1971). *J. Gen. Microbiol.* **63**, 211.

Stanbridge, T. N., and Preston, N. W. (1974). *J. Hyg.* **73**, 305.

Standfast, A. F. B. (1951). *J. Gen. Microbiol.* **5**, 531.

Sutherland, I. W., and Wilkinson, J. F. (1961). *J. Pathol. Bacteriol.* **82**, 431.

Vajdic, A. H., Goldner, M., and Wilson, R. J. (1966). *J. Gen. Microbiol.* **44**, 445.

Verwey, W. F., Thiele, E. H., Sage, D. N., and Schuhardt, L. F. (1949). *J. Bacteriol.* **58**, 127.

Wilson, R. J. (1963). *Can. J. Public Health* **54**, 518.

Problems Associated with the Development and Clinical Testing of an Improved Pertussis Vaccine

GEORGE R. ANDERSON

Division of Biologic Products, Bureau of Disease Control and Laboratory Services, Michigan Department of Public Health, Lansing, Michigan

I.	Do We Need an Improved Pertussis Vaccine?	44
	A. The Current Vaccines	44
	B. The Safety of the Current Pertussis Vaccines	45
	C. The Efficacy of the Current Vaccines	46
	D. The Cost of the Current Pertussis Vaccine	47
	E. The Acceptability of the Current Pertussis Vaccines in Clinical Use	48
II.	Why Don't We Have an Improved or Fully Acceptable Pertussis Vaccine Now?	48
III.	Where Do We Go From Here with the Development of an Improved Pertussis Vaccine?	51
	A. A Reexamination of the Current Vaccines	52
	B. Initiate Small-Scale Clinical Trials in Humans with Prototype "Purified" Vaccines	53
	C. Develop a Simple *in Vitro* Procedure for Detecting and Quantitating Protective Antibody	54
IV.	Can We Really Expect to Get New Pertussis Vaccines Approved, Marketed, and in General Distribution?	54
	References	55

Vaccines that are effective for the control of whooping cough have been used on large scale for over a quarter of a century and yet we know very little more about these vaccines now than was known 35 years ago when Kendrick and Eldering (1939) reported the first well-documented field trial in support of the efficacy of pertussis vaccine.

In the last 25 years vaccine manufacturers, government research funding agencies, and other groups have invested thousands of dollars in pertussis vaccine research and we still do not have a pertussis vaccine that would be considered by all responsible parties as being fully acceptable for human use.

This situation has to be disturbing to the users and recipients of pertussis vaccine, to the government regulatory agencies that approve the present vaccines, and to the hundreds of researchers and vaccine producers who have invested so much time and effort in attempting to develop a fully acceptable pertussis vaccine.

It is the purpose of this contribution to reassess the current status of pertussis vaccines, to describe the major problems that confront us in attempting to develop an improved vaccine, and to propose some approaches that might help bring some order to the chaos that now exists.

We can focus on the subject most clearly by asking four questions: Do we need an improved pertussis vaccine? Why don't we have an improved or fully acceptable pertussis vaccine now? Where do we go from here with the development of an improved pertussis vaccine? Can we really expect to get an improved pertussis vaccine to the marketplace?

I. Do We Need an Improved Pertussis Vaccine?

In order to answer this question it is necessary to briefly review the current pertussis vaccines.

A. THE CURRENT VACCINES

Three basic types of pertussis vaccine are being used in routine immunization. These include: (1) inactivated whole cell vaccines containing *B. pertussis* derived from solid culture media; (2) inactivated whole cell vaccines derived from liquid culture media; and (3) partially purified vaccines prepared from extracts of *B. pertussis* organisms derived from liquid culture media.

The inactivated whole cell vaccines containing organisms derived from solid and liquid culture media are similar to each other in many respects, with the prime differences being (1) the method of inactivation (heat or chemical), (2) the type of medium used for cultivation of the organisms (some are grown on agar media with animal blood, some on semisynthetic culture media, and some are produced from chemically defined media), (3) the strains of *B. pertussis* used in the vaccine, and (4) whether or not an adjuvant is included with the final vaccine.

Certain advantages and disadvantages have been claimed for each of these vaccines. The purported advantages and disadvantages of the prototype vaccines now in use in routine inmmunization are listed in Table I.

If we consider that in 40 years we have gone from a crude whole cell vaccine produced by a laborious discontinuous production procedure to the point where we can now prepare vaccines from *B. pertussis* grown in large-volume fermenters in a chemically defined medium, free of animal proteins, starch, charcoal, etc., we must admit that considerable progress has been made with the vaccine production method (Stainer, 1970).

If we consider only the vaccine production aspects we would have to say that improved pertussis vaccines have been developed and that these vaccines are now in use.

The introduction of defined media for growth of *B. pertussis*, the introduction of a more sophisticated vaccine production technique, and the application of more rigorous vaccine control measures (Minimum Requirements for Pertussis Vaccine, Code of Federal Regulations, 1975)

TABLE I
The Purported Advantages and Disadvantages of the Prototype Inactivated Whole Cell Pertussis Vaccines

Type of vaccine	Purported advantages	Purported disadvantages
Vaccines derived from solid agar media containing animal blood	Most of the valid data supporting efficacy have been obtained with these vaccines. Toxic substances released by the organism may be absorbed by or diffuse into the agar	Relatively high production costs. May contain small amounts of residual animal protein in the vaccine which may have sensitizing potential
Vaccines derived from semisynthetic solid agar media	Do not contain residual animal protein. Toxic substances released by *B. pertussis* may be absorbed by or diffuse into the agar base	Relatively high production cost. May contain additives such as charcoal which are difficult to exclude from the final vaccine
Vaccines derived from semisynthetic liquid media	Do not contain animal protein. More efficient to produce and lower cost than vaccines derived from solid agar. Some data available to support the efficacy of these vaccines	Greater lot-to-lot variability. May be more difficult to detoxify than vaccines derived from solid agar. Contain additives which may be difficult to exclude from final vaccines
Vaccines derived from chemically defined liquid medium	Large-volume fermenters are used in production process, thereby reducing production costs. Free of foreign protein and other nondesirable additives—greater lot-to-lot reproducibility	Do not have the body of experience in humans to properly evaluate these vaccines. May be more difficult to detoxify than vaccines derived from agar
Extract vaccines derived from semisynthetic liquid medium	Of lower toxicity in mice and lower reactivity in man than any of the whole organism vaccines	In animal studies these vaccines are of the same level of reactivity as the crude whole cell vaccines

have not stemmed the demand or the quest for an improved pertussis vaccine.

B. The Safety of the Current Pertussis Vaccines

In the last few years we have attempted to collect more precise information regarding the incidence of reactions in humans following the use of Diphtheria and Tetanus Toxoids and Pertussis Vaccine Adsorbed (DTP) produced by the Michigan Department of Public Health. These vaccines contain killed whole *B. pertussis* cells derived from a modified B–G medium. One of the limitations of this study has been that the follow-up has been handled for the most part by mothers, though a standardized report form has been used for recording reactions.

We have found that some degree of reaction occurs in about 60% of the recipients of our DTP vaccines. Fortunately, most of the reactions that do occur are mild (local reactions, fussiness, and/or a low-grade fever). As yet, no severe neurological or life-threatening anaphylactic reactions have been reported to us that were considered vaccine related. In fact, in the last 15 years we have distributed over 10,000,000 doses of DTP, DTP-polio, and single pertussis vaccine, and during this period of time we have not encountered any severe adverse reaction or resulting residual damage that could be attributed to the vaccine. This does not mean that such reactions have not occurred, but if they did occur it was not brought to our attention, and this is highly unlikely.

The reaction data that have been reported by other investigators for adsorbed DTP vaccines containing whole B. pertussis cells would appear to correlate with our own findings: a low incidence of the more severe reactions with a higher incidence of local and mild febrile reactions (Burland et al., 1968; Cohen, 1963; Muggleton, 1967).

There is one DTP vaccine containing an extract of B. pertussis cells that is being distributed in the United States at the present time (Weihl and Riley, 1963). This vaccine has been reported to induce fewer reactions in humans than a companion whole cell vaccine prepared by the same manufacturer. Data in well-controlled reaction-rate studies have not been reported to indicate how this vaccine would compare with whole cell vaccines prepared by the other manufacturers. However, the favorable lower reaction rates that were reported for this extract vaccine suggest that extract or partially purified vaccines may induce fewer reactions in humans than whole cell vaccines. One point must be made relative to the occurrence of reactions following pertussis vaccination. None of the reaction data that we have collected, nor any that we have seen reported by other investigators have been accompanied by matched data in nonimmunized controls. If this factor was taken into account, the reported reaction rate due to pertussis vaccination would probably be lowered. Because of the high incidence of mild reactions that is associated with the present pertussis vaccines there does appear to be a need for an improved pertussis vaccine.

C. The Efficacy of the Current Vaccines

The efficacy of whole cell pertussis vaccines, whether they be derived from solid agar media or liquid culture media, was confirmed in the clinical trials conducted by the British Medical Research Council during the period from 1951 to 1959 (Medical Research Council, 1951, 1959).

The protective efficacy of the whole cell vaccines was variable and appeared to be related to the potency of the vaccine as determined by the potency assay in mice. In general, the more potent vaccines afforded the highest degree of protection. The British field trial data, along with earlier data reported by Kendrick and Eldering (1939), constitute the direct evidence that is available to support the claims for efficacy of pertussis vaccine.

Admittedly then, there is only this evidence to support the protective efficacy of the pertussis vaccines that are being released for distribution in much of the world today. However, there is a great deal of indirect evidence, e.g., disease incidence data from national and local health agencies, to suggest that where pertussis vaccination has been carried out the incidence of clinical pertussis has decreased significantly (Cohen, 1963; Miller *et al.*, 1974; Warin, 1968).

By extrapolation from the British field trial data we can probably say that potent whole cell vaccines will afford protection to 75–90% of vaccinees. By extrapolation we might also assume (and have assumed) that any other pertussis vaccine, nonwhole cell, purified, or otherwise, that meets the current potency requirement, will probably provide a similar level of protection against pertussis, though this latter conclusion has not been fully substantiated.

Though the efficacy of pertussis vaccines has recently been questioned (Dick, 1974), we do not believe at this time that further work on an improved pertussis vaccine can be justified on the basis of the lack of efficacy of the present vaccines. In fact, with the decreasing incidence of pertussis in many parts of the world, it will be difficult to obtain meaningful data to support the efficacy of any new pertussis vaccine. It appears that when and if a new pertussis vaccine is introduced we will have to continue to depend on the mouse potency assay and disease incidence data to provide us with a partial answer regarding the efficacy of the new vaccine.

D.   The Cost of the Current Pertussis Vaccine

With the production techniques that are being used today by most manufacturers it probably costs less than 5¢ to produce a dose of pertussis vaccine. A significant part of this cost is accounted for by filling and packaging and testing costs. It must be emphasized that this cost figure does not include the costs for marketing or distribution of a product.

In the United States a dose of DTP vaccine can be purchased under large volume contract for less than 8¢. Even with inflation the cost of a dose of pertussis vaccine has risen by only about 20% over the last 5 years. It is doubtful that a highly purified vaccine can be made for any less than five times the cost of the present vaccines and the information we have collected in this laboratory would suggest that a purified vaccine may cost at least ten times as much as the vaccines that are in current use. From the cost angle there is every reason to continue with the present vaccines without pursuing further work toward improved pertussis vaccines.

E.   THE ACCEPTABILITY OF THE CURRENT PERTUSSIS VACCINES IN CLINICAL USE

The present vaccines have two major limitations from the standpoint of the user and the recipient. They cause some degree of discomfort and reaction in a high percentage of the recipients and the product must be administered by a parenteral route. Because of the reaction problem with pertussis vaccines there is some hesitation on the part of public health officials and practicing physicians to use vaccines containing pertussis.

In the United States at the present time, following the inoculation of a dose of single pertussis or DTP vaccine into an infant or a child, it is common practice for the physician to tell the mother that she should expect to see some degree of local reaction in the child, fussiness and fever, and that aspirin should be used to minimize these reactions. These types of comments do not necessarily help promote vaccination programs or provide great reassurance to the parents of the recipient regarding the safety of the product. Linnemann et al. (1974) have described some of the problems with pertussis that still persist. Foremost of these is the occurrence of pertussis in older children and adults. Many older children and adults apparently need booster doses of vaccine to reinforce their immunity to pertussis. Vaccine has been generally withheld from use in older children and adults because of a fear that the vaccine is more apt to induce severe adverse reactions in older children and adults than in younger children. It is not entirely clear whether or not pertussis vaccines present any greater reaction problem in the older age groups than in the younger age groups, but the fact that most of the product inserts contain the cautionary note "not recommended above the age of 6 years" tends to limit the use of the present vaccines to the younger age groups. Thus, current pertussis vaccines would have to receive a low rating in terms of acceptability by the public and by practicing physicians.

II.   Why Don't We Have an Improved or Fully Acceptable Pertussis Vaccine Now?

In the last 15 years a great deal of effort has been expended on the development of a purified pertussis vaccine, mostly on the premise that a purified vaccine may be less reactogenic. Two basic approaches have been used to isolate the protective antigen. In the first, the cells are disrupted and the active components are separated by physical and chemical means (Pusztai *et al.*, 1966; Sutherland, 1963; Wardlaw, 1966). In the second, the cells are treated with surface-active chemical agents, such as detergents or salts, to promote the release of the protective antigen from the cell wall (Ayme *et al.*, 1970; Nagel, 1967).

The purified preparations that have been produced by these techniques have been disappointing. Most investigators report at least one active component other than the protective antigen in the final vaccine. Another problem that has been of particular concern to us relates to the cost of preparing a purified vaccine. Most of the purification procedures that have been described are time consuming, difficult to adapt to large-scale production, and, with few exceptions, the yields of protective antigen have been extremely low: in the range of 10–15% of that in the starting material. In addition to the three problems already mentioned, a further problem that we have encountered is the lack of reproducibility in the quality and quantity of the final fractions obtained in the fractionation runs, even when consistently using the same starting material and fractionation procedure.

The development and introduction for clinical use of a purified pertussis vaccine has been greatly hampered by the lack of *in vitro* test procedures for both the detection and quantitation of the protective antigen (PA) in the vaccine itself, and the detection and quantitation of the protective antibody produced in the recipient.

At the present time the only reliable means for assaying for protective antigen is the *in vivo* potency assay in mice (Minimum Requirements for Pertussis Vaccine, Code of Federal Regulations, 1975). This test is subject to great variation and the cost of the mice alone that are used in a single test is in the range of $50. Furthermore, it takes 4 weeks to complete the test. When one considers the number of fractions to be assayed, from a single purification run, it becomes apparent that the need to use the costly, time-consuming mouse potency assay to identify PA is in itself counterproductive. Of equal or even greater concern is the lack of a reliable, simple test for the detection and quantitation of the protective antibody. A number of assays have been used for monitoring the response of an individual to pertussis vaccine (Winter, 1953). One of these tests, the passive protection test in mice, measures protective antibody, but this procedure, like the mouse potency assay, is extremely costly to perform and, obviously, is not practical for use in large-scale clinical trials or in mass antibody-screening programs.

Neither bacterial agglutinins nor complement-binding antibodies appear to be reliable indicators of protection against pertussis (Aftandelians and Connor, 1973). A bacterial antibody test was described by Ackers and Dolby (1972). However, there appeared to be a poor correlation between bactericidal antibody and protection.

Holt (1972) developed a tissue culture assay for the detection of a substance (probably antibody) that interferes with the adhesive or cell-binding properties of pertussis organisms. The antiadhesive antibody appears to correlate with mouse protective antibody, but the test, in its present form, could not be recommended for use in large-scale trials or in antibody-screening programs.

Another problem that has hampered pertussis vaccine development has been the difficulty in finding the ideal (highly potent, atoxic) strain of B. pertussis to use in vaccine production. In our department we have examined hundreds of strains of B. pertussis in an attempt to select strains that (1) represent the major antigenic types, (2) are highly immunogenic and easily and rapidly detoxified, and (3) grow well in culture media.

Selecting strains for antigenic type or for growth potential has not been a problem; however, selecting strains which produce potent vaccines with low toxicity (in mice) presents a much greater problem. So far we have been unable to find highly potent strains of B. pertussis with low levels of heat-labile toxin. The most potent strains are generally the most toxic strains.

Before any new pertussis vaccine can be considered as a viable alternative to any of the existing vaccines it must be tested extensively in humans. In the United States it is increasingly difficult to conduct vaccine trials in humans. It is certainly no longer desirable or ethical to use only institutional populations as the initial target for any new vaccine.

The trials should be conducted in those individuals most in need of the vaccine. Trials with experimental pertussis vaccines should primarily involve individuals in the 3 months to 1 year age group and the test subjects for any such trial will have to be recruited from the general population.

It has not been easy to find parents who are willing to volunteer their children for vaccine trials, nor has it been easy to find practicing physicians who are willing to administer experimental vaccines to infants and young children.

Possibly the greatest impediment to initiating clinical trials with any experimental or modified pertussis vaccine is the lack of a simple *in vitro* microassay for detecting and measuring protective antibody, requiring only minute amounts of blood for study. In attempting to initiate vaccine trials in Michigan we have found that physicians and the·parents of candidate test subjects may be willing initially to take part in testing a new experimental vaccine. When informed that, in addition to the vaccine injections, it will be necessary to collect a number of blood samples from each test subject (particularly if the test subject is an infant or child) they become reticent about participating in the trial. Fortunately, or unfortunately, pertussis has a low visibility in the public's eye. The disease has either been controlled so effectively for so long, or is so benign, that it is difficult to convince parents or physicians of the need for pertussis vaccine trials.

When contemplating vaccine trials, vaccine manufacturers and clinical investigators are immediately confronted with the problem of liability as a result of subjecting an individual to an unnecessary or an intangible risk. The development of new and improved vaccines is generally considered as being in the best interest of the consuming public; hence, vaccine development and clinical trial work should not be stifled unnecessarily by the fear of lawsuits, or other litigation. In order to facilitate and promote vaccine development and clinical trial work a need exists for a "no fault" insurance, such as that described by Krugman (1975), which would accommodate any serious problem yet would not relieve the manufacturer or user of responsibility for possible negligence.

Any individual who appears to be recruited for participation in a clinical trial must be fully informed of the type of procedure to be carried out. He or she (if under age, their parents) must be informed of the potential benefits and hazards of the procedure so the patient or parents can weigh the risks and benefits in making their decisions. Mills (1974) has reviewed the matter of "informed consent" in detail and has offered an approach for handling informed consent which should be applicable to those involved in and responsible for work with new experimental vaccines.

III.   Where Do We Go From Here with the Development of an Improved Pertussis Vaccine?

Is it realistic, based on past experience and with the data that we now have available, to believe that we can develop a pertussis vaccine that will be a significant improvement over the pertussis vaccines that are now in use in the United States and other parts of the world?

This is a very important question and we do not believe that it can be answered at the present time. It would appear to me that what we need to do is to direct our prime attention to data collection and to the initiation of small-scale field trials.

Our first priority should be to reassess the current risks associated with pertussis vaccination in humans. The main risk that has been clearly and consistently identified with the present vaccines is that of developing local and/or mild systemic reactions following vaccination.

The risk of brain damage following pertussis vaccination has been of concern but the risk of such an occurrence appears to be very low (Dick, 1966; Edsall, 1975; Griffith, 1974; Kulenkampff *et al.*, 1974). Nevertheless, with this concern and with the many and varied adverse responses that have been demonstrated in animals following the injection of pertussis vaccine, it is essential that we dispense once and for all with the question as to whether or not individuals who receive pertussis vaccine are at greater risk than the general population of developing a disease of the central nervous system, or a severe allergic disease, or some other disease syndrome that does not fit into either of these categories.

Pittman (1970) has presented an elegant summary of the complications and problems that may develop in man as a result of pertussis infection. Some of these same problems must be considered in any careful, long-term follow-up of recipients of pertussis vaccine.

In addition to long-term follow-up studies of the health status of pertussis vaccinees, there are other studies that can be conducted to provide us with the base-line information that will be needed to decide the course of future research work in the pertussis vaccine area.

These studies include: a reexamination of the current vaccines; the initiation of small-scale clinical trials with the more promising "purified" vaccines that have been described; and the identification or development of a simple *in vitro* procedure that can be used as a reliable indicator of protective antibody.

A. A Reexamination of the Current Vaccines

1. Small lots of potent whole cell vaccines should be prepared in which the opacity units (number of organisms) in the final vaccine have been reduced significantly from the number of opacity units being used in the present vaccines. We have found that by preparing vaccine pools from recent harvests of *B. pertussis* (harvested within the last 2 or 3 years) potent pertussis vaccines can be formulated with no more than 6–7 opacity units (OU) per dose. With this adjustment in OU, the number of organisms in the final vaccine will be less than half that being used in most of the current pertussis vaccines. Dick (1974) and Someya (1967) have reported that the reaction rate in humans is increased with vaccines containing higher numbers of OU. In light of the highly potent vaccines that can now be prepared there is absolutely no reason to continue to release vaccines that contain more than 9 or 10 OU per single dose. Small-scale trials should be initiated in humans in both children and adults to determine if reductions in OU result in a concomitant reduction in the incidence of local and mild generalized reactions.

2. The dose schedule should be reexamined to determine if one can be found that will significantly reduce the reaction rate without altering

the antibody response elicited. One dose of a highly potent vaccine may be sufficient to protect against clinical pertussis. Smaller booster doses (one-fourth to one-fifth the current dose), given at infrequent intervals, may be sufficient to maintain protection against disease. Linnemann *et al.* (1975) recently gave small booster doses of an adsorbed whole cell vaccine to adults at risk during an epidemic of pertussis without the appearance of severe disabling reactions in the vaccinees and with no greater problem with local and mild systemic reactions than is usually found in children.

3. Small lots of pertussis vaccine should be prepared and then varying quantities of pertussis immune globulin (human origin) should be added to each of these lots. In studies conducted in this laboratory it was found that human antipertussis globulin (PIG) added to potent pertussis vaccines destroyed residual toxicity without significant alteration in the apparent potency of the vaccine (in mice). The influence of PIG on the histamine-sensitizing factor (HSF), the lymphocytosis-promoting factor (LPF), the adjuvant factor (AF), etc., is not known at this time.

B.   Initiate Small-Scale Clinical Trials in Humans with Prototype "Purified" Vaccines

If we are going to continue to work on the development of "purified" vaccines we must get some human clinical trial data with some of the purified vaccines that have already been described. The following prototype purified vaccines would appear to be suitable candidates for use in clinical trial:

1. A detergent extract vaccine of the type described by Sutherland (1963);

2. A cell or cell fragment vaccine in which the bulk of the HSF has been removed in the supernate by a veronal wash, similar to that of Nagel (1967);

3. A sonicated extract vaccine with further purification of the PA by chromatographic procedures (S. Lehrer, personal communication, 1973). In this procedure the PA may be contaminated with HSF but the HSF activity can be reduced with heat treatment and formalin.

Prior to release for human clinical trial purified vaccines should be tested for AF, LP, HSF, and pyrogen factors. These vaccines should conform to the code of federal regulations (Minimum Requirements for Pertussis Vaccine, 1975). In addition, and as an added safety factor, we do not believe that a purified pertussis vaccine should be released for clinical trial if the average weight gain of mice injected with the test vaccine is less than 80% of that of an equal group of saline-injected controls; in addition, if any of the mice in the test vaccine group die during a 7-day period of observation, the lot should not be used. No less than 20 male mice and 20 female mice should be included in each group injected with test vaccine and in the group injected with saline.

C. DEVELOP A SIMPLE *in Vitro* PROCEDURE FOR DETECTING AND QUANTITATING PROTECTIVE ANTIBODY

Further exploration of the use of tissue culture as a substrate for detecting pertussis antigen-antibody interactions seems worthy of pursuit. A sensitive, reproducible, microtechnique has been developed in tissue culture for assaying for rabies antibody (Smith *et al.*, 1973). Some of the simple hardware that is used in this procedure would appear to be adaptable for use in a tissue culture technique for pertussis. Radioimmunoassay should also be investigated for use in detecting the protective antibody, particularly if we can prepare the protective antigen in essentially pure form and then bind it to a plastic or glass matrix.

IV. Can We Really Expect to Get New Pertussis Vaccines Approved, and in General Distribution?

The answer to this depends to a great extent on how much interest can be generated relative to the need for an improved pertussis vaccine. Someone will have to pay the bill for the conduct of the clinical trials and the follow-up that will be needed to confirm the suitability of any new pertussis vaccine. Will the vaccine manufacturers be willing to pay the bill for an improved pertussis vaccine? In a sense, they already have. Vaccine producers in the United States and Canada have committed a sizable amount of money to pertussis vaccine research and development. With pertussis vaccine prices being so low, with the economy being so uncertain, and with the lack of a clear-cut issue to challenge the safety and efficacy of the present pertussis vaccines, we would assume that manufacturers of biologics will probably assign a low priority to pertussis vaccine development.

If you ask the physicians if they want an improved pertussis vaccine they will generally say yes—if you ask them if they would be willing to pay for such a vaccine, they seem less interested. So, who pays the bill? It appears as though it may go by default to the federal government—who else? The type of priority to be given by federal agencies to pertussis vaccine research and development will have to be determined by someone else.

It would appear that the pertussis vaccine equation can be reduced to the following:

A. The acceptability of the present pertussis vaccines could be enhanced by reducing the number of organisms in the vaccine, by making adjustments in the immunization schedule, and by adhering more carefully to the known contraindications to exclude high-risk individuals from the vaccination procedure, e.g., children from families with a history of neurological disease.

B. We need to carefully reassess the potential for development of an acceptable purified pertussis vaccine and, if this potential does exist, how such a vaccine might best be achieved.

C. Clinical trial, field trial, and immunization and disease surveillance programs must be given the highest priority in terms of funding and work commitment. There is absolutely no reason to spend additional time and money for basic research on pertussis if we cannot do a more adequate job of using and interpreting the information that is now available.

The development and clinical testing of an improved pertussis vaccine will be costly and may incur some risks. If we want a better pertussis vaccine we must be willing to pay the price. The question then still remains, what price are we willing to pay for an improved pertussis vaccine?

REFERENCES

Ackers, J. P., and Dolby, J. M. (1972). *J. Gen. Microbiol.* **70**, 371–382.

Aftandelians, R., and Connor, J. D. (1973). *J. Infect. Dis.* **128**, 555–558.

Aymes, G., Mynard, M. C., Donikian, R., and Triau, R. (1970). *Symp. Ser. Immunobiol. Stand.* **13**, 89–98.

Burland, W. L., Sutchliffe, W. M., Voyce, M. A., Hilton, M. L., and Muggleton, P. W. (1968). *Med. Off.* **119**, 17–19.

Cohen, H. H. (1963). *Antonie van Leeuwenhoek* **29**, 183–201.

Dick, G. (1966). *Can. J. Public Health* **57**, 435–446.

Dick, G. (1974). *Proc. R. Soc. Med.* **67**, 371–372.

Edsall, G. (1975). *Practitioner* **215**, 310–314.

Griffith, A. H. (1974). *Proc. R. Soc. Med.* **67**, 372–374.

Holt, L. B. (1972). *J. Med. Microbiol.* **5**, 407–424.

Kendrick, P. L., and Eldering, G. (1939). *Am. J. Hyg.* **29**, 133–153.

Krugman, R. D. (1975). *Pediatrics* **56**, 159–160.

Kulenkampff, M., Schwartzman, J. S., and Wilson, J. (1974). *Arch. Dis. Child.* **49**, 46–49.

Linnemann, C. C., Partin, J., Perlstein, P., and Englander, G. (1974). *J. Pediat.* **85**, 589–591.

Linnemann, C. C. Jr., Perlstein, P. H., Ramundo, N., Minton, S. D., Englender, G. S., McCormick, J. B., and Hayes, P. S. (1975). *Lancet* **2**, 540–544.

Medical Research Council. (1951). *Br. Med. J.* **1**, 1464–1472.

Medical Research Council. (1959). *Br. Med. J.* **1**, 994–1000.

Miller, C. L., Pollack, T. M., and Clewer. A. D. E. (1974). *Lancet* **2**, 510.

Mills, D. H. (1974). *J. Am. Med. Assoc.* **229**, 305–310.

Minimum Requirements for Pertussis Vaccine, Code of Federal Regulations (1975). 21, Part 600, Subpart A, § 620. 1-620.7.

Muggleton, P. W. (1967). *Public Health* 81, 252–263.

Munoz, J., and Hekstein, B. M. (1966). *J. Bacteriol.* 91, 2175–2179.

Nagel, J. (1967). *Nature (London)* 214, 96–97.

Pittman, M. (1970). *In* "Infectious Agents and Host Reactions" (S. Mudd, ed.), pp. 239–270. Saunders, Philadelphia, Pennsylvania.

Pusztai, Z., Eckhardt, E., and Jaszovszky, I., Joó, I. (1966). *Symp. Ser. Immunobiol. Stand.* 3, 119–126.

Smith, J. S., Yager, P. A., and Baer, G. M. (1973). "Techniques in Rabies (WHO)," 3rd ed., Appendix 5, pp. 254–257. World Health Organization, Geneva.

Someya, S. (1967). *Bull. Inst. Public Health, Tokyo* 16, 23–32.

Stainer, D. W. (1970). *Symp. Ser. Immunobiol. Stand.* 13, 89–98.

Sutherland, I. W. (1963). *Immunology* 6, 246–254.

Wardlaw, A. C. (1966). *Symp. Ser. Immunobiol. Stand.* 3, 99–118.

Warin, J. F. (1968). *J. R. Soc. Health* 88, 21–26.

Weihl, C., and Riley, H. D. (1963). *Am. J. Dis. Child.* 106, 210–215.

Winter, J. L. (1953). *Proc. Soc. Exp. Biol. Med.* 83, 866–870.

Problems Associated with the Control Testing of Pertussis Vaccine

JACK CAMERON

Connaught Laboratories Ltd., Willowdale, Toronto, Canada

I.	Introduction	57
II.	Colonial Variation in *Bordetella pertussis*	58
III.	Opacity Standards	65
IV.	Potency Assay	66
V.	Role of Adjuvant	69
VI.	Agglutinins	72
VII.	Toxicity	73
VIII.	Summary	77
	References	79

I. Introduction

The responsibility of a manufacturer of vaccines is to satisfy himself first and thereafter the appropriate control authority, whether it be Canadian, American, European, or other, who may choose to examine his product subsequently, that the product meets the requirements specified by that authority. When the product is diphtheria or tetanus toxoid the manufacturer's responsibility is almost nominal in that the gamut of tests is well established, well correlated with performance in the field, and such that a competent technical staff can both perform and interpret the assays with a considerable degree of assurance. This is not so with pertussis (whooping cough) vaccine in spite of the existence of standards covering the same major parameters.

Pertussis vaccine seems to come into a special category in that, although the incidence of the disease has been strikingly reduced with the development of well-standardized vaccines, and although it is well recognized that there are cyclical upsurges in incidence—triennially in Ontario, quinquennially in the United Kingdom (why? an interesting problem in itself)—there still seems to be a nagging fear in many people's mind either that vaccines are not doing as good a job of prevention as might be expected or that they are too reactive to be as widely used as they are.

Potency and toxicity are the major points of concern and, unfortunately, one tends to be defined in terms of the other: there is an upper limit set to the number of organisms that may be contained in a single dose of vaccine, although a greater number, on occasion fourfold, could still yield a nontoxic vaccine which, of course, would be four times as potent. It is such a balancing of risks between potency and toxicity that causes so much concern.

Both potency and toxicity will be discussed in this presentation. There are established requirements for both, but these are not always met without difficulty. It would also be true to say that the toxicity requirement is not accepted without demur as being either definitive or even necessarily meaningful.

The other aspects of control testing in general to be discussed are: colonial variation in *Bordetella pertussis,* an almost wholly unexplored area of investigation; opacity standards, the yardstick by which the number of organisms per dose of vaccine is determined; the role of adjuvant in vaccine; and the suggestion, gaining ground in some areas, that the production of agglutinins following injection of vaccine might be developed as an alternative to the mouse protection test as a more convenient form of potency assay.

II. Colonial Variation in *Bordetella pertussis*

The WHO requirements (Report, 1964) begin by describing the strains of bacteria to be used in the manufacture of pertussis vaccine. For interest, the following two quotations indicate the requirements for strains of *B. pertussis* (Report, 1964) and for *Vibrio cholerae* (Report, 1969a) or use in the manufacture of vaccine.

3.1.1. Strains of *Bordetella pertussis*
Strains of *Bordetella pertussis* used in preparing vaccine shall be identified by a full record of their history, including their origin, characters on isolation, and particulars of all tests made periodically for verification of strain characters.
They shall be maintained by a method that will retain their ability to yield potent vaccine.

3.1.1. Strains of *Vibrio cholerae*
Strains used for the production of vaccine shall be (a) *Vibrio cholerae* classical biotype, Inaba serotype, and (b) *Vibrio cholerae* classical biotype, Ogawa serotype, and shall be identified by historical records.
They shall be in the smooth phase of growth and shall the properties described in section 3.1.2 below.
Vaccines prepared from *V. cholerae* strains National Institutes of Health, USA, 41 Ogawa and 35A3 Inaba, have proved effective in controlled field trials . . . .

3.1.2. Characterization of strains
All strains selected for vaccine production shall have the following properties:
(a) *Morphological properties.* Smears made from cultures after growth at $36° \pm 1°C$ for not more than 24 hours on a suitable nutrient medium adjusted to pH 7.4–8.0, when stained by Gram's method, give Gram negative rods, the majority being slightly curved.
(b) *Cultural properties.* When examined after 18–24 hours' growth on a nutrient agar that has been adjusted to pH 7.4–8.0 and may contain 0.5% of bile salt, the strains give colonies that are typically smooth and translucent.

Stereomicroscopic examination using oblique light is helpful in recognizing typical smooth colonies on agar plates.

(c) *Biochemical properties.* After 18–24 hours' growth in a set of peptone–water media containing different sugars, the strains ferment mannose and saccharose with the production of acid but not gas and fail to ferment arabinose.

(d) *Serological properties.* The strains are agglutinated by nondifferential cholera O group 1 antiserum. Such strains, when tested against monospecific cholera-agglutinating serum Ogawa, are agglutinted to titre if they are Ogawa serotype, and show no significant agglutination if they are Inaba serotype.

The strains are not agglutinated by anticholera O rough serum.

(e) *Properties of stability in suspension.* The strains grown for not more than 24 hours at 37°C on nutrient agar adjusted to pH 7.4–8.0 and then suspended in isotonic saline show no signs of clumbing or aggregation when held for at least 5 hours at 37°C.

(f) *Other properties.* The strains conform to the reactions of *V. cholerae* classical biotype given in the following table . . . .

Tests	Reactions
Sensitivity to Mukerjee's choleraphage Group IV	Sensitive
Sensitivity to polymyxin B (50 IU/disc)	Sensitive
Voges–Proskauer reaction	Negative
Agglutination of chicken erythrocytes	No agglutination
Tube haemolysis of sheep erythrocytes	No haemolysis

Why the differences? It is doubtful if this question can be answered other than in general terms. One explanation may be that so much more is known about the behavior of the cholera vibrio as compared with *B. pertussis,* although ironically this knowledge has not yet led to the preparation of a wholly satisfactory cholera vaccine! Since definitive criteria cannot be given for the manufacture of a vaccine, perhaps it does no harm to define rigidly the strains of bacteria from which it is to be made. Certainly, in this respect, a great deal is known about *V. cholerae.*

Bordetella pertussis, on the other hand, tends almost to be defined in the negative sense in that it neither ferments nor does it produce gas from the usual range of sugars (lactose, maltose, sucrose, glucose, mannite, dulcite); it does not produce indole, nor does it reduce nitrates to nitrites (Bergey, 1974). The author is aware of an incident where a strain of *Brucella abortus,* similarly inert biochemically, was actually accessioned to a culture collection as *B. pertussis* on the basis of such a negative identification. Fortunately the error was discovered and rectified!

Even in purifying cultures the almost negative attitude toward *B. pertussis* is exemplified in that, where faster growing organisms are purified by selecting single colonies that appear after 24–48 hours growth, pertussis growth is allowed to proceed for at least 72 hours, usually longer, and it is assumed that the culture taking this time to grow is likely to be *B. pertussis.*

Certainly no systematic examination of the colonial morphology of pertussis has ever been carried out comparable to that done on *Brucella abortus, Salmonella typhi,* and *Vibrio cholerae.* Possibly the work of Leslie and Gardner (1931) on the "phases" of pertussis comes closest, although this is more concerned with the degradation of laboratory strains than with the colonial morphology of fresh isolates. The type of description given by Wilson and Miles (1975) might even be considered romantic rather than scientific; it certainly does not compare with the considerably more perceptive description given of *V. cholerae* in the same volume.

Why are such different standards for production strains of bacteria, as those indicated in the WHO requirements, allowed to persist at such a critical stage of the production process? Later in the same requirements there are the usual additional standards for nontoxicity, potency, safety, and sterility, but without a more meaningful definition of the strains of *B. pertussis* to be used in the manufacture of the vaccine. This is almost to confuse the shadow with the substance.

What if single-colony isolates are made? Figures 1–4 show a number of colonial variants isolated on the same lot of Bordet–Gengou (1906) medium. All strains from which isolates were made are so-called phase I strains and wholly acceptable for the manufacture of vaccine.

This variation is not the same as the phenomenon of "modulation" described by Lacey (1960) who showed that the colonial form of *B.*

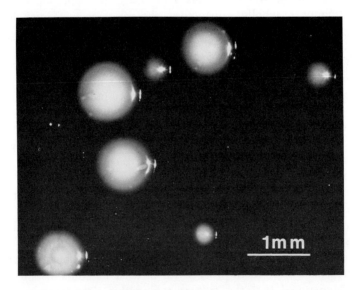

FIG. 1. Single colonies of *Bordetella pertussis.* Note different sizes and papillate colony, bottom left (×28).

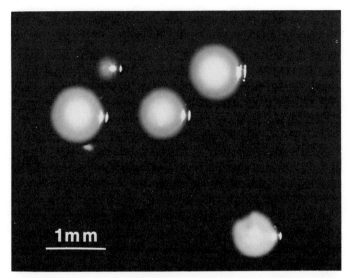

Fig. 2. Single colonies of *Bordetella pertussis*. Note different sizes, papillate colony, bottom left, and concentric appearance of others (×28).

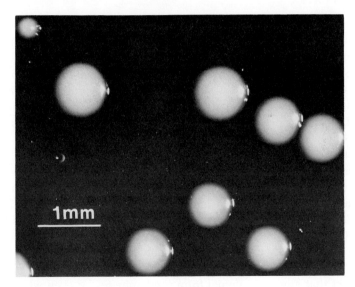

Fig. 3. Single colonies of *Bordetella pertussis*. Note relatively homogeneous white colonies and different types of concentric marking, also small opaque colony, top left (×28).

pertussis can be modified by altering the composition of the growth medium.

Currently, fresh isolates submitted to the author's laboratory for sero-

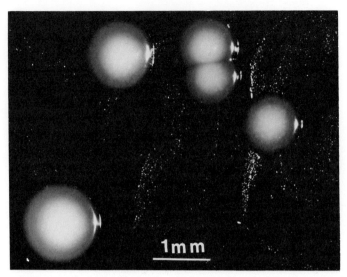

Fig. 4. Single colonies of *Bordetella pertussis*. Note large, translucent colony, right side, and generally more translucent appearance of colonies compared with those in Figs. 1–3 (×28).

typing—all coming from a single hospital—are examined to determine homogeneity of colonial form. A pictorial record is maintained, isolates being photographed daily once single colonies begin to develop, to insure that apparent differences observed in colonial form are real and not time-related artifacts. One of the simplest forms of differentiation, not colonial, is the existence of hemolytic and nonhemolytic colonies, sometimes within the same isolate. Serologically these colonies may or may not be homogeneous.

Observations to date show that of 12 isolates examined the predominant isolate is type 1,2,3 (9 isolates). This observation is at variance with an earlier Canadian study (Chalvardjian, 1965) and with the findings in the later and more extensive survey carried out by the Public Health Laboratory Service in the United Kingdom (Report, 1969b). The colonial form, while generally though not invariably consistent for each individual isolate, may differ from isolate to isolate. So far there are insufficient data to suggest that any one colonial form predominates, or even what properties, if any, different colonial forms signify.

In the work reported by Cameron (1967) a number of laboratory strains of *B. pertussis,* some used in the manufacture of vaccine, some variants of the original mouse pathogenic W18-323 strain introduced by Kendrick *et al.* (1947) were shown to be both colonially and serologically heterogeneous. This type of investigation was extended to fresh isolates in the course of the Public Health Laboratory Service survey in the United Kingdom and it was found that even fresh isolates were not serologically homogeneous, a finding confirmed by Stanbridge and Preston (1974). They also reported the isolation of serotypes 1, 1,2 and 1,3 from the W18-323 challenge strain; no comment was made on colonial morphology.

What is the relevance of such observations to vaccine manufacture? There is no answer to this question at present but one point is certain: were strains of *V. cholerae* and *S. typhi* to display such heterogeneity of colonial form they would simply not be acceptable for the manufacture of vaccine. Why are there not the same rules for *B. pertussis?*

An examination of some of the properties of clones derived from such an investigation yields interesting data.

Table I indicates the heterogeneity that can exist within a pure culture of *B. pertussis:* colonial variation, presence and absence of hemolytic activity, and serological variation. Nor is this confined to old laboratory cultures: fresh isolates too may be heterogeneous in all of these features. Table II, on the other hand, indicates differences in the ability of clones to produce protective antigen, to sensitize mice to histamine

TABLE I

Colonial Variants Isolated from Strains of *Bordetella pertussis*[a]

Strain	Description of variant	Serotype
2992[b]	Hemolytic, umbonate, crenated, purple	1,2,3
	Hemolytic, convex, entire, white	1,2(weak),3
	Hemolytic, convex, entire, purple	1,2,3
	Hemolytic, convex, entire, cream	1(weak),2(weak),3
	Nonhemolytic, convex, entire, purple	1,2(weak),3
	Nonhemolytic, convex, entire, cream	1,3
4132[b]	Hemolytic, convex, entire, white	1,2
	Hemolytic, umbonate, entire, white	1,2
	Nonhemolytic, convex, entire, white	1
	Nonhemolytic, umbonate, entire, white	1
5438[c]	Hemolytic, convex, entire, gray	1,3
	Nonhemolytic, convex, entire, gray	1,3
5476[c]	Hemolytic, convex, entire, gray	1,3
	Nonhemolytic, convex, entire, gray	1,3
1262[d]	Hemolytic, umbonate, entire, white	1
	Hemolytic, convex, entire, white	1,2
5618[d]	Hemolytic, convex, entire, white	1,2,3

[a] From Cameron (1967).
[b] Laboratory strains.
[c] Freshly isolated strains.
[d] Variants of the Kendrick 18-323 strain.

TABLE II

TOXICITY, PROTECTIVE CAPACITY, AND HISTAMINE-SENSITIZING
ACTIVITY OF COLONIAL VARIANTS OF *Bordetella pertussis*[a]

Strain	Serotype of variant	Mouse-toxic dose (number of organisms $\times 10^9$)	Protective capacity (IU per 20×10^9 organisms)	HSD 50 (number of organisms $\times 10^6$)	Ratio of HSD 50 to protective capacity ($\times 10^6$)
2991	1,2	10	10	0.4	0.04
	1,2	10	11	12	1.1
	1	35	4	100	25.0
2992	1,2,3	10	4	5	1.25
	1,2(weak),3	10	52	3	0.06
	1,2(weak),3	10	59	8	0.14
	1,2,3	10	23	6	0.23
4132	1,2	10	16	1	0.06
	1	40	2	100	50.0
	1,2	10	7	2	0.29
	1	40	2	20	10.0

[a] From Cameron (1967).

(Munoz and Bergman, 1968), and to cause loss of weight in mice. No suggestion is made, at this stage, that any particular colonial or serological variant is associated with any particular property except possibly that the type I variants appear to produce relatively little protective antigen and histamine-sensitizing factor and to be considerably less toxic. The emphasis at this point, however, is simply on the fact that such variation exists and should be controlled, if possible, in the interests of producing better-defined vaccines.

Even at this relatively superficial level there are exciting possibilities: selection of clones for the predominance of a particular property, i.e., protective antigen in the case of vaccine manufacture or histamine-sensitizing or lymphocytosis-promoting factor (Morse, 1965) if the particular interest concerns these factors; or investigation of colonial morphology per se in order to be able to interpret visual differences in colonies.

To read of pertussis being used as an adjuvant (Emmerling *et al.*, 1969) when almost certainly no two investigators are likely to be using the same material, other than in the broadest sense, is somewhat disconcerting. Certainly more meaningful control of the strains of *B. pertussis* used in vaccine manufacture and related investigations is necessary in order to insure greater reproducibility of results.

III. Opacity Standards

The toxicity of pertussis vaccines is controlled by setting an upper limit to the number of organisms that may be contained in a single dose of vaccine, usually 20×10^9 in the case of fluid or nonadsorbed vaccines and, in the United States, 16×10^9 in the case of adsorbed or adjuvant vaccines. For this purpose a reference, Opacity Standard, is available through both the World Health Organisation in Copenhagen and the Bureau of Biologics in Bethesda. The opacity of this preparation, by definition, is considered to be equivalent to 10 opacity units and to be the same as that of a suspension of B. pertussis containing 10×10^9 organisms per ml. Since vaccines are required to contain no more than 20×10^9 organisms per dose, they are standardized by adjusting the opacity so that where the recommended dose is 1 ml, a 1 in 2 dilution of the vaccine has the same opacity as the reference preparation, i.e., it has an opacity of 10 opacity units or contains 10×10^9 organisms per ml; where the recommended dose is 0.5 ml the vaccine will have an opacity of 40 opacity units and in this case a 1 in 4 dilution should yield a suspension of the same opacity as the reference preparation.

In practice, suspensions of B. pertussis for vaccine manufacture are usually held in a concentrated form and the appropriate dilution factor calculated by measuring the opacity of the concentrated suspension. The opacity of the final vaccine is checked. In the case of adsorbed vaccines such a final check is not possible because of the presence of the adsorbent and is determined simply by calculation from the concentrated suspension.

To count cells by opacity can, of course, be faulted: it does not take into account differences in the size or shape of cells brought about by growth conditions or differences in size of cell from strain to strain. Direct counts are probably more accurate; other parameters suggested are DNA, nitrogen, and dry weight. However, two points must be borne in mind: we do not know what we really are measuring so that all of the methods may be ill-chosen. More important, standardization by opacity serves the purpose in being a simple, readily performed operation that insures a good measure of uniformity from laboratory to laboratory without the need for expensive equipment. This is an important consideration for developing countries trying to undertake the production of a vaccine that presents problems to more sophisticated laboratories.

The problem facing manufacturers is that although both reference Opacity Standards are arbitrarily regarded as equivalent to 10 opacity units, or 10×10^9 organisms per ml, the United States standard is about twice as opaque, in absolute terms, as that of WHO. Since both the United States and WHO authorities require vaccine to contain a minimum of 4 protective units in the equivalent of 20×10^9 organisms per dose, it follows that United States vaccines can contain almost twice as many organisms as WHO vaccines for the same number of protective units. More important, since the toxicity of vaccines is controlled by limiting the number of organisms per dose, United States vaccines by this criterion can, in theory, be almost twice as toxic as they need be.

Since this report was prepared, a meeting of the International Association of Biological Standardisation discussed the problem and recommended to the appropriate Expert Committee of the World Health Organization that the permanent Opacity Standard described by Perkins *et al.* (1973) be adopted as the new International Reference Preparation. If the present WHO standard is accepted as having a value of 1, the current United States preparation has a value of about 1.8 and the new standard a value of 1.1. The new standard will still be said to be equivalent to 10 opacity units and to be equivalent to the opacity of a suspension containing 10×10^9 organisms per ml.

IV. Potency Assay

The potency of pertussis vaccines is measured by developing a standard, 3-point regression curve, comparing the response of groups of mice immunized with serial dilutions of the vaccine being assayed and that of groups of mice immunized with corresponding dilutions of a reference vaccine (U.S. Standard Pertussis Vaccine) and, given that the other parameters are satisfied—homogeneity of response, parallelism of response curves, lethality and level of challenge—by assigning a numerical value, in mouse-protective units, to the vaccine. This should, in general, be no less than 4 and no more than 12 protective units per single human dose in the United States and no less than 4 protective units per dose elsewhere.

The methodology of the potency assay was largely developed in the United States and is described in detail by Kendrick *et al.* (1947). The current recommended dose of 4–12 mouse-protective units contained in no more than 20×10^9 organisms in fluid, i.e., nonadjuvant or nonadsorbed vaccines (16×10^9 organisms in adsorbed vaccines), was developed through the investigations of Felton and Verwey (1955), Jaffe (1955), and M. Pittman (unpublished). The number of organisms in a single dose is, of course, determined by means of the Opacity Standard.

Potency testing makes heavy demands on mice in terms of numbers: a test involving 5 samples and a reference preparation, usually the maximum number that can be handled comfortably by operators since at challenge the mice have to be anesthetized and challenged within a fixed time, requires 350 to 400 mice of one sex and weighing 14 to 16 gm at immunization. The test lasts 4 weeks, 2 weeks immunization and 2 weeks postchallenge for observation of mortality; if 10 samples are tested per week regularly, not an unusual number, this means the number of mice permanently on test is about 3000. These figures take no account of any concurrent research program which, in terms of priorities, usually has to be fitted around production testing. In spite of almost pedantic attention to the detail of the assay, it is sometimes difficult to produce consistent results. It is perhaps understandable why an alternative form of assay would be welcome.

Recent work by Finney *et al.* (1975) suggests that the problem may be quite fundamental. In an analysis of one year's data (1970), 32 assays involving groups of 32 mice in a range of 3,5-fold serial dilutions, Finney and his colleagues found that there was as much as an 8-fold range in the ImD_{50} values. When 95% confidence limits were derived for the 32 assays it was found that 13 of the sets of data even fell outside these limits. The data for the years 1971, 1972, and 1973 confirmed the 1970 results. The investigation had stemmed from the fact that "A large part of the expenditure of time, effort and materials in a series of assays is . . . devoted to work with the reference preparation, this expenditure being as much as half of the total if there be only one test preparation in each assay" (Finney *et al.*, 1975, p. 1). In spite of the authors' desire to be able to pool data on a series of assays to provide a single and definitive dose–response curve that could be used, without further observations on the reference, for all subsequent assays, they concluded that this was inadvisable and that it was desirable to retain "the usual statistical practice: relative potency should be estimated from simultaneous trials of the reference and test vaccines, in order to eliminate variations in responsiveness of subjects from one occasion to another (Finney *et al.*, 1975, p. 9).

It might seem unlikely that there can be such a variation in mice from assay to assay, bearing in mind their dependability in so many other biological assays. Nevertheless, examination of corresponding data in my own laboratory involving a different strain of mouse essentially confirms these findings (J. D. Sparkes and J. Hatcher, personal communication). One seems to be driven, albeit unwillingly, to the conclusion that there is an inherent unsoundness in the assay, perhaps in the use of the intracerebral challenge, but clearly much investigation is needed to validate such an assertion.

The assay was first developed with 10 days between immunization and challenge: this has now been extended to 14–17 days. Several groups of workers, Evans and Perkins (1954a), Andersen (1957a,b) and Yoshioka *et al.* (1962), have shown not only that immunity is well developed within hours of immunization—promunity or proimmunity, as it has

been called—but also that there is a critical period around day 14 when
protection tends to wane, only to rise again to a maximum by day 17 and
to stay at this level for at least 40 days. When a 3-point assay was examined over such a period by Evans and Perkins (1955) it was found
that although survival at the top dose was fairly constant from days 10
to 40, with possibly a slight reduction at day 14, survival with the lower
doses increased with time. As shown in Fig. 5, protection with lower
doses of vaccine clearly increased with time.

If the response curve at day 14 consists of two parts, as was suggested
by Evans and Perkins (1955), do we know that the rate of development
of the two parts is constant for all vaccine strains? Perhaps this is an
area which should be reexamined.

Results of assays can also be affected by the strain of mouse (Pittman,
1966). It is disconcerting to find the requirements for the challenge in
assays defined in terms of both colony-forming units, i.e., viable count,
and LD_{50}, but such is the case for both the pertussis and typhoid assays
in the United States (Report, 1975a), which ultimately does more to define the mouse than the challenge. When the required values are mutu-
ally exclusive, as may be the case, what should be done? Should the
rules or the mouse be changed?

In our own experience the ImD_{50} of the U.S. Standard Pertussis Vaccine in the Connaught mouse has increased from 0.1 protective units in
1955 to 0.32 units today: the ImD_{50} in the NIH mouse is about 0.16 units
(C. H. Manclark, personal communication). To assume that, in spite
of these differences, the use of the same reference vaccine will put mat-
ters right is quite wrong! At the end of the 14-day observation period
postchallenge there are usually a number of "sick" mice and a judgment
has to be made as to whether these mice should be regarded as "dead"

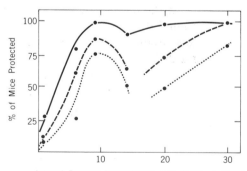

FIG. 5. Increase in protection in mice with time, related to amount of pertussis
suspension injected. Solid line: pertussis suspension undiluted. Long dashes: suspension diluted 1 part in 2.5; short dashes: suspension diluted 1 part in 6.25.

or "alive." In experiments with Connaught mice, such mice show remarkable tenacity and may well be alive from 14 to 28 days after they
have been pronounced dead! They lose weight and are undoubtedly
badly affected by the challenge, so much so that they appear to be permanently "runted," but they are certainly not dead. Such a situation
makes the assessment of results extremely difficult since it can affect
either overall mortality or the parallelism of response curves, which are
both grounds for rejecting, or for having to repeat, assays. This may
mean a delay of 4 weeks and the use of possibly 400 more mice! Experiments with NIH mice, on the other hand, confirmed the results noted at
the Bureau of Biologics, Bethesda, for both ImD_{50} and LD_{50}. This suggests that while it may not be possible to have access to the NIH mouse,
it may well be essential to standardize one's own mouse against the NIH
mouse both with regard to performance of the reference vaccine made
available by the Bureau of Biologics and of material representative of
one's own production.

In a detailed examination of the role of the challenge in the assay,
Murata et al. (1971) found no indication that a uniform method for the
preparation of the challenge reduced heterogeneity of results. In earlier
work, Cameron (1967) had reported the existence of two serotypes in
a culture of strain W18-323, types 1 and 1,2. Assayed separately, each
had the same LD_{50}. An ensuing investigation of the importance of serotype composition in relation to both challenge and immunization strains
suggested that protection is quite independent of serotype composition
of both vaccine and challenge strain. At that time further investigation
led to the finding of derivatives of strain W18-323 of serological composition 1; 1,2; and 1,2,3; but not of 1,3. The nondependence of protection
on serotype composition of vaccine was confirmed by Saletti and Genna
(1969).

On the other hand, with a vaccine such as pertussis where the levels of
many of the biological activities of the organism other than protective
antigen are not controlled, and where the organism, B. pertussis, is accepted in many areas of research as an adjuvant (Emmerling et al., 1969)
and as a nonspecific immunostimulant (Finger et al., 1967), is it surprising that there are problems with reproducibility in potency assays?
Perhaps the wonder is that the lack of reproducibility is not greater.

V. Role of Adjuvant

Usually an adjuvant is introduced into a vaccine to enhance the immune response of the host: references to this are legion. Adjuvant may
also be introduced to localize injected material and to prevent severe
systemic reactions otherwise observed with nonadjuvant (nonadsorbed)

vaccines as well as to enhance the immune response. This is particularly relevant in veterinary practice, e.g., in the case of vaccines against hemorrhagic septicemia (cattle plague, pasteurellosis) (M. P. Sterne, personal communication).

When an adjuvant is introduced into a vaccine it is necessary to prepare a corresponding adjuvant or adsorbed reference preparation. This is because the immune response is greatly enhanced and the slope of the response curve is so different from that observed with a nonadjuvant preparation as to make it almost impossible to compare responses to nonadjuvant and adjuvant vaccines in the same assay.

In the case of diphtheria and tetanus toxoids, preparation of reference adjuvant vaccines has presented technical problems. When freeze-dried —the usual way to preserve reference preparations—the physical structure of the adjuvant is affected, causing some concern about reproducibility of response in the reconstituted preparations. Nevertheless, differences in response to nonadjuvant and adjuvant diphtheria and tetanus toxoids are so great, about 100-fold, that different forms of assay are necessary. In the case of nonadjuvant or fluid toxoids the assay is based on immunization of guinea pigs followed by challenge with toxin. For adjuvant vaccines the guinea pigs are immunized, bled, and their antibody titers measured; no less than 2 units of antibody per ml of serum is acceptable—usually 4–7 are observed—compared with a hardly measurable 0.01 unit in the case of nonadjuvant vaccines.

As opposed to this situation, experimental adjuvant salmonella vaccines, particularly those against *Salmonella dublin* and *Salmonella typhimurium,* give rise to remarkably high agglutinin titers in the host, but still the vaccines do not protect. It is important, therefore, to know that an adjuvant enhances the immune response rather than to assume that it does. This kind of misleading assumption is currently apparent in the United Kingdom where pertussis vaccines came under fire first for lack of potency (Report, 1969b), then for unacceptable reactivity. One of the grounds for suggesting that later vaccines are more effective is that they now contain adjuvant (Miller *et al.,* 1974).

Does adjuvant indeed improve the immune response to pertussis vaccine? So small, in fact, is the effect of adjuvant on the immune response in mice that the same nonadjuvant reference vaccine can be used to assay adjuvant vaccine. More than this, a careful examination of the results of mouse protection tests shows that, if anything, adjuvant, in this case aluminum hydroxide, actually depresses the immune response (Cameron, 1972; see Table III). This has recently been confirmed by Novotny and Brooks (1974). The other widely used adjuvant, aluminum phosphate, appears to enhance the immune response by a factor of 2–3 (L. Higy Mandic, J. D. Van Ramshorst, personal communications). Bear-

TABLE III

COMPARISON BETWEEN THE POTENCY OF THE PERTUSSIS
COMPONENT OF 10 LOTS OF PLAIN (NONADSORBED) AND
ADSORBED DTP VACCINE[a]

Potency parent pertussis component (PU per dose)	Plain (nonadsorbed) DTP vaccine		Adsorbed DTP vaccine	
	Potency (PU per dose)	Potency ratio[b]	Potency (PU per dose)	Potency ratio[b]
15.2			5.7	0.38
			2.6	0.17
31.0			3.6	0.12
			8.1	0.26
4.3	6.1	1.42	5.1	1.19
4.2	4.2	1.00	2.1	0.50
3.9			2.8	0.72
			3.2	0.82
4.5	5.9	1.31	2.5	0.56
6.9	8.4	1.22		
	13.0	1.88		
9.2	7.1	0.77		
	4.9	0.53		
5.7	3.8	0.67	14.5	2.54
6.1	4.7	0.77		
	5.0	0.82		

	Plain DTP	Adsorbed DTP	Difference
Arithmetic mean potency:	1.04	0.73	Not significant
Geometric mean potency:	0.92	0.52	$P = 0.05$

[a] From Cameron (1972).

[b] The potency ratio is obtained as: potency of plain or absorbed DTP vaccine/potency of parent pertussis component.

ing in mind the confidence limits in a pertussis assay, a considerable number of assays is required to establish this almost marginal difference. The fact that the same reference preparation can be used for nonadjuvant and adjuvant vaccines tends to confirm the almost negligible effect of adjuvant on the immune response.

Pertussis vaccine, of course, is rarely used by itself, but is more commonly used in conjunction with diphtheria and tetanus toxoids and, in Canada, with poliomyelitis vaccine. It is in conjunction with diphtheria and tetanus particularly that adjuvant comes into its own. This is of major importance in immunization programs, particularly in developing countries, but does not justify the rather uncritical attitude shown in more sophisticated laboratories toward the negligible effect of adjuvant on pertussis. In being relatively unaffected by adjuvant, pertussis vaccine is in a similar category to cholera and typhoid vaccines as well as to experimental enterobacterial vaccines in general where adjuvant does little or nothing to enhance protective responses.

VI. Agglutinins

The mouse protection test, the standard assay procedure for measuring the potency of pertussis vaccines, is time-consuming to perform and subject to much variation. Most workers in the field would welcome a dependable, alternative form of assay but to date none is available. It should also be remembered that the mouse protection test is one of the few assays where correlation with performance of the vaccine in the field has been established (Report 1951, 1956, 1959); any alternative must, therefore, meet this additional requirement of being correlated with field performance. One such alternative, under serious consideration during the extensive Medical Research Council (MRC) trials in Britain, was the measurement of the production of agglutinins in mice. Most active in this field at the time were Evans and Perkins (1953, 1954b): they were about to recommend the adoption of this procedure (F. T. Perkins, personal communication) until it was learned that Pillemer's antigen (Pillemer et al., 1954), an extract of B. pertussis adsorbed on human red cell stromata and one of the preparations included in the MRC trial, protected children but was nonagglutinogenic. In the trial, it proved too reactive in children to be recommended for clinical use as a vaccine but its nonagglutinogenic nature ruled out the use of agglutinins as an alternative to the mouse protection test and, in a sense, paved the way for the adoption of the mouse protection test in the United Kingdom.

Of late, however, there has been a tendency to look toward agglutinins again as an alternative form of potency assay (Preston, 1966). Since Evans and Perkins (1955) conclusively demonstrated that a protective vaccine is not necessarily agglutinogenic, such an approach more or less amounts to a rejection of the mouse protection test.

Such responses cannot be considered seriously as an alternative to the mouse protection test until they are subjected to the same rigorous criteria as the mouse protection test. At the moment there is the rather circular argument that because adjuvants can be shown to enhance the agglutinin response to pertussis in rabbits and infants (Butler et al., 1969; Preston et al., 1974) and because adjuvants generally enhance the immune response, the immune response in pertussis can be assumed to be enhanced because the agglutinin response is demonstrably enhanced.

As indicated in the previous section, experience of the effect of adjuvant on other gram-negative organisms should suggest caution in assessing its effect in the case of pertussis. The only yardstick is protection and this has not to our knowledge been shown to be enhanced in parallel with enhanced agglutinin responses.

VII. Toxicity

Discussions of the toxicity of pertussis vaccine and its measurement can be emotive in the extreme. The different degrees and types of reaction observed are well reviewed by G. S. Wilson in his excellent monograph "The Hazards of Immunisation" (1967). There is much concern about this problem in Europe at the moment, particularly in the United Kingdom, and this has led to a much reduced uptake of vaccine, resulting also in a reduced uptake of diphtheria and tetanus toxoids, since DPT vaccine (diphtheria, tetanus, and pertussis) is by far the most common source of pertussis vaccine. The concern has become so serious that the Joint Committee on Vaccination and Immunisation felt constrained to reaffirm in June, 1975 (Report, 1975b) the opinion they had given in 1974 that

. . . the Joint Committee upholds the view that the hazard to children from contracting whooping cough exceeds the hazard associated with immunisation. The Committee's unanimous view is that the policy of offering pertussis vaccine in infancy should not be changed . . .

They go on further to state

. . . that currently used vaccines are offering a high degree of protection and that whooping-cough is still severe, especially in infants. Severe adverse reactions to the vaccine are rare, and evidence on this point has come from the P.H.L.S. [Public Health Laboratory Service] and from the Committee on Safety of Medicines.

Reactions certainly must be a matter of concern and one particular concern must be the interpretation of the case histories of serious reactions referred to by Strom (1969), Dick (1974), and Ehrengut (1974). Bearing in mind that tens of millions of doses of vaccine are released annually in North America and Europe, it is almost inconceivable that, with a reasonable "reporting back" system, a problem of the magnitude implied by some authors, serious enough, for instance, to warrant stopping the use of pertussis vaccine, could go undetected.

The toxicity of vaccines is controlled by the mouse toxicity or mouse weight-gain test, although the test is essentially concerned with weight loss. Mice weighing 15 gm are injected with half of a human dose of vaccine: they lose weight over the first 24 hours postinjection and, for a vaccine to be nontoxic, they are required to regain their starting weight within 72 hours of injection and thereafter to gain a further 3 gm by the 7th day postinjection compared with their starting weight.

Much of the development of this test must be attributed to the well-documented work of Pittman and Cox (1965). Even in this report, however, some of the problems with the test are evident: the weight gain for normal saline-injected control mice increased from 4 gm in 7 days in 1955 to 6 gm in 7 days in 1963. It was also noted that when adjuvant vaccines were assayed the adjuvant itself caused weight loss—some adjuvants more than others—presumably for different reasons, compared with pertussis. Allowance had to be made for this in the test, although there was no means of distinguishing which part of the weight loss was attributable to pertussis and which to adjuvant.

The strain of mouse is also important. Some mice, including the Connaught mouse, are almost unaffected by the injection of vaccine—they just go on gaining weight (see also Van Ramshorst, 1969); the absolute weight gain suggested of 3 gm is only relevant in the case of the NIH mouse—some mice are basically heavier and gain more weight, others are lighter and gain less. To set an absolute weight gain of 3 gm fails totally to take into account both the general nonavailability of the NIH mouse and the differences in growth characteristics of different mice. Van Ramshorst (1969) has also elaborated on this point. If manufacturers are to be required to perform this test it is essential that a reference preparation be made available, not so much to assist in monitoring toxicity as to calibrate other lines of mice against the NIH mouse. Van Ramshorst (1969) has already suggested a weight gain of 50% of that of controls, based on the ratio of 3 gm required by United States regulations to the 6 gm gained by normal saline-injected control mice. This assumes a parallelism not yet demonstrated experimentally between strains of mice. The author prefers the use of a reference "toxicity" preparation to be used to calibrate mice against the NIH mouse.

If pertussis vaccine can be judged to be toxic by this test, is it uniquely toxic? Figure 6 shows what happens when cholera and typhoid vaccines are assayed similarly.

It is fairly clear that all vaccines induce a similar weight loss in mice, cholera and typhoid, more so than pertussis and DTP. Therefore the test must lose some of its credibility as a unique test for detecting toxicity in pertussis vaccines. At best it can possibly be regarded as an "in-process" test for lot-to-lot control. It is doubtful if, statistically, it can detect quantitative differences in toxicity and there are no data whatsoever to suggest that qualitative differences in toxicity are even likely to be reflected in loss of weight in the mouse. Since pertussis is the only gram-negative vaccine injected into children just weeks old, is not its reputation one that might be equally shared with other gram-negative

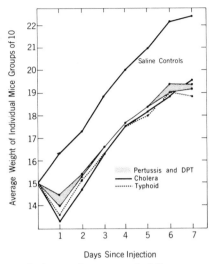

FIG. 6. Comparison of the performance of different vaccines in the pertussis mouse toxicity test.

vaccines such as cholera and typhoid? If we use the logic of the mouse toxicity test, our answer must be in the affirmative.

If we ask if the in-process control test is appropriate, we are avoiding the question of what is being controlled. Oddly enough, nearly all the factors associated with *B. pertussis* and considered at one time or another as possible candidates for the role of "toxic" factors are protein in nature and heat-labile. In our experiments, however, temperatures up to 100°C only marginally reduced the capacity of pertussis to cause weight loss in mice. Presumably the heat-labile entities are destroyed at this temperature. What remains? Since pertussis is a gram-negative organism, the almost automatic (and fashionable?) thought is lipopolysaccharide. This immediately recalls the similarity of response in the weight-gain test caused by all three gram-negative vaccines, cholera, typhoid, and pertussis.

In the course of manufacture of vaccine, various means and agents are used to inactivate the organism heat such as formalin, thimerosal, and combinations of these. Where the weight-gain test has been thought to control toxicity, has this been toxicity readily removable by conventional methods (Nelson and Gottshall, 1967), such as heat treatment as opposed to the more fundamental toxicity of pertussis as a gram-negative organism?

The test itself is subject to many vagaries, as Tables IV and V show. In the course of the test mice are affected by heat, light, and size of cage (Cameron, 1969). Moreover, as already indicated, some mice, including

TABLE IV

WEIGHT GAIN OF MICE INJECTED WITH SALINE UNDER
DIFFERENT CONDITIONS OF ILLUMINATION AND
TEMPERATURE[a,b]

Illumination	Temperature (°C)	Average mouse weight gain (days after injection)		
		1	3	7
Total darkness	11	0.77	1.00	2.65
	23	0.56	0.90	2.49
8 Hours artificial light	11	0.54	0.67	3.59
	23	0.22	1.28	2.45
16 Hours natural light	23	0.53	2.03	3.99
24 Hours artificial light	11	0	1.06	4.05
	23	1.78	2.81	5.44

[a] From Cameron (1969).
[b] Size of cage: 40 × 13 × 13 cm throughout.

TABLE V

EFFECT OF SIZE OF CAGE ON WEIGHT GAIN OF MICE[a]

Length of cage[b] (cm)	Material injected	Average mouse weight gain (days after injection)		
		1	3	7
15	Saline	0.11	1.36	3.84
	Reference preparation	−0.82	1.48	2.87
	BA 3685	−0.93	0.95	2.72
30	Saline	0.12	1.40	3.42
	Reference preparation	−0.95	1.57	3.46
	BA 3685	−2.21	−2.61	1.94
40	Saline	0.17	1.50	3.43
	Reference preparation	−0.06	2.27	3.48
	BA 3685	−2.06	−3.29	−0.68
53	Saline	−0.10	1.00	3.30
	Reference preparation	−0.48	1.86	4.18
	BA 3685	−1.30	0.60	2.80

[a] From Cameron (1969).
[b] All cages were 13 cm wide and 13 cm deep.

the Connaught mouse, gain weight on injection instead of losing weight.

Supposing it were possible to extract the protective antigen, would the present problems disappear? For one, the mouse toxicity test would almost certainly become redundant since it reflects an effect of heat-stable material. There is also much evidence to suggest that the protective antigen is heat-labile. The requirement that 4 protective units be recovered from no more than 20×10^9 organisms would also have to be revised since this requires, quite literally, 100% extraction of protective antigen. The antigen itself will have to be characterized in its own right for mouse protection, capsular antigens (Evans and Adams, 1952), hemagglutinin (Keogh and North, 1948), heat-labile toxin (Evans and Maitland, 1937), histamine-sensitizing activity, lymphocytosis-promoting activity, and, in fact, all activities of B. pertussis that at one time or another have been regarded as possible manifestations of toxicity. Fortunately, there are laboratory models for most of these but not, of course, for the major problem, the source of the delayed reactions in children. If lipopolysaccharide is responsible for these reactions in conjunction with a sensitized child—which would, to some extent, explain the unpredictability of reactions—there is more reason than ever to work on the isolation and characterization of protective antigen, and to demonstrate its noninvolvement in the reactivity of pertussis vaccine.

VIII. Summary

This contribution deals solely with the problems in control testing of pertussis vaccine. One cannot work with pertussis without being involved in all its facets: epidemiology of the disease—a killer disease in many parts of the world—efficacy of vaccines, potential toxicity of vaccines, and the role of its many metabolic products identified and characterized in the course of time, intensively investigated, then no longer fashionable. Bordetella pertussis is a challenging organism that has not yet yielded its innermost secrets. It is therefore easier to list problems than to suggest more useful alternatives.

An intensive study of colonial morphology should at least lead to a better standardization of strains of B. pertussis for use in the manufacture of vaccines. There are also indications that such a study may lead to a better fundamental understanding of the organism.

To compare the requirements listed by WHO for the production of cholera and typhoid vaccines with those listed for whooping-cough vaccine simply leaves one asking "Why the differences?" Although the question probably cannot be answered it is certainly one which should be asked and which needs an answer, preferably in the form of further intensive investigation. Since different colonial forms can be identified by their ability to produce greater or lesser amounts of protective antigen, histamine-sensitizing factor, and lymphocytosis-promoting factor, it might be possible, at a later date when the roles of such factors are better understood, to select clones for desirable combinations of these factors or even for their absence.

The outstanding problem in determining opacity—the existence of two standards—is well on the way to being resolved. The use of opacity, a relatively simple way of controlling the number of organisms in a dose of vaccine, must only be regarded as a means to an end. When more definitive parameters become available for controlling toxicity it will have served its purpose.

The potency assay has always been tedious to perform and there seems little prospect of improvement. In spite of meticulous attention to the detail of the test and rigid control of the parameters that can be controlled, problems involving strain of mouse, weight of mouse, and preparation and performance of challenge occur with disturbing regularity. The variability demonstrated in the performance of the reference preparation, i.e., performance of the mouse, suggests an inherent unreliability in the assay—reference preparations are not generally as variable as such data suggest. Perhaps a comparison with other assays which use mice as the test animals could give some indication of what can be expected of mice in terms of reproducibility. It is also clear that the use of a reference preparation does not compensate for differences in the performance of mice; it simply emphasizes them. The route of intracerebral challenge is unusual but, at present, is the only effective one available. Although the challenge strain, W18-323, can be shown to exist in a number of different serological forms, this feature seems to play no role in the results of assays.

Pertussis vaccine today is most generally available as a component of adsorbed DTP vaccine (diphtheria and tetanus toxoids and pertussis vaccine). We should not assume that the adsorbent (or adjuvant) does for pertussis what it does for diphtheria and tetanus toxoids. It possibly reduces or localizes reactions at the time of injection, a benefit to some extent counterbalanced by the fact that the injection nodule may persist longer; more than this cannot, however, be claimed for adjuvant.

The difficulties and expense of potency assay in mice make the possibility of an alternative form of assay attractive. The production of agglutinins was considered in the 1950s but rejected when Pillemer's nonagglutinogenic antigen was shown to protect children. Of late the question has been resurrected, particularly since it can be shown that adjuvant vaccines give better serological responses than nonadjuvant vaccines. Although this is true, it does not detract from the fact that these differences are not reflected in the mouse protection test which, having already been correlated with the field performance of vaccines in children, must remain the yardstick of comparison until displaced by an equally well-validated alternative assay.

Now that the toxicity and potential risks associated with the use of smallpox vaccine in developed countries have been acknowledged, and smallpox vaccination has been stopped in some of these countries, pertussis vaccine has to some extent become the next target for the antivaccination lobby. This is to entirely miss the point that in general the risk from whooping cough far exceeds the risk from vaccination. Nevertheless, those in the field are well aware of the potential toxicity of the vaccine and of the lack of appropriate parameters to detect and control it. The mouse toxicity or weight-gain test is at present used to control toxicity but, at best, it can only be regarded as a lot-to-lot in-process control test. By its very nature it can only dependably detect quantitative differences in toxicity. It has not even been suggested that qualitative differences are reflected in the weight gain of mice. An added complication is that different strains of mice behave differently in the assay and, in the performance of the assay, mice are affected by heat, light, and size of cage. The nature and infrequency of reactions suggest more a type of sensitivity or hypersensitivity, in which the infant plays a contributing role. It is unlikely in the extreme that a complete lot of vaccine, a reactive lot *capable of eliciting such reactions in all children injected with it,* has ever been released. Nevertheless, identification of the factor or factors capable of precipitating such a reaction in a receptive host should be one of the goals in research on the toxicity of pertussis vaccines.

REFERENCES

Andersen, E. K. (1957a). *Acta Pathol. Microbiol. Scand.* **40**, 227–234.
Andersen, E. K. (1957b). *Acta Pathol. Microbiol. Scand.* **40**, 235–247.
Bergey, D. H. (1974). "Bergey's Manual of Determinative Bacteriology," 8th ed. Williams & Wilkins, Baltimore, Maryland.
Bordet, J., and Gengou, O. (1906). *Ann. Inst. Pasteur, Paris* **20**, 731–741.
Butler, N. R., Voyce, M. A., Burland, W. L., and Hilton, M. L. (1969). *Br. Med. J.* **1**, 663–666.
Cameron, J. (1967). *J. Pathol. Bacteriol.* **94**, 367–374.
Cameron, J. (1969). *Prog. Immunobiol. Stand.* **3**, 319–323.
Cameron, J. (1972). *Proc. Symp. Comb. Vaccines,* Zagreb, 1972, pp. 55–67.
Chalvardjian, N. (1965). *Can. Med. Assoc. J.* **92**, 1114–1116.
Dick, G. (1974). *Proc. R. Soc. Med. B* **67**, 15–16.
Ehrengut, W. (1974). *Dtsch. Med. Wochenschr.* **99**, 2273–2279.
Emmerling, P., Finger, H., and Bruss, E. (1969). *Z. Med. Mikrobiol. Immunol.* **155**, 48–64.
Evans, D. G., and Adams, M. O. (1952). *J. Gen. Microbiol.* **7**, 169–174.
Evans, D. G., and Maitland, H. B. (1937). *J. Pathol. Bacteriol.* **45**, 715–731.
Evans, D. G., and Perkins, F. T. (1953). *J. Pathol. Bacteriol.* **66**, 479–488.

Evans, D. G., and Perkins, F. T (1954a). *Br. J. Exp. Pathol.* **35**, 322–330.
Evans, D. G., and Perkins, F. T. (1954b). *J. Pathol. Bacteriol.* **68**, 251–257.
Evans, D. G., and Perkins, F. T. (1955). *Br. J. Exp. Pathol.* **36**, 391–401.
Felton, H. M. and Verwey, W. F. (1955). *Pediatrics* **16**, 637–651.
Finger, H., Emmerling, P., and Schmidt, H. (1967). *Experientia* **23**, 591–592.
Finney, D. J., Sheffield, F. W. and Holt, L. B. (1975). *J. Biol. Stand.* **3**, 1–10.
Jaffe, V. R. (1955). *J. Pediatr.* **47**, 716–719.
Kendrick, P. L., Eldering, G., Dixon, M. K., and Misner, J. (1947). *Am. J. Public Health* **7**, 803–810.
Keogh, E. V., and North, E. A. (1948). *Aust. J. Exp. Biol. Med. Sci.* **26**, 315–322.
Lacey, B. W. (1960). *J. Hyg.* **58**, 57–93.
Leslie, P. H., and Gardner, A. D. (1931). *J. Hyg.* **31**, 423–434.
Miller, C. L., Pollock, T. M., and Clewer, A. D. E. (1974). *Lancet* **2**, 510–513.
Morse, S. I. (1965). *J. Exp. Med.* **121**, 49–68.
Munoz, J., and Bergman, R. K. (1968). *Bacteriol. Rev.* **32**, 103–126.
Murata, R., Perkins, F. T., Pittman, M., Scheibel, I., and Sladky, K. (1971). *Bull. W.H.O.* **44**, 673–687.
Nelson, E. A., and Gottshall, R. Y. (1967). *Appl. Microbiol.* **15**, 590–593.
Novotny, P. and Brookes, J. E. (1975). *J. Biol. Stand.* **3**, 11–30.
Perkins, F. T., Sheffield, F. W., Outschoorn, A. S., and Hemsley, D. A., (1973). *J. Biol. Stand.* **1**, 1–10.
Pillemer, L., Blum, L., and Lepow, L. H. (1954). *Lancet* **1**, 1257–1260.
Pittman, M. (1966). *Symp. Ser. Immunobiol. Stand.* **5**, 161–166.
Pittman, M., and Cox, C. B. (1965). *Appl. Microbiol.* **13**, 447–456.
Preston, N. W. (1966). *J. Pathol. Bacteriol.* **91**, 173–179.
Preston, N. W., Mackay, R. I., Bamford, F. N., Crofts, J. E., and Burland, W. L. (1974). *J. Hyg.* **73**, 119–125.
Report. (1951). *Br. Med. J.* **1**, 1463–1471.
Report. (1956). *Br. Med. J.* **2**, 454–462.
Report. (1959). *Br. Med. J.* **1**, 994–1000.
Report. (1964). *W.H.O., Tech. Rep. Ser.* **274**.
Report. (1969a). *W.H.O. Tech. Rep. Ser.* **413**.
Report. (1969b). *Br. Med. J.* **4**, 329–333.
Report. (1975a). "Code of Federal Regulations." U.S. Govt. Printing Office, Washington, D.C.
Report. (1975b). *Lancet* **2**, 544.
Saletti, M., and Genna, G. (1969). *Symp. Ser. Immunobiol. Stand.* **13**, 126–131.
Stanbridge, T. N., and Preston, N. W. (1974). *J. Hyg.* **73**, 305–310.
Strom, J. (1969). *Symp. Ser. Immunobiol. Stand.* **13**, 157–160.
Van Ramshorst, J. D. (1969). *Prog. Immunobiol. Stand.* **3**, 324–326.
Wilson, G. S. (1967). "The Hazards of Immunisation." Univ. of London Press (Athlone), London.
Wilson, G. S., and Miles, A. A. (1975). "Topley and Wilson's Principles of Bacteriology, Virology and Immunity," 6th ed. Arnold, London.
Yoshioka, M., Nahase, Y., Takatsu, K., and Kasuga, T. (1962). *Kitasato Arch. Exp. Med.* **35**, 13–20.

Vinegar: Its History and Development

HUBERT A. CONNER

*Department of Physical Sciences,
Northern Kentucky University, Highland Heights, Kentucky*

AND

RUDOLPH J. ALLGEIER

Wheaton Place, Catonsville, Maryland

I.	Introduction	82
II.	In the Beginning	83
	A. Origin of Vinegar	83
	B. Production and Food Uses of Vinegar	83
III.	Nonfood Uses of Vinegar through the Centuries	87
	A. Medicinal Uses	87
	B. Solvent Properties	89
	C. Cosmetic Uses	89
IV.	Vinegar in Food Processing	89
	A. Ancient and Current Uses	89
	B. Vinegar Substitutes	90
V.	Production Methods	90
	A. Raw Materials	90
	B. Ancient Methods	91
	C. Orleans Process	91
	D. Quick Process	92
	E. English Production	94
	F. Modern Processes	95
VI.	Major Types of Vinegar	100
	A. White Distilled Vinegar	100
	B. Cider Vinegar	103
	C. Wine Vinegar	104
	D. Malt Vinegar	105
VII.	The Industry	105
	A. United States and British Production	105
	B. European Economic Community	108
	C. Manufacturers' Associations	109
VIII.	Microbiology of Vinegar Production	110
	A. Early Observations	110
	B. Recognition of Biological Origin of Vinegar	111
	C. Confirmation of Biological Nature	111
	D. Bacterial Flora of Generators	112
	E. Taxonomy	113
	F. Nutrition and Metabolism	117
IX.	Mechanism of the Biological Conversion of Ethanol to Acetic Acid	120
	A. First Observations: Abbé Rozier and Lavoisier	121
	B. Döbereiner's Development of the Equation for the Reaction	121
	C. Presently Accepted Mechanism	121

      X.   Composition of Vinegar 124
     XI.   Summary 126
            References 127

I.   Introduction

Vinegar is one of several fermented foods prepared and used by early man; and like others, wine, beer, bread, and certain foods from milk, its discovery predates our earliest historical records. Its first uses were as a beverage, a condiment, a preservative, and, possibly not until later, a household cleansing and medicinal agent.

Nearly any unfortified wine or beer when exposed to the atmosphere will develop an acidity which, in many cases, results from the action of *Acetobacter* in converting the ethanol to acetic acid. Under normal household conditions, the conversion occurs quite slowly and there are no sharp lines of demarcation between the categories of wine, old wine, sour wine, and vinegar. Thus the question whether wine and vinegar were to be considered as one was discussed in the Halakah (Löw, 1901). While vinegar was more stable and had a greater number of applications in the early households than its progenitors wine and beer, its formation was not desirable, since it entailed the replacement of a pleasant beverage with a commodity having more mundane merits and having no capacity for stimulating the mood of the consumer.

The manufacture and use of vinegar has continued uninterrupted from the earliest times to the present. Precise figures are not available to show the amounts of vinegar produced and utilized throughout the world today, but in the United States alone the consumption exceeds 140,000,000 gallons a year. The manufacturing process has developed from an uncontrolled, naturally occurring sequence of fermentations to a carefully monitored and automated fermentation. The important uses, the preservation and enhancement of foods, are the same today as those noted in the earliest records.

This chapter is an expansion and updating of a review of the history and development of vinegar which has recently appeared (Allgeier *et al.*, 1974). Additional historical references are included, and the technology, biochemistry, bacteriology, reaction mechanism of the conversion of ethanol to acetic acid, and vinegar industry are discussed in more detail than in the earlier publication.

Wüstenfeld's (1930) book contains a good review of the history of vinegar production and advances in equipment to 1929 and·is well documented in comparison to the earlier books on the subject by Brannt (1900), Lafar (1898), and Dussance (1871). These contain a great deal of information on vinegar but, for the most part, lack references to the literature, which makes it difficult to follow developments on the subject. Mitchell (1926) also presents a detailed discussion of the history of vinegar, and particularly its development in England. A Technical Information Service Report by Schierbeck (1951) contains details on the practical aspects of vinegar manufacture. Attention is called to several recent reviews by Vaughn (1942, 1954), Llaguno (1964), and Joslyn (1970). These papers summarize in considerable detail important research results and production improvements.

II. In The Beginning

A. ORIGIN OF VINEGAR

The word "vinegar" is derived from two French words, *vin* and *aigre* meaning sour wine, but the term is now applied to the product of the acetous fermentation of ethanol from a number of sources. Duddington (1961) states that wine-making is an art dating back at least 10,000 years; therefore it is safe to assume the existence of vinegar for at least that period of time. The factors involved in the conversion of wine to vinegar were not known; hence the spoilage of wine was a common occurrence. Huber (1927a) notes that the Talmud contains accounts of large quantities of wine and beer that were lost by turning to vinegar at the time of the forced emigration of the Jews from Palestine to Babylon. There are many records which refer to vinegar, attesting to its importance in ancient civilizations. Apparently the earliest are those from Babylon about 5000 B.C.

B. PRODUCTION AND FOOD USES OF VINEGAR

1. Babylonians

Huber (1927a,b) has published two interesting papers on the preparation and uses of vinegar by the Babylonians. They used vinegar both as a flavor enhancer and as a pickling agent or preservative. Thus, while they may not have been the first to discover these attributes, credit must be given to the Babylonians for first recording what are still today the most important uses of vinegar in foods.

Vinegar was prepared by the Babylonians from the juice or sap of the date palm, from date wine and raisin wine, and from beer. Palm trees were pierced or tapped near the influorescence or spathe surrounding the flowers. The sweet sap which exuded was conducted into vessels bound to the tree (Huber, 1927a; Moldenke and Moldenke, 1952). This palm juice or sap had a low sugar content and yielded a wine which was not very stable and turned to vinegar within three or four days after the fermentation. A more stable wine was prepared by the fermentation of "date honey" (Huber, 1927a). The latter was not obtained from bees but from the fruits. Ripe dates were collected and exposed to sunlight and subsequently processed by an unrecorded method to yield a thick syrup resembling honey. It is this material which is referred to as honey so many times in the Bible and other writings (Moldenke and Moldenke, 1952). The fermentation of date honey led to a wine having a higher concentration of ethanol than that from palm sap; hence vinegar prepared from date honey wine contained a higher concentration of acid than did palm sap vinegar. A third type of vinegar was prepared by adding water to the residues from the date wine preparation, yielding first a dilute fruit juice which soon turned to vinegar.

Vinegar also was made from beer brewed from barley malt and barley, or from a mixture of dates and these cereals. The beer made from cereal was reserved for festive occasions while that for daily consumption was made from dates or cereal and dates, the latter being much less expensive than were cereals in Babylon.

It was in Babylonian times about 3000 B.C. that commercial vinegar production was taken over by the beer brewing industry. During the last 2000 years B.C. breweries changed to increasingly large date-processing establishments and utilized the by-products of their operations for the preparation of vinegar.

Heads of large Babylonian households were not dependent upon breweries for their vinegar but were able to produce what they needed. Economics and the quality and stability of the product determined what source was used for vinegar. The home-produced vinegar was referred to as Edomitish vinegar in the Talmud, the name being derived from the word "Edom," the biblical name for the Arabian desert.

Desert tribes did not have acess to commercial beer or vinegar but made their own from a malt or travel bread. The latter was produced from germinated cereals which were crudely ground and made into bread. After baking on both sides the bread was stable for many months and was a main staple of travelers and desert nomads. Beer was produced by cutting or crushing the bread to yield a crude grist and moistening the latter wtih water and/or beer. Sugars in the bread, produced by the malted cereals, soon fermented to yield ethanol which in turn fermented to yield vinegar. Surprisingly enough, this process for beer and vinegar was not invented by the desert nomads but was an ancient cultural tradition developed by the Babylonians as early as 4000 B.C. Raisin wine, mentioned above, was a preferred beverage in Babylon households and therefore was an important source of vinegar.

Vinegar was used by the Babylonians in cooking, along with spices, to enhance what at times undoubtedly was a monotonous diet. Thus they prepared spiced lentils and Polenta, a millet porridge, with vinegar. Other grains, barley and wheat, also were roasted or parched and served in a similar manner. A large number of spices, including tarragon, ruta, neander leaf, capers, portulaca, and, most important of all, saffron, were used. Huber (1927b) states that these probably were used with vinegar to add variety to their meals rather than to prepare spiced vinegars. It is well known that seasoned foods are regarded highly in the Orient.

Meat, fish, fruits, and vegetables were preserved by pickling in vinegar, usually with the addition of spices, herbs, and pepper. While meat and fish were not plentiful, vegetables and fruit grew in abundance. Thus melons, cucumbers, pumpkins, peaches, apricots, apples, and pears were preserved by cutting them into pieces and pickling them in wine or fruit vinegar to which spices and herbs were added. Some of the latter include fennel, tarragon, cardamon, anise, carroway, pepper, saffron, and sage. Therefore vinegar played an important role in the Babylonian household and economy by allowing the preservation of fresh food items and consequently extending the time period when these were available for consumption (Huber, 1927b).

Mention should be made of cider and of cider vinegar, since the etymology of the word connotes something ancient. Cider, that is, what we call fermented cider, was the wine or strong drink "Shekkar" of the Phoenicians and was well known to the Aryan race populating northern Europe before the beginning of recorded history. The souring of apple cider to yield vinegar was an ancient art (Alwood, 1903). Löw (1901) states that in the biblical period vinegar was prepared from either wine or cider.

2. Romans

The Romans manufactured vinegar and used it in a number of ways. Mixed with water and sometimes with water and eggs, it was the "Posca," a common drink of the Roman soldiers and of slaves. Both the Greeks and Romans made extensive use of vinegar both in cooking and at their meals, as indicated by some of their writers. Cato (234–149 B.C.) was the first of a number of Romans to write on the art of husbandry (Cato and Varro, transl. by Hooper and Ash, 1934; Cato, transl. by Brehaut, 1933). His De Agri Cultura contains a wide assortment of observations and instructions on many facets of agriculture. Included are a few precepts regarding the use of vinegar. Wine for the hands (slaves) was made by fermenting 12 parts of must (grape juice) mixed with 2 parts of sharp vinegar and 50 parts of boiled water. Instructions were given for the use of vinegar in the preservation of lentils, for the preservation and seasoning of green olives, and for the preparation of a confection of green, ripe, and mottled olives using vinegar, oil, and spices.

Columella lived from about 4 B.C. to 65 A.D. His volume, *Res Rustica De Arborubus* (*On Agriculture and Trees*), contains 18 references to vinegar (Columbella, transl. by Forester and Heffner, 1955). One reference gives explicit directions for the preparation of vinegar from wine that has gone flat. The latter was fortified with yeast, dried figs, and honey (sources of sugar) and diluted with vinegar. The added vinegar served to supply a heavy inoculum of *Acetobacter* and also to make conditions less suitable for the development of contaminating organisms. Thus he includes the same ingredients used for vinegar production today: ethanol, sugar, yeast, and a viable culture of *Acetobacter*.

For those regions ". . . in which wine and therefore vinegar also, is lacking . . ." he gives detailed instructions for preparing the latter from figs and for obtaining additional vinegar from the figs by adding water. He also gives methods for the preservation of foods by pickling. Included are olives, elecampane, stalks of cabbage, unopened flowers of wild or cultivated parsnip, and many herbs. "All of these are conveniently preserved by one method of pickling, that is to say, a mixture of two thirds of vinegar and one third of hard brine" (Columella, transl. by Forster and Heffner, 1955). Clark and Goldblith (1974) point out that pickling may have come into increased popularity in Rome during the years intervening between those of Cato and of Columella. The former refers only occasionally to vinegar but Columella gives specific instructions for pickling a number of vegetables and herbs.

Pliny the Elder (23–76 A.D.) in his *Natural History* noted that "It is a peculiarity of wine among liquids to go mouldy or else to turn into vinegar; and whole volumes of instructions how to remedy this have been published" (Pliny, transl. by Rackham, 1945). Nearly 2000 years later Pasteur (1862a) discusssed the action of "la fleur du vin et la fleur du vinaigre." Pliny regarded vinegar very highly and referred to it many times: ". . . in vinegar, on the other hand, notwithstanding its tart and acid taste, there are very considerable virtues, and without it we should miss many of the comforts of civilized life" (Pliny, transl. by Bostock and Riley, 1856). A similar thought was expressed by Huber (1927b) with respect to the Babylonians (see above). Rackham quotes Pliny as follows: "No other sauce serves so well to season food or to heighten a flavour"; "I have often said, and shall often have to say, how often it is a beneficial ingredient with other things" (Pliny, transl. by Rackham, 1945).

3. Biblical References

The King James Version of the Bible contains five references to vinegar in the Old and four in the New Testament. In the Revised Standard Version, vinegar is mentioned only four times in the Old Testament. Ruth is invited by Boaz to dip her morsel in vinegar in the King James

Version, but in the Revised Standard Version the word "vinegar" has been changed to "wine." The word vinegar as used in the Old Testament passages comes from the Hebrew word "Chomets," which refers to a liquid such as wine or other strong drink that has turned sour (Moldenke and Moldenke, 1952).

All references to vinegar in the King James Version of the New Testament refer to the offering of it to Christ before his crucifixion and while he was on the Cross. Specific references are given in the bibliography. The following is from John (19:29–30): "Now there was set a vessel full of vinegar: and they filled a spunge with vinegar, and put it upon hyssop, and put it to his mouth. When Jesus therefore had received the vinegar, he said, 'It is finished'; and he bowed his head, and gave up the ghost."

4. Medieval and Modern Uses

Tannahill (1973) cites several references to the use of vinegar in food from the time of the Greeks to the seventeenth century. She quotes a recipe from one of the earliest cookbooks of modern Europe, *Le Viandier de Taillevent*, compiled in 1375 by Guillaume Tirel. The recipe was for the preparation of cameline and listed vinegar as one of the ingredients. Platina (1475) gives innumerable recipes for preparing many foods with vinegar, including those made with herbs, vegetables, meat, and fish. Digbie (1669) recommends vinegar for fewer dishes than does Platina and, except for the preparation of horseradish and mustard, all are for the cooking of meat of one kind or another.

According to Kromer (1972), vinegar made in Modena, Italy has been famous since the Renaissance and, to this day, it is made there in old barrels (see the Orleans process, Section V,C). Enthusiastic users of this unusual beverage include the composer Rossini and Sir Winston Churchill.

III. Nonfood Uses of Vinegar through the Centuries

Since vinegar resulted from the refermentation of beverages of one type or another, its first prehistoric use probably was in foods. However, its ready availability led to many early nonfood uses.

A. MEDICINAL USES

There are many references dating back to antiquity to the use of vinegar in medicine. Possibly vinegar was the first antibiotic known to man, a bacterial product antagonistic to other microorganisms. Thus the principle of antibiosis has been used since the earliest civilizations. Hippocrates, known as the father of medicine, applied it to his patients in 400 B.C. (Hippocrates, transl. by Adams, 1849). Löw (1901) mentions medicinal uses of vinegar in biblical times, including its application as a dressing for wounds.

While Huber (1927a,b) records no medicinal uses for vinegar by the Babylonians, Columella recommends it for treatment of scab, the bite of a mad dog, and treatment of wounds in ox hooves (Columella, transl. by Forster and Heffner, 1955). Pliny lists 28 maladies he claims are relieved by wine vinegar, 17 by squill vinegar, and an equal number by vinegar lees (Pliny, transl. by Bostock and Riley, 1856). Vinegar has been applied to the ear, as described in an Assyrian medical text translated by Thompson (1931). Dussance (1971) refers to medicinal vinegars and presents several formulations, among which are: antiseptic vinegar of the four thieves, antiscorbutic vinegar, camphorated vinegar, colchicum vinegar, squill vinegar, black drop vinegar, and syrup of vinegar. The four thieves of Marseilles used a concoction containing vinegar to ward off infection by the plague. While the effectiveness of their prophylactic may be questioned, it has been reported that all of the thieves died of old age (Kromer, 1972). Fetzer (1930) reports that vinegar was employed in the Civil War with considerable success to prevent an incipient attack of scurvy among soldiers. It was employed during the First World War in the treatment of wounds. More recently, Ochs (1950a) has made use of the Assyrian remedy to treat chronic middle ear diseases when other methods failed. A second paper by Ochs (1950b) gives a good historical background and also describes his treatment of acute external otitis, the therapeutic technique he employed, and several case histories. Jarvis (1959) discusses the use of apple cider vinegar in formulations for the following: liniment to treat lameness, poison ivy, shingles, night sweats, burns, varicose veins, impetigo, and ringworm.

Vinegar finds but little use in modern medicine and one is tempted to dismiss the claims of ancient and some recent authors regarding the medicinal value of vinegar. While the ancient applications of vinegar for medicinal purposes are of historic value only, it is of interest to note that some *Acetobacter* produce biologically active compounds. Gilliland and Lacey (1966) found a strain of *Acetobacter rancens* which was lethal to yeasts. The active principal could not be isolated and was very unstable, being destroyed by chemicals or heat. Antoniani *et al.* (1958) and Yamamoto *et al.* (1959) have studied a biologically active material in vinegar that inhibited the growth of human tumor cells *in vitro*. Lanzani and Pecile (1960) showed that the material inhibited tumor growth in mice injected with Ehrlich's ascites tumor fluid. While the early writers may have overestimated the medicinal efficacy of vinegar, their observations are quite valid with respect to the merits of vinegar for flavor enhancement and food preservation. Methods they describe are used today with only minor modifications.

B. Solvent Properties

It is noteworthy that the solvent effects of vinegar were understood by the ancients. Olives were sprinkled with vinegar to release them from their pits (Löw, 1901). A well-known anecdote concerning Cleopatra is related by Pliny: "To gain a wager that she would consume at a single meal the value of a million sisterces, she dissolved pearls in vinegar which she drank" (Pliny, transl. by Holland, 1964). Titus Livius reported that as Hannibal marched over the Alps to Rome, his soldiers used vinegar to crack rocks blocking his way:

> The soldiers being set to make a way down the cliff, by which alone a passage could be effected, and it being necessary that they should cut through the rocks, having felled and lopped a number of large trees which grew around, they made a huge pile of timber; and as soon as a strong wind fit for exciting the flames arose, they set fire to it, pouring vinegar on the heated stones, they render them soft and crumbling. They then open a way with iron instruments through the rock thus heated by the fire, and softened its declivities by gentle windings so that not only the beasts of burden, but also the elephants, could be led down it (Titus Livius, transl. by Spillman and Edmonds, 1895).

Fee discredits this account and considers it to be a fiction invented by the Romans to rationalize their defeat by Hannibal (M. Fee, cited in Pliny, transl. by Bostock and Riley, 1856).

The alchemists experimented extensively with vinegar for which they used the symbols $+$ and $\times$, while the characters ✚ and ✳ were used for distilled vinegar. Acetic acid was prepared from vinegar by neutralization, concentration, and acidification followed by distillation (Mitchell, 1926). Their distilled vinegar does not correspond to what we call white distilled vinegar today (The Vinegar Institute, 1974).

C. Cosmetic Uses

Vinegar was used early, and even today, as a cosmetic aid. Kromer (1972) records the use of vinegar as a beautifier. Lucrezia Borgia is said to have bathed in vinegar daily to keep herself fresh and well-groomed. For some time women have been known to finish a shampoo with a vinegar rinse (Fowler and Schwarz, 1973).

IV. Vinegar in Food Processing

A. Ancient and Current Uses

Vinegar was, in Babylonian and undoubtedly in pre-Babylonian times, and still is used in cooking. Nomadic people use it to prepare foods such as leguminous vegetables.

The attributes of vinegar which made it valuable in the past, such as its preservative action and its flavor-enhancing properties, are those which make it valuable today.

French cooking has long been noted for the quality of its sauces; many of these are flavored with vinegar (Gogúi, 1856). Vinegar is used extensively in the United States in a number of ways. Both the British and the Americans use it on fish and chips. Crawford (1929) describes vinegar as a wholesome food condiment.

Vinegar can reduce the pH of certain food products to a level inhibiting the growth of microorganisms, including some spore-forming bacteria (Heid and Joslyn, 1963). It is used as an additive to a large variety of foods, including pickles, pickled vegetables, spiced fruits, aspics, salad dressings, mayonnaises, mustards, catsups and other tomato products, relishes, fish products, barbecued poultry, marinated and pickled meats, breads, sauces, cheese dressings, and soft drinks. New uses for vinegar are devised as new foods are developed. Seasoned vinegars, prepared by infusion of cider or white vinegar with herbs, spices, shallots, or garlic are used for condiment purposes. A significant amount of vinegar is sold through retail outlets for use in the home. While distilled vinegar finds many nonfood uses in homes, most of it and the bulk of wine and cider vinegars are used in food preparation.

B. VINEGAR SUBSTITUTES

Several synthetic acids are becoming fairly important as acidulants in food processing. Among those now produced for use in foods are adipic, fumaric, and succinic acids. Vinegar is superior to these materials in food processing (Heid and Joslyn, 1967). Its benefits include the production of a superior flavor and texture in the final product and longer retention of these and other attributes.

V. Production Methods

A. RAW MATERIALS

1. Sources of Ethanol

Vinegar can be made from a variety of fermentable substances, the essential requirements being that they are satisfactory, economical sources of ethanol. The use of palm sap, expressed date juice, date honey, and cereal has been noted (see Section II,B,1). Grape and apple juice also have been discussed (see Section II,B,2), these being the most important sources of vinegar after distilled ethanol. The commonly used substrates have been fruits and their juices, cereals, sugar syrups, and ethanol. Vinegar can be and has been made from nearly any fermentable material containing sugar or starch. Among these materials are molasses, sorghum, and maple syrups, fruits, berries, melons, coconut, honey, potatoes, beets, malt, grains, and whey.

2. Factors Determining Choice of Raw Materials

The climate and available crops of each country determined what raw materials were suitable for ethanol and, therefore, vinegar production. During the last two centuries sugar cane, beets, and potatoes have been produced where soil, climate, and economic conditions permit their growth, and vinegar has been produced from these sources as well as from cereal grains. Through the centuries, sources of sugar and starch have moved from the flowers and fruits downward through the stem, as in sugar cane, to the underground parts such as beets and potatoes (Hildebrandt et al., 1967).

B. ANCIENT METHODS

Preparation of vinegar as a household commodity and the subsequent commercial manufacture by Babylonian breweries has been mentioned (see Section II,B,1). Possibly because of the availability of grape wine and its ready conversion to vinegar, the latter was manufactured in France on a commercial scale and exported long before vinegar manufacturers were established in England.

The first method used is what Cruess (1948) called the "let alone process." Wine, left in open containers, was converted to vinegar. Bacteria occurring naturally on the fruits, or which were carried by the containers or in the air, converted the ethanol to acetic acid. Later, in the old "fielding process," casks of wine were laid out in open fields (Mitchell, 1926), and the fermentation was initiated by transferring small amounts of vinegar from cask to cask: The acetic acid thus introduced tended to inhibit the development of lactic bacteria which would yield an inferior product. No doubt the quality of vinegar thus produced varied considerably from batch to batch. Nevertheless, this process was carried out for several hundred years. The earliest published description of a method of vinegar manufacture appears to be that of Oliver deSerres in 1616 (cited by Bitting, 1928, 1929).

C. ORLEANS PROCESS

The Orleans process, named after the French city, is one of the oldest commercial vinegar processes for which we have detailed information,

as described by Wüstenfeld (1930) and Mitchell (1926). This is a re-
finement of the old fielding process and is still in use today in some parts
of the world.

The equipment consists of a large cask with a 2-inch hole bored just
above the center in one end of the cask, and in the other end another
hole the same size was drilled about an inch below the stave containing
the bung hole. These holes were covered with screening to prevent the
entrance of fruit flies. A spigot was provided for withdrawal of the
product, and a tube passed through the bung to the bottom of the cask
for the purpose of adding wine or other alcoholic liquor. This tube was
arranged as described so that the surface film of bacteria would not be
disturbed. Inoculation was usually made from a previous batch. The
liquid in the cask was generally a low-grade wine mixed with about 20%
of its volume of fresh vinegar, which served as the inoculum and increased the acidity, thus preventing the growth of contaminating microorganisms. Instead of being laid out in fields, the casks were stored in
a heated building or in an underground cellar.

The process was slow and the efficiency low by present standards (77
to 84%) (Wüstenfeld, 1930; Joslyn, 1970). Objectionable features of this
method were the gradual filling of the barrel with slime or "mother of
vinegar" and the slow rate of reaction (Mitchell, 1926).

Despite its disadvantages, the Orleans process yielded a preferred
product and was used for the manufacture of wine vinegar for many
years after the introduction of the "quick process." Although attempts
were made to improve the Orleans process, there was little change in
manufacturing methods (Mitchell, 1926). Pasteur (1862a) devised an
acetator consisting of a large shallow-covered trough with holes at the
sides for the admission of air. Claudion (Mitchell, 1926) superposed a
number of shallow fermentation vessels in such a manner that the bottom
of one formed the top of the one below. Air was admitted to the surface
of the liquid through ports located in the sides of the fermentation vessel,
just above the surface of the liquid.

D. Quick Process

Surprisingly enough, the second earliest published description of a
method of vinegar manufacture was not that of the slow or Orleans
processes but of a forerunner of the quick process. Mitchell (1926)
quotes almost in its entirety a description of a French method using
packed generators, published anonymously (1670) by "An Ingenious
Physitian of that Nation." Two large casks were used, each fitted with a
"trevet" (false bottom) and a cock for drawing off the liquid. The casks
were packed first with a layer of grape vine twigs and then with "rape,"

the latter being the stems left after removing grapes from the clusters. One cask was completely filled with wine; the other only half-filled. At appropriate intervals one-half of the contents of the filled cask was transferred to the half-full cask. During hot weather it was necessary to transfer wine from cask to cask at least twice daily in order to keep the temperature within proper limits for vinegar production.

The Dutch technologist Boerhaave (1732) is usually given credit for constructing the first generator using packing and applying the trickling principle for vinegar production; although, as mentioned above, generators packed with rape were used as early as 1670. Boerhaave used materials such as pumice, branches of vines, and grape stems for packing generators. Wüstenfeld (1930) reports that Kastner conducted experiments in 1823 to improve the old Boerhaave process, and at about the same time Schüzenbach trickled spirits through beechwood shavings by gravity flow to obtain vinegar. Specific details of Boerhaave's and Kastner's contributions are the subject of much confusion and speculation in the literature (Wüstenfeld, 1930). Mitchell (1926) states that Schüzenbach introduced the use of a vat instead of a cask for the acetification process and provided mechanical means for the repeated distribution of the acidic liquor over the packing. He provided holes for ventilation near the bottom of the generator and is given the credit for inventing the so-called German process (Wüstenfeld, 1930).

A British patent issued to Ham (1824) for the production of vinegar describes:

1. A revolving pump located at the top of the generator and a sprinkling device composed of two perforated pipes attached to the pumps;

2. A false bottom "full of large holes" for supporting the packing and located to allow space for the vinegar-in-process below;

3. A hole in the side of the generator just above the surface of the liquor and fitted with a pipe connected to a bellows or air pump to extract the air. Provision was made to connect the air pump to an aperture in the cover of the generator so that air could be forced through the packing from above or below; and

4. A steam pipe for keeping the vinegar-in-process at a temperature between 80° and 100°F.

Ham's contributions may not have received the recognition they deserved. His generator incorporated many features which were not widely adopted for many years. Although he did not use the word "generator" there seems to be little doubt that his device was the forerunner of the Frings circulating generator.

The following is a direct quotation from the patent:

> . . . I protest against the use or introduction of any mechanical or other
> method, contrivance, or principle whatever, by which twigs or other materials
> presenting an extensive surface can be wholly and constantly kept wetted by
> the vinous liquor to be acidified for the purpose of forming vinegar (and I claim
> the exclusive right to this process for no other purpose) so that a current or
> shower of the said liquid, and also a current of atmospheric air, can and shall
> be simultaneously passing over, in, or through all parts and interstices of the
> said twigs or other materials in any direction . . . .

Although the sequence of events and the precise contributions of each
investigator are not clear, the works of these four men (Boerhaave,
Kastner, Schüzenbach, and Ham) were responsible for the widespread
adoption of the quick vinegar process.

Many types of generator packing materials have been employed successfully, including coke, rattan, charcoal, corncobs, ceramics, grape
stems, and various types of wood such as beechwood, fir, cypress, redwood, oak, and some pines (Wüstenfeld, 1930). The choice of packing
material depends mainly on availability and cost. Generally beechwood
shavings or curls are the preferred packing, and in some plants they
have been in continuous use for as long as 50 years.

Feed or mash for the generator consists of fermented cider or cereal
extract, wine, or a solution of dilute ethanol. The concentration of the
latter may vary from 10 to 14% by volume of ethanol, depending on the
manufacturer. In addition to the ethanol, it is necessary to add nutrients
for the *Acetobacter* (see Section VI,A). Numerous patents summarized
by Bitting (1928, 1929) illustrate the many innovations which advanced
the state of the art during the 1880s and early 1900s.

E. English Production

Since the English consumed beer and ale, their vinegar was obtained
by the souring of those beverages rather than from wine. The earliest
product was simply ale converted partially into vinegar by long exposure
to the air. By analogy, the product became known as *alegar* which stood
in the same relationship to ale as did vinegar to wine. Thus over a number of centuries, English vinegar has been produced from cereals. The
vinegar industry evolved out of the brewery, since the production of
vinegar was a convenient way of utilizing surplus beer or ale.

The Rush family owned a "vinegar yard" in Castle Street, Southwark
as early as 1604 (G. F. Bond, personal communication, 1975). This plant
was purchased by R. and W. Pott in 1790 and it remained in their family
until 1902, when it was sold to Beaufoy and Co. The latter was combined
with two other firms in 1932 to form British Vinegars, Ltd. (Slater, 1970).

Dating from the seventeenth century, vinegar became a source of revenue for the British. Mitchell (1926) gives a detailed account of the early

taxation of vinegar products. Taxation of vinegar began in England in 1673 during the reign of Charles II. Vinegars obtained by the souring of beer or ale were taxed a duty of sixpence per gallon. It is interesting that revenue suffered from the evasion of this duty, for in the year 1696 William III imposed a penalty of 40 shillings a barrel for vinegar concealed from the gaugers or sent from the brewhouse without notice to the excise officers. By an act passed in 1710, the duty on vinegar increased to ninepence per barrel and remained at that rate throughout the following century. During the reign of George III (1796) vinegar makers were prohibited from having distilleries upon their premises. A license was required indicating whether they intended to make vinegar from grain or from molasses or sugar. These are a few examples of the laws which affected the operation and taxing of vinegar-producing plants and their products. Mitchell states changes continued to be made in the tax laws on vinegar in England even through the last century.

F. MODERN PROCESSES

Improvements in vinegar manufacturing methods developed slowly until the twentieth century. During the last 50 years, paralleling advances in other fermentation processes, significant improvements have been made in production methods as a result of unique equipment and pure culture techniques.

1. The Circulating Generator

A major improvement in the quick vinegar process was made when Frings (1932, 1937) introduced the circulating generator. The process was similar in principle to previously described methods but had many advantages over the latter. For example, vinegars of higher acid concentration were produced, requiring less tankage, and an increased production rate was achieved. Forced aeration, temperature control, and semicontinuous operation have produced higher efficiencies. Good descriptions of the process are given by Hansen (1935), Prescott and Dunn (1949), and by Slater (1953). In the United States, and presumably in other countries, Frings generators rapidly replaced many of the older trickling generators which were not equipped with forced ventilation and adequate process controls. Many of the Frings generators are still in operation throughout the United States but the trend is to replace these with the Frings Acetator (see next subsection).

The circulating Frings generator is simply a large tank constructed in a variety of dimensions, generally of wood, including redwood and fir but preferably cypress. The vertical timbers are held in place with steel hoops made from rods and fitted with threaded lugs for tightening the staves. Some stainless steel generators are in use, but in the United States wood is the material of choice for circulating generators. A false bottom supports the beechwood curls above the collection chamber. Air for the oxidation of the ethanol is supplied by a simple fan-type blower and is distributed to the generator by a number of equally spaced inlets located in the sides of the generator just beneath the false bottom. The generator is equipped with a pump for circulating the feed-vinegar mixture from the holding section beneath the false bottom, through a heat exchanger to the top of the generator. A spray mechanism distributes the liquid over the surface of the packing. Cooling water to the heat exchanger is regulated to control the internal temperature of the generator below 35°C.

Meters and regulators monitor three parameters in the operation: (1) the circulation rate of the mash; (2) the flow of cooling water to the mash cooler or heat exchanger; and (3) the quantity of air delivered to the generator. Thermometers are installed at several different heights on the tank to indicate the temperature of the mash as it trickles through the generator.

The circulating generator does not require close attention other than that necessary to maintain airflow, temperatures, and recirculation rates and to prevent the ethanol concentration in the mash from falling below 0.3–0.5%. Its performance is quite consistent over long periods of time, but long-time trends do occur in which production rates may decline perceptibly, level off, and eventually increase to their original high values. Reasons and corrective measures for these fluctuations in production rates are not known. Temperature variations do not appear to be a factor, since the fluctuations appear to be independent of the season of the year. Variation in the flora of the generator or mutation of the predominating organism may be responsible, but unequivocal data are not available to support this concept. The difficulties inherent in the isolation of vinegar bacteria from circulating generators and in the comparison of the efficiency of strains in the laboratory mitigate against the development of definitive data on the cause of fluctuations in production rates in circulating generators.

From the 1930s to the present, there has been continued interest in improving the design and production efficiencies of the process. Martinez (1963) has prepared a comprehensive bibliography of the patent literature from 1937 to 1962.

2. Submerged Culture—Frings Acetator

Tait and Ford (1876) and Fowler and Subramaniam (1923) studied what we now refer to as submerged culture for vinegar production. The latter investigated the preparation of pure cultures of *Acetobacter,* the effects of amount of inoculum, concentration of ethanol, forced aeration rate, temperature, stimulants such as activated silt, and the action of extraneous bacteria on submerged acetification. Despite these studies, no practical applications were derived from them. It was not until the late 1940s that the technology used in the production of antibiotics led to the new method for the production of vinegar. The submerged culture technique in use today is the result of studies by Hromatka and co-workers (Hromatka and Ebner, 1949, 1950, 1951, 1955; Hromatka *et al.,* 1951; Hromatka and Kastner, 1951). Concurrently, Haeseler (1949) described submerged acetification in a separate paper. These investigations led to the development of the Frings Acetator. This highly automated fermentor makes possible shorter production cycles and higher yields than were obtained with the circulating generator and eliminates much of the unexplained variation in production rates and efficiencies which were inherent in the circulating generator.

The Acetator consists of a stainless steel tank fitted with internal cooling coils and a bottom-entering, high-speed agitator. The latter is designed to pull air from the room to the bottom of the tank, where it is finely dispersed by the agitator and distributed throughout the liquid in the tank. Automatic devices control the aeration rate and the flow of cooling water.

The equipment is operated batchwise for white distilled vinegar. When the ethanol content of a batch falls to 0.3 or 0.2%, by volume, about 35–40% of the contents of the tank are removed as finished product. Fresh feed is pumped in to restore the original level and the cycle starts again. It is imperative that the agitator and air supply operate continuously during the change. Cycles usually are from 25 to 35 hours in length.

The production rate from the Acetator may be 10-fold higher than that from a trickling unit, depending on the size of the units. The yield from the Acetator is higher than that from the trickling generator. Values of 94 and 85% respectively, have been reported by manufacturers.

Advantages of the Acetator over the trickling type of generator include:

1. Superior performance (noted above)
2. Freedom from variations in production rates (noted above)
3. Small size
4. Ease and economy of change from one type of vinegar to another
5. Ease of start-up to full production after a long shut-down
6. Adaptability to continuous production of cider vinegar

Some disadvantages of the Acetator are:

1. Susceptibility to complete production stoppage as a result of a one minute power failure. Some manufacturers install stand-by generators which supply instantaneous power in the event of a failure.

2. High rate of power consumption. Total power consumption for a given amount of vinegar is no greater than for the trickling generator.

3. Production of cloudy vinegar which requires some type of fining treatment before it can be filtered.

Suggested modifications in the process of vinegar manufacture have appeared in the literature. Ebner (1967) has described the operation of the Acetator to produce 13% acetic acid. Ebner *et al.* (1967) have described a self-priming Acetator which operates without a compressor. Mori *et al.* (1970) studied the kinetics of the submerged acetic fermentation, and the behavior of *Acetobacter rancens* toward oxygen; and Enenkel and Ebner (1970) described the effect of improved gas dispersion and gas distribution on gas–liquid transfer.

3. The Cavitator

The success of the Frings Acetator prompted others to develop a less expensive model. One such modification was that described by White (1966). Liquor from an acetifier is circulated under high pressure through an aspirator where intimate mixing of liquor and air takes place. The system is easily installed in existing wooden vats and is adapted to continuous production. White (1966) states that it is believed to be the first continuous submerged culture carried out for the manufacture of vinegar and still is the only continuous process used in Great Britain. A 98% conversion efficiency of ethanol to acetic acid has been achieved in commercial operations. Another modification was the use of the Cavitator, which had been developed for the biological oxidation of sewage (Burgoon *et al.*, 1960; Cohee and Steffens, 1959; Beaman, 1967). Despite attempts to use the Cavitator for vinegar production on a commercial scale, technical difficulties have forced its abandonment.

4. The Tower Fermentor

A recent novel improvement in aeration systems applied to vinegar production has been described by Greenshields and Smith (1971, 1974) and Imrie and Greenshields (1973). It makes use of the tower fermentor (sometimes called the column fermentor) which consists of a column fitted at the bottom with a perforated plate. A patent has been granted recently to Greenshields (1972).

A 3000-liter pilot-scale production unit has been in use for more than a year. Both malt and spirit (white distilled) vinegar have been produced in the unit. The fermentor is constructed of polypropylene rein-forced with fiber glass and is 2 ft in diameter and approximately 20 ft tall with an additional expansion chamber 4.5 ft in diameter and 6 ft high. Aeration is accomplished through a plastic perforated plate cover-ing the cross section of the tower and holding up the liquid (Green-shields and Smith, 1974).

Two production units, each having a capacity of 10,000 liters, are planned by a British firm. Cost of the tower fermentor is said to be ap-proximately half that of a Frings Acetator of equivalent productive capacity (R. N. Greenshields, personal communication, 1975). Data are not available which would allow a comparison of production rates and efficiencies with those of the Frings Acetator.

5. Processing of Vinegar

Regardless of which manufacturing process is used, vinegar made for retail merchandising requires careful clarification. This is accomplished by filtration, usually with the use of filter aids and/or fining agents such as diatomaceous earth or bentonite, respectively. Masai and Yamada (1974) have been granted a patent for a process for the clarification of vinegar broth using bentonite and alginic acid, or salt thereof.

After clarification vinegar is bottled, sealed tightly, pasteurized at 140°–150°F for 30 minutes, and then cooled to 70°F (Weiser et al., 1971). Some manufacturers pasteurize the vinegar before bottling and bottle it hot. While this procedure is more economical, it is less efficient from a bacteriological standpoint. Carton dust found in all bottles can recontaminate the vinegar after pasteurization.

6. Concentration of Vinegar

Concentrated (200° grain or 20% acetic acid w/v) vinegar was de-veloped to alleviate two problems involved in its use. In the pickle processing industry, cucumbers are treated first with a salt brine and then with vinegar which soon becomes ineffective as the water from the cucumbers is extracted. In the past the diluted vinegar was replaced with a fresh charge and the former discarded, resulting in a loss of solids such as salt, sugar, and other flavoring material. A concentrated vinegar allows the restoration of the original acidity of the mixture without replacement and reduces the loss of flavor ingredients. The sec-ond motivating factor for developing a concentrated vinegar was the high cost of transporting a solution which contained 90% water. By con-centrating the vinegar to 200° grain, a substantial reduction was made in transportation costs. A process described by Dooley and Lineberry (1965) and developed by Girdler Process Equipment Division, Chemetron Corp., Louisville, Kentucky, yields 200° grain from 120° grain vinegar at a rate of 750–1000 lb/hr.

The concentration process involves slush-freezing 170° grain vinegar (obtained by blending 120° with 200° grain material) in a scraped-surface heat exchanger and then centrifuging the ice from the slush to produce the concentrate. A portion of the ice from the centrifuge is delivered to a feed-stream cooler where it is used to reduce the temperature of the fresh incoming vinegar, thereby reducing the heat load of the heat exchanger. Condenser water from the refrigeration system is used to melt the remaining ice. The melted ice is recycled to the vinegar generators since it contains 1.0–2.5% acid, depending on the centrifuge spin time, the grain of the original vinegar, and other factors. Vinegars of 300° grain can be produced by blending 200° and 120° grain vinegar to yield intermediates of higher than 170° grain. Vinegars of higher grain than 300° are seldom produced, since the cost of concentration increases disproportionately with the increase in grain strength. A bulletin published by the Chemetron Corporation (undated) gives details on the process.

VI. Major Types of Vinegar

A. White Distilled Vinegar

While white distilled vinegar can be made by the acetous fermentation of ethanol from any source, almost the entire production of this vinegar in the United States is derived from synthetic ethanol. In England the term "distilled vinegar" is applied to a distillate of malt vinegar, whereas in the United States the word "distilled" refers to the ethanol used as a raw material; the vinegar itself is not distilled.

Production data for white distilled vinegar in the United States are given in Section VII,A. This vinegar is used primarily in the pickle, salad dressing, tomato products, and mustard industries and in shelf packs for sale through retail outlets.

Specifications for white distilled vinegar have been developed by The Vinegar Institute (1974); they cover acid content (4% calculated as acetic acid), appearance (clarity), color, odor, trace metal content (copper and iron), and residual ethanol content (0.5% by volume, maximum).

While the first production of vinegar from fermented beverages occurred before any existing record, the use of an alcoholic distillate for vinegar must have been developed comparatively recently. But even this development does not appear to be well documented. Wüstenfeld (1930) mentions that in 1823 Schüzenbach trickled wine spirits through beechwood shavings to obtain vinegar.

Probably the first spirits used for vinegar were, by present standards, crude distillates containing traces of ethyl acetate, acetaldehyde, and fusel alcohols. The latter would of course be partially oxidized and esterfied and thus yield a more flavorful product than the present white distilled vinegar.

For many years, ethanol from the fermentation of molasses was used in the United States to produce distilled vinegar; neutral grain spirit was more expensive than ethanol from molasses and hence was little used. Beginning in the early 1950s, increasing amounts of synthetic ethanol from ethylene (derived from natural gas) were substituted for fermentation ethanol for vinegar production in the United States and other parts of the world.

Generally denatured ethanol is used in the United States for vinegar production. Several specially denatured alcohols (SDA) may be used, but of these SDA-29 is used by most vinegar manufacturers. It contains one gallon of ethyl acetate in 100 gallons of 190° proof ethanol. Another formulation, SDA-18, is prepared by adding 100 gallons of vinegar containing not less than 9% of acetic acid to 100 gallons of 190° proof ethanol. SDA 35-A is made by adding 5 gallons of ethyl acetate to 100 gallons of 190° proof ethanol (U.S. Industrial Chemicals Company, 1969).

For the preparation of distilled vinegar a feed or mash is made up to contain 10–14% ethanol v/v. A source of nutrient for the *Acetobacter* is added and the mixture delivered to the generator. If trickling generators are used, the new charge may be added to the holding tank at the bottom of the generator. Compositions of the nutrients, or vinegar foods, vary according to the manufacturer or compounder. Some vinegar producers prefer to compound their own nutrient rather than to purchase a prepared product. The usual ingredients are a source of ammonia nitrogen, phosphate [apparently Pasteur (1862b) was the first to suggest the use of diammonium phosphate to supply both of these ingredients], glucose, autolyzed yeast or pantothenate, and salts of organic acids. The addition of these organic salts is a logical development from the studies of Brown and Rainbow (1956). A quite satisfactory nutrient has been developed for both the circulating generator and the Frings Acetator (Nickol *et al.*, 1964). The minimum essential ingredients and their optimum amounts for maximum production rates have not been established.

As a result of the replacement of fermentation with synthetic ethanol, studies were initiated on analytical methods designed to differentiate between vinegars made from ethanols from the two sources. Also a need developed to detect the adulteration of cider and wine vinegars with the much cheaper distilled vinegar.

Edwards and Nanji (1938) and White (1971) discuss at length analytical methods and results which might yield a pattern to possibly detect the adulteration of malt vinegar with acetic acid or with distilled white vinegar. The methods included oxidation value (the amount of $KMnO_4$ reduced by a specified amount of distillate from the vinegar), the iodine value, and the ester value. The data have ranges from 2- to more than 8-fold for the three attributes in different samples. Despite the authors' views, the large inherent variations in samples make it virtually impossible to detect moderate adulteration of a malt vinegar, for example, with a cheaper spirit vinegar or with acetic acid.

Faltings (1952) was the first to show that fermentation ethanol and acetic acid derived from this ethanol have a measurable radioactivity, and he suggested that this phenomenon could be useful in differentiating biogenic from synthetic materials. Studies on this subject have been presented by Kaneko et al. (1973) and Masai et al. (1973). Hildebrandt et al. (1969) have reviewed a paper by Simon et al. (1968) who determined ^{14}C in ethanol and acetates of different origins by using a liquid scintillation counter and reported the suitability of this method for analyzing vinegars. Mecca et al. (1969, 1970) described similar studies on a number of wines and foods. Masai et al. (1973) also estimated ^{14}C in various vinegars and vinegar-containing foods, using a liquid scintillation counter, and detected synthetic acetic acid in several products purchased in retail markets in Japan. Their technique appears capable of distinguishing between foods and liquids containing synthetic acetic acid derived from fossil sources of carbon and those containing vinegar made from fermentation ethanol, which contains carbon from recently photosynthesized carbohydrates.

Vinegar or acetic acid, derived by any method, from an intermediate produced from a fossil fuel contains a low concentration ^{14}C and can be distinguished from vinegar or acetic acid derived from compounds containing recently photosynthesized carbohydrates. These sources carry relatively high concentration of ^{14}C and include wood and ethanol produced from carbohydrates by fermentation. This method cannot be used to determine whether the acetic acid in a food was produced chemically from a fossil fuel or whether it is vinegar produced from synthetic ethanol by Acetobacter. Likewise, the acetic acid in vinegar produced from fermentation ethanol can not be distinguished by this method from so-called synthetic acetic acid made by the destructive distillation of wood. Inherent variations in the materials undergoing analysis and in the method of analyses make it impossible to detect a moderate adulteration of, say, wine vinegar with acetic acid or with white distilled vinegar from fossil fuel.

Meinchein et al. (1974ab,c) have shown that the carboxyl carbon of acetic acid manufactured from apple cider is enriched in ^{13}C to the extent of about 20%, compared to 2% enrichment for the methyl carbon. The distribution of ^{13}C in acetic acids might serve as a means of distinguishing between samples of acetic acid from synthetic or from fermentation sources. However, it may not be adequate to detect the adulteration of vinegar, since wide variations are observed in the isotopic compositions of vinegars (W. G. Meinschein, personal communication, 1975). Thus the $\Delta\delta$ ^{13}Cs (δ ^{13}C for —COOH minus δ ^{13}C for —CH$_3$) for two commercial vinegars were 19% and 12%, from Michigan and Connecticut, respectively.

B. CIDER VINEGAR

This material may be made (in the U.S.) by ". . . the alcoholic and subsequent acetous fermentation of the juice of apples or concentrate thereof" (The Vinegar Institute, 1974). Specifications for cider vinegar include the same attributes as those for white distilled vinegar except that higher values are allowed for copper (5 ppm maximum vs. 1 ppm for white distilled vinegar) and iron (10 ppm maximum vs. 2 ppm), and allowance is made for 1 ppm of lead.

Beech (1972) has described in detail the making of apple juice and fermented apple juice (which the English call cider) as it is carried out in England, and Vaughn (1954) discusses the preparation of cider vinegar.

Modern methods of producing apple juice from the peelings of flumed apples may yield a juice having a lower sugar content than that from apples conveyed dry. In addition, most manufacturers use a type of peeling equipment which utilizes a jet of water to remove the peelings from the blades. While some of this water can be removed by passing the peelings over a shaker screen, the process can be expected to yield a juice having a Brix of 9° to 11° or about 1.0° Brix lower than that from whole dry-conveyed apples. If vinegar is produced by the submerged culture method, a yield of about 5 grains of acetic acid can be expected from each 1.0° Brix. Thus apple juice having a Brix of 10° will yield a vinegar of approximately 50 grain or 5% acetic acid after the yeast and *Acetobacter* fermentation. Some manufacturers blend apple juice concentrate with the apple cider to insure the desired concentration of ethanol in the fermented stock. Alcoholic fermentation of the apple juice may be allowed to proceed of its own accord, without the addition of yeast or nutrient. Addition of yeast is, of course, necessary when apple juice concentrate is used; but additional nutrients are not required. Storage of the fermented cider allows the settling of the yeast and finely suspended solids and the occurrence of a secondary fermentation in which malic acid is converted to lactic acid.

Fermented cider may cause a "sliming up" of circulating generators to a much greater extent than does the dilute ethanol used for white distilled vinegar. The sliming is attributed to the growth of *Acetobacter xylinium* and may interfere seriously with the operation of the generator. By blocking the passages in the shavings the slime may produce channeling, which causes a drop in conversion rates and efficiencies. Some control of the sliming may be accomplished by the proper regulation of the circulation and aeration rates. Dave and Vaughn (1969) state that sliming can be prevented by adjusting the stock so that, when acetified, it contains 9.0–10.0% acetic acid. This simple technique is not available to cider vinegar manufacturers who ferment unfortified apple juice. In extreme cases of sliming, steaming and/or washing of the packing or even repacking the generator may be required. No such difficulties are experienced with the Frings Acetator (see Section V,F,2).

Cider vinegar is bottled for retail sales and hence must be filtered to yield a sparkling, clear product. Filtration of this vinegar is not accomplished as easily as that of white distilled vinegar. Even after an effective filtration, cider vinegar may form a cloudy or colloidal precipitate. The latter may eventually settle and form a sediment in the bottle. The precipitate and its sporadic occurrence bear much resemblance to the phenomena which occur in unfermented cider. A long storage of maturation of the cider vinegar before bottling reduces the incidence of precipitation.

Monties and Barret (1965) analyzed the chill hazes, formed in apple juice, cider (fermented apple juice), and cider vinegar, and found them to contain tannins (67%), amino acids (19%), and sugars (4%). They postulated that the hazes were formed by the polymerization of tannin precursors during the processing and that oxidation of the pulp by air would prevent the formation of haze. Their premise leaves unexplained why haze occurs in cider vinegars, all of which have been subjected to extreme conditions of oxygen tension during their production.

Precipitation in unfermented cider has been attributed by Johnson *et al.* (1968) to the formation of complexes by the reaction of polymeric phenolic compounds (derived from leucoanthocyanidins and catechins) with proteins. From the appearance of the precipitates and their mode of formation in the clarified liquids, a similarity in composition may be anticipated in the sediments from apple cider and apple cider vinegar. Factors initiating the precipitation have not been defined. Certainly all samples of apple juice contain leucoanthocyanidins, catechins, their decomposition products, and protein; yet not all samples of cider nor cider vinegar show the phenomenon of reclouding in the bottle.

C. Wine Vinegar

This is made by the alcoholic and acetous fermentation of grape juice or by the acetous fermentation of grape wine, produced according to federal regulations, in a manner quite similar to that for cider vinegar. The same general specifications apply to wine vinegar as to that from cider except that for the former a tolerance of 120 ppm, maximum, of sulfur dioxide is allowed (The Vinegar Institute, 1974).

Wine vinegar is manufactured primarily in those countries that are important wine producers; namely, Italy, France, West Germany, Spain, Portugal, South Africa, and various South American nations. Lower quality wines may be used, or those that have soured and hence are undesirable for table use but nevertheless yield excellent vinegar.

D. MALT VINEGAR

This product is made by the alcoholic and subsequent acetous fermetation of an extract of a malt mash or of a mash containing malt, corn, and/or barley. Amylolytic enzymes in the malt convert the starch in the grains to sugars. The mash is crudely clarified by means of a brewer's slotted clarifier and fermented with yeast to yield an ethanolic solution. The latter is then acetified in the customary manner. Details of the method of manufacture are given by White (1966, 1970) and by Slater (1970).

Very little malt vinegar is manufactured in the United States but in England it is one of the major types produced. Much of the English malt vinegar is exported to the United States where it is used on fish and chips.

VII. The Industry

A. UNITED STATES AND BRITISH PRODUCTION

No records are available that give the number or productive capacity of operating vinegar plants in the United States or in the world. In the United States, approximately 50 companies operate plants located primarily in the Middle West in a band extending from Michigan to Texas. Other plants are located in California and in the northeastern states. As in many other industries, there has been a trend for the large companies to increase their production at the expense of the smaller ones.

The Robert H. Kellen Company, Atlanta, Georgia has conducted surveys of the United States Vinegar Industry since 1967. Questionnaires are sent annually to all major American manufacturers. For the 1974 survey, conducted in 1975, it was estimated that the data received represented 88% of all vinegar domestically manufactured and bottled. Forty-eight firms participated in the survey and of these 36 manufactured and bottled vinegar, 9 manufactured only, while 3 bottled only. These production data, as supplied by the Robert H. Kellen Company (personal communication, 1975) for all types of vinegars, are given in Table I. Table II

TABLE I

UNITED STATES PRODUCTION OF
VINEGARS OF ALL TYPES[a]

Year	Volume (gallons)
1967	117,653,945
1968	128,909,076
1969	132,786,465
1970	140,469,806
1971	144,604,472
1972	141,135,842
1973	141,653,234
1974	146,334,630

[a] Data supplied by the Robert H. Kellen Company (personal communication, 1975).

TABLE II

VOLUMES OF THE MAJOR TYPES OF VINEGARS SOLD IN 1974[a]

Types	Volume (gallons)	Grain strength[b]
White distilled vinegar	111,826,363	100
Cider vinegar	24,679,813	60
Wine vinegar	5,042,668	50
Malt vinegar	1,149,988	50
Others	3,635,630	—
Total reported	146,334,630	

[a] Data supplied by the Robert H. Kellen Company (personal communication, 1975).

[b] Grain strength is 10 times the concentration of acetic acid w/v.

contains data on the volumes of the most important types produced in 1974, and Table III contains the estimated volumes of vinegar sold for the most important uses in that year.

It is interesting to note that while the present production of vinegar is used almost entirely in foods, an attempt was made during World War I to manufacture industrial acetic acid by the acetification of ethanol. In 1915 the U.S. Industrial Chemicals Company of Baltimore, Maryland constructed and operated two of the largest vinegar plants ever built. The acetic acid was converted with lime to calcium acetate and the latter pyrolyzed to acetone.

One plant contained 200 generators 18 ft in height and 25 ft in diameter; the second contained 960 generators measuring 18 ft in height and

TABLE III

MAJOR USES AND DISPOSITIONS OF VINEGAR
MANUFACTURED IN 1974[a]

Use	Volume	
	Gallons[b]	% of reported production[c]
Pickle industry	30,909,517	21.1
Shelf pack	30,507,372	20.8
Salad dressings industry	25,319,274	17.3
Tomato products industry	17,651,360	12.1
Mustard industry	15,600,178	10.7
Other processed foods	9,062,367	6.2
Nonmanufacturing bottler	9,232,839	6.3
Other vinegar manufacturers	6,343,738	4.3
Carry-over	741,459	0.5
Nonfood usage	636,032	0.4
To export	320,548	0.2
Total	146,324,684	
Total reported production	146,334,630	

[a] Data supplied by the Robert H. Kellen Company (personal communication, 1975).

[b] Gallonages attributed to various end uses or industries were calculated from estimates supplied by contributing firms and based on the total industry production data. The discrepancy between the two totals resulted from omissions and inaccuracies in the survey data.

[c] Based on the total of column 2.

10 ft in diameter. Douglas fir and cypress woods were used to construct the generators. The latter differed from most conventional generators in that they were constructed with two false bottoms. No provision was made for aeration other than that of a natural draft through holes provided in the sides below the lower false bottom. A freely rotating device at the top of each generator served to distribute the liquid evenly over the beechwood shavings. Fermented malt extract was used as the nutrient for the *Acetobacter* during the acetification of cane molasses fermentation-ethanol. The plant operated for a number of years, until competition from acetone produced by direct fermentation made the process uneconomical.

Consistent production figures are not available for the amount of vinegar produced annually in the United Kingdom. Estimates supplied by the Vinegar Brewers Federation (G. F. Bond, personal communication, 1975) are as follows (expressed as gallons of vinegar containing 5% acetic acid):

Malt	16,000,000
Alcohol (spirit)	7,000,000
Others (wine, cider, etc.)	500,000
Total	23,500,000

The development of the industry has been somewhat different in England than in the United States. Large vinegar-manufacturing firms have been formed by the amalgamation of smaller ones. From 46 firms in 1900, the number was reduced to 7 by 1968. These include one large combine which occupies a dominant position in the industry, three other large independent firms, and three smaller ones.

B. European Economic Community

Vinegar production data for the European Economic Community were supplied by the Verband der deutschen Essigindustrie e. V. (J. Mürau, personal communication, 1975). Table IV contains production data, in hectoliters of 10% vinegar, for the years 1958 to 1974, inclusive. With respect to the total volume of vinegar manufactured, the major producing nations in the Western world are in this order: United States (first), Germany, France, The United Kingdom, and Italy. The American pro-

TABLE IV

Production of Vinegar by Nations of the European Economic Community[a]

Year	Belgium	Germany	France	Italy	Holland	United Kingdom	Denmark	Total production
1958	120,700	788,100	622,700	400,300	62,830			1,994,630
1959	112,630	779,400	638,015	400,300	61,730			1,992,075
1960	116,282	748,000	625,800	440,450	58,332			1,988,864
1961	114,441	758,435	713,200	445,000	60,128			2,091,204
1962	112,147	814,580	681,297	442,500	57,546			2,108,070
1963	118,460	849,945	677,358	440,000	61,842			2,147,605
1964	98,800	909,519	695,318	450,000	59,696			2,213,333
1965	124,992	871,255	688,212	457,500	58,130			2,200,089
1966	120,439	913,091	744,386	460,000	89,486			2,327,402
1967	108,672	968,382	802,916	462,500	91,780			2,434,250
1968	113,911	982,930	791,899	470,000	103,397			2,462,137
1969	110,881	918,517	813,410	416,000	100,154			2,358,962
1970	110,204	1,035,378	891,840	408,000	102,393			2,547,815
1971	124,386	1,054,092	905,250	400,000	105,250			2,589,679
1972	122,197	1,001,392	921,290	425,000	111,855	554,050		3,135,784
1973	119,856	1,082,409	964,560	450,000	119,606	564,800	115,142	3,416,373
1974	130,095	1,024,045	922,609	468,000	128,893	611,800	102,317	3,387,759

[a] Data are expressed in hectoliters of vinegar containing a 10 % concentration of acetic acid. Data supplied by Verband der deutschen Essigindustrie e.V. (J. Mürall, personal communication, 1975).

duction of 10% vinegar for 1974 was calculated to be 5,048,000 hectoliters, a figure substantially larger than that of the entire European Economic Community.

Production volumes for the United Kingdom given by the Verband der deutschen Essigindustrie e. V. do not agree with the estimate supplied by the Vinegar Brewers Federation. The values are 13,458,000 and 23,500,000, respectively. Possibly some of the discrepancy arises from the captive use of the vinegar by its producers.

Table V contains comparative data on the number of plants, average production per plant, and per capita consumption of vinegar for 1974. The inordinately large production per plant in the United Kingdom and the United States results from the trend of the larger plants to increase their production at the expense of the smaller ones. Average production per plant varies from 5.5×10^3 hectoliters for Italy to 112×10^3 for the United States. There is a significant correlation ($P > 0.95$) between the average plant production and the per capita consumption of vinegar for the eight nations. Consumption varies from 0.85 liter in Italy to 2.4 liters per person in the United States.

C. Manufacturers' Associations

In 1945 a number of United States vinegar manufacturers formed an organization to serve as a forum for the discussion of problems common to the industry. The original association was reorganized in 1955 as The Vinegar Institute and in 1967 its management was taken over by the Robert H. Kellen Company, Atlanta, Georgia. At present it represents

TABLE V

Number of Operating Plants, Average Plant Production, and Vinegar Consumption per Capita in the European Economic Community for 1974

Country	Number of producing plants	Vinegar production per plant[b]	Per capita consumption[c]
Belgium	16	8,131	1.33
Denmark	3	34,106	2.03
Germany	50	20,481	1.65
France	34	27,136	1.76
Italy	84	5,571	0.85
Holland	9	14,321	0.95
United Kingdom	7	87,400	1.09
United States[a]	45	112,175	2.40

[a] Figures calculated from data supplied by The Robert H. Kellen Company and based on an estimated population of 210×10^6.

[b] Hectoliters of 10% vinegar.

[c] Liters of 10% vinegar.

companies producing over 80% of the vinegar manufactured in the United States.

A primary objective of The Vinegar Institute is to coordinate activities among the member companies in order to improve the quality of vinegar, and to increase its acceptance by the consumer and its profitability to the manufacturer and bottler. Some activities of The Vinegar Institute include the development of specifications and standards of identity for different types of vinegar, and the dissemination of information regarding pending legislation which may affect the manufacturers. It also conducts discussions and negotiations with various state and federal governmental agencies regarding restrictive regulations and laws applicable to vinegar manufacturers. Some topics which have been so discussed include labeling requirements, assurance of an adequate supply of raw materials, interpretation of existing regulations, and compliance with guidelines of the Environmental Protection Agency.

In England an association of vinegar companies existed as early as 1756. The purpose of this organization and of several others which followed it, in 1816, 1818, and 1838, was to stabilize the price of vinegar at a point where the manufacturer could realize a profit. A London and County Vinegar Brewers Association formed in 1901 was superseded from 1916 to 1917 by the United Vinegar Brewers Association (Slater, 1970). In 1930 the name was changed to the Malt Vinegar Brewers' Federation and in 1955 the name was changed again to the Vinegar Brewers' Federation, its present name.

The Verband der deutschen Essigindustrie e. V. was organized on June 23, 1949 to serve as a forum for the solution of problems common to the industry. Topics which have been considered include means of obtaining adequate supplies of raw materials (spirit and wine), problems resulting from increased manufacturing costs, negotiations with the acetic acid industry, labeling requirements, and the definition and standards of identity for vinegar. The organization presently includes 47 companies which operate 71 vinegar plants. Its effectiveness since 1958 may be judged by the increased sales (up 73.3%), increased price (up 158%), the increase in per capita consumption of vinegar (up 49.6%), and an increase in the portion of the market shared with the acetic acid industry (up to 81.8% from 55.7%).

VIII.   Microbiology of Vinegar Production

A.   EARLY OBSERVATIONS

Vaughn (1954) points out that the living nature of "mother of vinegar" may have been suspected before it was suggested by Boerhaave (1732).

Persoon (1822) described pellicle formation on various liquids and gave mother of vinegar the name *Mycoderma*, i.e., "mucinous skin." He examined the pellicle and observed that it was made of up cells resembling those of yeast, but he did not recognize the relationship between the formation of a pellicle and the production of acetic acid.

B. RECOGNITION OF BIOLOGICAL ORIGIN OF VINEGAR

Kützing (1837) was the first to recognize the microbial nature of vinegar. He showed that mother of vinegar contained minute dotlike organisms arranged in chains and was the first to isolate acetic acid bacteria from vinegar. He believed the organism to be an alga and named it *Ulvina aceti*, and he asserted quite positively that acetic acid was formed from ethanol by the vital activities of *U. aceti*. His concept attracted little or no attention at the time because very shortly Liebig (1839) expressed his theory that mother of vinegar was devoid of life and was a structureless precipitate of albuminous matter. Liebig believed that the mode of action of mother of vinegar was identical to that of platinum black in oxidizing ethanol to acetic acid.

C. CONFIRMATION OF BIOLOGICAL NATURE

Not until 1862–1864 were Kützing's views proved to be correct. Pasteur (1862a,b, 1864) showed that the acetic fermentation was a biological process whose inception and maintenance were the result of the metabolism of a living organism. He applied the name *Mycoderma aceti*, which, according to Lafar (1898), was first employed by R. Thompson. In the beginning Pasteur may not have understood that acetic acid was a product of the metabolism of the organism. He believed that the organism possessed a peculiar property of being able to retain the oxygen of the air and condense it in a manner similar to the action of platinum black (Lafar, 1898). Von Knieriem and Mayer (1837) showed beyond doubt that the production of acetic acid was a biological process and could not be classed with the oxidation of ethanol with platinum black. The latter oxidized both dilute and concentrated ethanol and had no critical temperature beyond which it did not take place. The production of acetic acid by fermentation, on the other hand, did not take place in solutions containing more than 14% ethanol, nor above 40°C.

It should not be assumed that there was an immediate and universal acceptance of the role of *Acetobacter* (*M. aceti*) in the conversion of ethanol to acetic acid. As late as 1891 Dittmar refused to accept the concept that a few grams of *M. aceti* could oxidize the many pounds of ethanol that passed through a Schützenbach tower. He believed that the wood shavings acted exactly as does platinum black, condensing oxygen in their pores and transferring it to the ethanol. He also believed that the *Mycoderma* converted a small portion of the ethanol to acetaldehyde, which in turn attracted and activated oxygen atoms that became available for the oxidation of the ethanol.

D. BACTERIAL FLORA OF GENERATORS

Although there still are many trickling or circulating generators in operation, no adequate laboratory or plant scale studies have been made comparing the efficiencies or conversion rates of isolates from different generators. One obstacle to such a study is the lack of a suitable simple technique for making the comparison. High concentrations of acetic acid do not develop in shake flasks containing mash for white distilled vinegar.

A second obstacle is the difficulty of knowing whether an organism isolated from a generator is the one primarily responsible for the oxidation of the ethanol. This is particularly true of those circulating generators producing white distilled vinegar. The difficulties of isolating active acetifying bacteria from generators have been discussed by only a few investigators (see Wiame, 1949; Razumovskaya and Loitsyanskaya, 1956; Shimwell, 1954; Wiame and Lambion, 1950; Wiame *et al.*, 1959). Direct plating of organisms from vinegar frequently is unsatisfactory since there are so few bacteria to be found in vinegar as it issues from a circulating generator. Special enrichment techniques are required in many cases to reveal the presence of the *Acetobacter* (H. A. Conner, G. B. Nickol, and F. M. Hildebrandt, unpublished data, 1960–1970). Wiame *et al.* (1959) obtained good results using plates containing deionized silica gel acidified with acetic acid.

Another difficulty in comparing strains may be their instability. Shimwell (1956) published results of a study in which he purportedly obtained four widely differing cultures of *Acetobacter*. Starting with a pure culture of *Acetobacter mesoxydans* (isolated from malt vinegar) and by using two selective methods of cultivation, he isolated mutants resembling *A. xylinum*, *A. xylinoides*, *A. orleanse*, and *A. rancens*. The first three were cellulosic types and were isolated from a laboratory acetifier which was packed with wood wool and through which was circulated a culture of the *A. mesoxydans*. The *A. rancens* was isolated from a submerged culture of the *A. mesoxydans*. The instability of *Acetobacter* cultures is discussed in more detail in the next section.

A few investigators have used or attempted to use pure culture techniques for vinegar production; among these were Henneberg (1926), Manteïfel (1939), Shaposhnikov *et al.* (1940), Wiame and Lambion (1951), and Suomalainen *et al.* (1965). However, Shimwell (1954) seems to be the only investigator to carry out all of Koch's postulates in connection with the acetic acid fermentation. An *Acetobacter* culture was isolated from a vinegar generator and characterized as belonging to Frateur's *mesoxydans* group of *Acetobacter* (see Section VIII,E). When used to seed laboratory acetifiers of both the recirculating and submerged aeration types, the organism produced acetic acid at a rate equal to or better than that achieved in industry. An isolate from the laboratory recirculating acetifier, after two months' operation of the latter, was identical morphologically, culturally, and physiologically to the strain originally isolated from the working acetifier.

Vinegar prepared by a method other than that using the Frings Acetator (see Section VII,B) involves the development of mixed cultures of bacteria. This is particularly true of the preparation of malt, wine, and cider vinegars where the substrates provide nutrients suitable for the development of a number of organisms. The conditions under which the conversion is carried out preclude a pure culture fermentation, but they are such that only specialized types of bacteria develop.

Possibly a succession of different organisms develops during the "let alone" or Orleans process, but no studies are available to support this conjecture. Similarly, few studies are available to show the number of kinds of bacteria that develop in circulating generators. Gamova-Kayukova (1950) isolated 12 strains of *Acetobacter* from vinegar plants located in Moscow, Lenigrad, and Odessa. He assigned the 12 strains to four species which he named *Bacterium schüzenbachii* (6 strains), *Bacterium xylinum* (3 strains), *Bacterium xylinoides* (2 strains), and *Bacterium ascendens* (1 strain). Frateur and Simonart (1952) isolated from a Frings generator strains of *Acetobacter* having biochemical properties corresponding to those of *A. rancens, A. mesoxydans,* and *A. xylinum.* Wiame *et al.* (1959) isolated from an experimental generator an organism differing from other *Acetobacter* in that it requires an acidic medium for growth. The bacterium, named *A. acidophilum,* did not grow above pH 4.3. Hromatka and Leutner (1963a,b) studied the bacterial flora of submerged vinegar fermentations.

E. Taxonomy

1. Conventional Taxonomic Schemes

Acetobacter strains are widely distributed in nature and have been isolated from many sources, including vinegar generators, beer, wine, yeast, wort, apple juice, cane juice, flowers, vegetables, and many fruits. The early workers undoubtedly did not realize that a generator might contain more than one species or strain of *Acetobacter* or that different generators might contain different strains or species.

Apparently Hansen (1879b) was the first to recognize that acetic acid bacteria were pleomorphic. His observations were confirmed by Brown (1886a) using a culture isolated from beer, which he named B. aceti.

Hansen (1879a) also was the first to distinguish between species of Acetobacter. He isolated three species which were later named Bacterium aceti, Bacterium pasteurianum, and Bacterium kützingianum. These were distinguished from one another by their morphology, nature of the pellicles, colony form on solid media, and the reaction of their capsules toward iodine. The latter reacted with the capsular material of B. pasteurianum and B. kützingianum to yield a blue color, showing the presence of starch in the cell envelope.

Subsequent to Hansen's work many strains or species were isolated by other workers. Lafar (1895) conducted a series of experiments to prove the existence of various types of microorganisms which produced acetic acid from ethanol. The organisms differed not only morphologically but physiologically. Asai (1968) has published a comprehensive treatment of the classification and metabolism of the acetic acid bacteria including a listing of the important strains or species of Acetobacter which had been isolated from 1837 to 1962. Several earlier, also excellent, publications discuss the physiology and classification of the acetic acid bacteria. Among these communications are those of Vaughn (1942), Shimwell (1948), Brown and Rainbow (1956), Razumovskaya and Loitsyanskaya (1956), Rao (1957), Cooksley and Rainbow (1962), and Rainbow (1966).

The first publication on the classification of acetic acid bacteria was that of Hansen (1894) who recognized four strains. These included the three organisms that he had isolated previously (Hansen, 1879b) and A. xylinum (Brown, 1886b). The latter was distinguished by the thick, leathery pellicle of cellulose formed on liquid media. Three other investigators, Beijerinck (1898), Henneberg (1898), and Rothenbach (1898), published classification schemes and a year later Hoyer (1899) modified the Beijerinck system. All of these investigators used differences in habitats as a primary, and morphological and physiological properties as supplementary bases of classification. A comprehensive classification was published by Janke (1916) which was based on the metabolism of nitrogen and carbon sources as well as pellicle structure, mobility, and acid formation from carbohydrates.

The continual isolation of organisms by a number of investigators resulted m a proliferation of the number of species described in the literature and increased the difficulty of including them in a rational classification.

Physiological and biochemical properties have been used to classify the acetic acid bacteria by a number of investigators, including Kluyver and de Leeuw (1925), Herman and Neuschul (1931), Asai (1934), and

Frateur (1950). Visser't-Hooft (1925) reduced the number of species. His study was followed by that of Vaughn (1942) who proposed a simplified scheme based on the ability or inability of the organisms to oxidize acetic acid. He recognized only 7 types of species. However, Shimwell (1948) reversed the trend started by Visser't Hooft and Vaughn and included several more species in Vaughn's scheme.

Frateur (1950) has published a comprehensive study of *Acetobacter* in which he lists all of the important strains or species that have been described from the time of Pasteur (1868) to Shimwell (1948). He recognized 10 species and divided these into 4 groups on the basis of their biochemical properties: peroxydans, oxydans, mesoxydans, and suboxydans. The properties used for classification purposes were (1) the presence or absence of catalase, (2) and (3) the ability to oxidize acetic and lactic acids, respectively, (4) the presence or absence of ketogenic activity, and (5) the ability to produce gluconic acid from glucose. Two organisms used for vinegar production, *A. aceti* and *A. rancens*, were placed in the mesoxydans and oxydans groups, respectively. *A. xylinum*, which sometimes is responsible for the sliming up of vinegar generators, also is a member of the mesoxydans group.

Not until 1954 was it discovered by Leifson that mobile *Acetobacter* carried polar, multitrichous and peritrichous flagella. Prior to this finding it was believed that the mobile strains of *Acetobacter* had only polar monotrichous flagella. Shimwell (1958, 1959) confirmed Leifson's observation and published the first electron micrograph of an *Acetobacter* strain. The micrograph was taken by James M. Shewan of Torry Research Station, Aberdeen, Scotland. Leifson suggested that the term *Acetobacter* be applied only to peritrichously flagellated species and nonflagellated species which can oxidize acetate and lactate to CO_2 and H_2O. This characterization of *Acetobacter* was accepted by Shimwell (1959).

2. Present Views

Shimwell (1959) rejected Frateur's (1950) classification on the basis of the variability of *Acetobacter* species. While Asai (1968) gives Tošić and Walker (1946) credit for first observing variability in *Acetobacter*, Shimwell and Carr (1959) point out that Beijerinck (1911) observed the loss of pigmentation and an increase in slime formation in certain strains of *Acetobacter*.

During the careful examination of *Acetobacter* cultures Shimwell (1956, 1957, 1959) found that the majority of his cultures were mixtures of two or more species derived by spontaneous mutation (see Section VIII,D). After a vigorous "purification" by plating and replating, the mutant strains would again mutate or sometimes back-mutate to yield a culture resembling but not necessarily identical to the original. All but one of five biochemical attributes of *Acetobacter* were mutable. The one exception was the ability to grow in a medium containing ethanol as the sole source of carbon and ammonium salts as the sole source of nitrogen.

By plating out a culture of *A. aceti* on unhopped beer agar, Shimwell (1957) obtained five different colony forms; one, a smooth circular type did not form a coherent film on Hoyer's medium as did the parent culture. Pleomorphism of acetic acid bacteria was first observed by Hansen (cited by Lafar, 1898) and was confirmed by Shimwell (1959). The latter found the "abnormal" forms observed by Hansen and also both coccoid and crescent-shaped cells in the same culture. Thus Shimwell found biochemical, cultural, and morphological mutants in his cultures. He states that the classification and identification of *Acetobacter* strains are virtually impossible since a given property may diminish until it has been lost, or at other times a culture may gain a biochemical or other property of another species. In either case a culture of *Acetobacter* may no longer correspond to the species originally assigned. After an extended study of many strains representing all available classical species of *Acetobacter* and some new strains isolated from various materials, Shimwell (1959) concluded that *Acetobacter* strains cannot be classified in any manner, regardless of what criteria are used.

De Ley (1961) also rejected the species concept for *Acetobacter* but for a different reason than did Shimwell (1959). In contrast to the results of Shimwell (1957, 1959) and Shimwell and Carr (1959), De Ley observed no mutation or change in the cultures studied (two of which were the same as those studied by Shimwell and Carr in 1959) but rejected the species concept on the basis of the continuous spectra of variations in biochemical properties from one strain to another. He suggested the elimination of species names since he found no abrupt changes in the sequence of *Acetobacter* strains and that within each phylogenetic line the strains can be arranged in a continuous gradation from the most complex to the one having the minimal enzyme system. Differences between closely related strains consist only of the presence or absence of one or two enzymes. A similar view was expressed by Asai *et al.* (1964). De Ley (1961) did not find that strains of acetic acid bacteria were as variable as claimed by Shimwell and rejected the latter's view of the abnormally high mutability of *Acetobacter*.

De Ley's views are supported by physiochemical studies by Scopes (1962). The latter found that the infrared spectra of 9 strains of *Acetobacter* formed a smoothly graduated series, showing no marked differences that could be correlated with species categories or that could be attributed to the presence or absence of specific biochemical properties.

While other investigators have observed variations in *Acetobacter*

(Beijerinck, 1911; Tošić and Walker, 1946), apparently none has observed the facile mutation found by Shimwell (1957, 1959). Suomalainen *et al.* (1965) found no variations in 100 isolated colonies of a strain of *A. rancens* used for the commercial production of vinegar in Finland.

De Ley (1961) and Schell and De Ley (1962) did not find that *Acetobacter* were as mutable as claimed by Shimwell (1959). De Ley (1961) observed no obvious changes in the properties of 39 strains of *Acetobacter* in studies carried out over a 3-year period and rejected Shimwell's view of the abnormally high mutability of acetic acid bacteria. Schell and De Ley (1962) reinvestigated some of the strains studied by De Ley (1961) and observed that occasional aberrant colony types were not biochemical but merely colony variants. Three biochemical mutants of *A. aceti* were found. One was a spontaneous mutant lacking catalase; the other two mutations were induced by 0.04% $MnCl_2$ and lacked the particulate glucose-oxidative system.

Shimwell "purified" his cultures by plating and replating, but no investigator has studied comprehensively the effectiveness of this technique for purifying *Acetobacter* strains. No investigator has isolated single-cell cultures of *Acetobacter* by means of a micromanipulator. If such single-cell isolates mutated readily, Shimwell's (1959) observations would be confirmed. An obvious explanation of his results is that he was studying contaminated cultures, and Schell and De Ley (1962) did find Shimwell and Carr's (1960) strain of quasi-*Acetobacter* to be a mixed culture, but there is no other evidence to support this view. Possibly unrecognized differences in cultural techniques and conditions in the separate laboratories might be responsible for the mutability observed by Shimwell and not by others. For the present, the differences between the results of Shimwell and others remain unresolved.

Strains of *Acetobacter*, other than *A. aceti* which have been used commercially or which have been isolated from vinegar generators, are referred to as *Bacterium curvum*, now named *A. curvum* (Henneberg, 1926); *Bacterium schüzenbachii*, now named *A. schüzenbachii* (Gamova-Kayukova, 1950); *A. acidophilum* prov. sp. (Wiame *et al.*, 1959); and *A. rancens* (Suomalainen *et al.*, 1965). A very efficient strain of *Acetobacter* has been found (Hromatka and Leutner, 1963) for use in the Frings Acetator. H. A. Conner, G. B. Nickol, and F. M. Hildebrandt, unpublished data (1960–1970), used this organism with good results in laboratory circulating-type generators.

F. NUTRITION AND METABOLISM

Acetobacter strains used for vinegar production are characterized by their ability, under proper conditions of high aeration rate and high acidity, to convert ethanol rapidly and with high yields to acetic acid. While other organisms have been found to oxidize ethanol (Stanier, 1947) none of these is capable of the transformation under the extremely acid conditions tolerated by *Acetobacter*. The latter utilizes glucose but does not form acetic acid from it or other carbohydrates. Ammonia nitrogen can serve as the sole source of nitrogen for most strains. While neither vitamins nor growth factors are required by *A. aceti* in laboratory cultures, it is common practice in industry to add autolyzed yeast (Hildebrandt *et al.*, 1961) or pantothenate (Nickol *et al.*, 1964) to the diluted ethanol for the production of white distilled vinegar.

Rainbow (1966) and Asai (1968) have reviewed in detail the extensive studies made comparing the nutrition and metabolic processes of a large number of *Acetobacter* strains. Most investigators in this field have chosen to select strains from culture collections for their metabolic studies. Little emphasis has been given to cultures now used for vinegar production; i.e., relatively few strains of *Acetobacter* recently isolated from vinegar generators and presently used for vinegar production have been included in the studies. Hence, categorical statements are not possible regarding the nutrition and metabolism of the organisms in commercial use for vinegar production throughout the world.

Some strains such as *A. suboxydans* have been studied because of their unusual metabolic products. Other cultures investigated are sufficiently related to organisms used commercially to allow extension of certain general concepts to the latter.

1. Carbon Metabolism

The comparative carbohydrate metabolism of acetic acid bacteria has been studied by De Ley (1961), Kondo and Ameyama (1958), and Ameyama *et al.* (1965). De Ley and Schell (1959) found that *A. aceti* converted about 80% of the consumed glucose into gluconate and used the remaining 20% as a source of carbon and energy. Glucose was metabolized by two independent, competing oxidative pathways; one led to the formation of gluconate, the other was the hexokinase pathway followed by the hexose monophosphate cycle. Mannose, galactose, xylose, 1-arabinose, and ribose were oxidized to the corresponding sugar acids but were unsuitable carbon sources for the growth of most acetic acid bacteria. The strain of *A. aceti* used was able to oxidize mannitol, fructose, mannose, galactose, xylose, glycerol, erythritol, sodium *d*-lactate, ethanol, and sodium acetate.

De Ley (1961) showed that earlier experiments of Neuberg and Simon (1928a,b) were invalid. The latter claimed that *A. ascendens, A. pasteurianum,* and *Glucobacter suboxydans* ferment glucose with the formation of about equal amounts of ethanol and CO_2. The absence of the glycolytic cycle in *Acetobacter* is now recognized. King and Cheldelin (1952a,b) showed that A. *suboxydans* did not contain the enzymes of the tricarboxylic acid (TCA) cycle, but Rao (1955) showed the presence of the TCA cycle enzymes in A. *aceti* and King *et al.* (1956) demonstrated the presence of all these enzymes in a strain of A. *pasteurianum*.

Stouthamer (1959) studied the carbohydrate metabolism of 20 *Acetobacter* strains belonging to the oxydans, mesoxydans, and suboxydans groups. Enzymes of the pentose or hexose monophosphate oxidative cycle were present in all groups. The TCA cycle was present in the oxydans and mesoxydans strains but was absent in the suboxydans group. These results on the whole were confirmed by De Ley (1961). Simplified maps of the carbon metabolism of *Acetobacter* presented by both investigators were in good agreement. Nakayama (1960) showed that a "crude extract" (centrifuged at 10,000 rpm for 10 minutes) of an organism used for commercial vinegar production showed oxygen consumption in the presence of β-hydroxybutyrate, citrate, α-ketoglutarate, succinate, malate, oxaloacetate, and pyruvate; thus he demonstrated the presence of enzymes involved in the TCA cycle. Williams and Rainbow (1964) found 7 enzymes involved in the TCA cycle in extracts of 5 lactaphilic strains of acetic acid bacteria.

2. Nitrogen Metabolism

Rainbow and Mitson (1953) and Brown and Rainbow (1956) characterized as lactaphiles those *Acetobacter* which grew relatively poorly in amino acid–glucose media, but which grew well when lactate provided the source of energy and carbon. The classification of lactaphiles corresponds to Vaughn's (1942) overoxidizing, and Leifson's (1954) *Acetobacter* categories and includes A. *aceti* and A. *rancens* strains which have been used for vinegar production. The ability of lactaphilic *Acetobacter* to utilize inorganic nitrogen has been noted and the supposition has been advanced that most vinegar bacteria are included in this group.

Few studies have been made relative to the nitrogen metabolism pathways of *Acetobacter*. Brown and Rainbow (1956) showed that lactaphiles grew readily on a substrate containing ammonium sulfate as a nitrogen source and lactate as a source of energy and carbon. Poor growth was obtained on substrates containing ammonium salts and only glucose as a source of energy. Growth occurred in the presence of organic acid salts if glucose also was present, succinate being the most effective; malate, citrate, and fumarate were less efficient in promoting growth. Binary mixtures containing citrate, fumarate, malate, or succinate and glucose, glycerol, ethanol, or acetate promoted the growth of lactaphiles in media containing inorganic nitrogen.

Glutamate, proline, aspartate, and alanine were rapidly assimilated by lactaphiles when grown on a substrate containing acid-hydrolyzed casein (CAH). Good growth was obtained when L-proline, L-glutamate, or L-aspartate, ninhydrin-reacting compounds were formed which were different from those in the control suspension. The proline, glutamate, and aspartate systems yielded new materials having the chromatographic mobilities, respectively, of glutamate, aspartate, and alanine. Cooksley and Rainbow (1962) found that alanine arose through enzymatic decarboxylation of L-aspartate by a β-decarboxylase.

While the mechanism of ammonia assimilation is not known, Rainbow (1966) has suggested that the initial step is the synthesis of glutamate, catalyzed by glutamic dehydrogenase, from ammonia and 2-oxyglutarate generated in the TCA cycle. He postulated the formation of L-aspartate from L-glutamate and oxaloacetate by transamination. Other amino acids also may arise in the same manner by transamination from glutamate.

3. Growth Factors

Rainbow and Mitson (1953) found that *A. mobile* required only *p*-aminobenzoic acid as a growth factor. Two other members of the lactaphilic group had no vitamin requirement when grown on vitamin-free CAH. Brown and Rainbow (1956) extended this study to an additional 8 identified lactaphilic strains of *Acetobacter*, including *A. aceti* and *A. rancens*, and 8 untyped lactaphilic isolates. None of the lactaphiles tested had any requirement for added growth factors. Those studies were supported by the results of Ameyama and Kondo (1966) who also found that *Acetobacter* strains required no vitamins for growth.

IX. Mechanism of the Biological Conversion of Ethanol to Acetic Acid

The ancients recognized that exposure to air was required for the conversion of wine or beer to vinegar, and that it was necessary to prevent the free access of air to wine in order to preserve it. Thus Pliny reports: "Districts with a milder climate store their wine in jars and bury them in the ground entirely, or else up to a part of their positions, so protecting them against the atmosphere . . ." (Pliny, transl. by Rackham, 1945, p. 273). At the time of Pliny (Pliny, transl. by Bostock and Riley, 1856, p. 267) wine vessels were coated inside with pitch or with resin. While the primary purposes may have been to flavor the wine and to prevent losses by leakage and evaporation, the technique doubtless helped to delay the conversion to vinegar.

A. First Observations: Abbé Rozier and Lavoisier

The Abbé F. Rozier (Lafar, 1898; Vaughn, 1954) recognized in 1786 that air was absorbed by the wine in the process of being converted to vinegar. In 1793 Lavoisier showed that oxygen was the only component of the air involved in the conversion (Lafar, 1898). In 1823 Edmund Davy (cited by Pasteur, 1862a) observed that when platinum black was moistened with spirits of wine it became hot and the odor of acetic acid became apparent.

B. Döbereiner's Development of the Equation for the Reaction

Döbereiner (cited by Mitchell, 1926) studied the reaction and observed that the ethanol reacted with the oxygen to yield water and acetic acid, but no carbon dioxide. He expressed the equation for the reaction as

$$C_4H_6O_2 + 40 \rightarrow C_4H_4O_4 + 2(HO)$$

Carbon was considered to have an atomic weight of 6 and oxygen of 8.

C. Presently Accepted Mechanism

Apparently Liebig (1851) was the first to note that acetaldehyde is an intermediate in the oxidation of ethanol to acetic acid (Mitchell, 1926). Pasteur (1862b) was the first to recognize that acetaldehyde was an intermediate in the formation of acetic acid by the so-called *Mycoderma aceti*. He also realized that the conversion of ethanol to aldehyde was a dehydrogenation reaction. Henneberg (1897) showed that the first step in the conversion of ethanol to acetic acid was the formation of acetaldehyde according to the equation

$$\underset{\text{ethanol}}{CH_3CH_2OH} + O \rightarrow \underset{\text{acetaldehyde}}{CH_3CHO} + H_2O \tag{1}$$

The second step is the formation of acetic acid from acetaldehyde. The latter first reacts with water to yield hydrated acetaldehyde which, in turn, is oxidized, or (properly) dehydrogenated, to yield acetic acid in accordance with the following equations:

$$\underset{\text{acetaldehyde}}{CH_3{-}CHO} + H_2O \rightarrow \underset{\substack{\text{hydrated} \\ \text{acetaldehyde}}}{CH_3C\overset{H}{\underset{OH}{-}}OH} \tag{2}$$

$$CH_3C \overset{H}{\underset{OH}{\diagup}} -OH + [O] \xrightarrow[\text{dehydrogenase}]{\text{aldehyde}} CH_3COOH + H_2O \tag{3}$$

hydrated
acetaldehyde

Two protons of the hydrated acetaldehyde are activated and transferred to the oxygen atom.

Neuberg and Nord (1919) confirmed the production of acetaldehyde as an intermediate by blocking the reaction with the addition of calcium sulfite and/or calcium bisulfite to the acetic fermentation. Bisulfites react with aldehydes to yield an addition product as follows:

$$2CH_3CHO + Ca(HSO_3)_2 \rightarrow 2CH_3 \overset{O-H}{\underset{H}{\overset{|}{-}C\overset{|}{-}}}SO_3Ca_{1/2} \tag{4}$$

The addition product is not oxidized by acetic acid bacteria and accumulates in the fermentation liquor.

Later Neuberg and Windisch (1925) proposed a second mechanism by which acetaldehyde is converted to acetic acid. Two molecules of acetaldehyde, formed as in Reaction (1), may react with each other in the so-called Cannizzaro reaction to yield acetic acid and ethanol:

$$CH_3CHO + CH_3CHO \xrightarrow{H_2O} CH_3COOH + CH_3CH_2OH \tag{5}$$

acetaldehyde   acetic   ethanol
acid

The ethanol then undergoes reoxidation according to Reaction (1) and the cycle is repeated until all but a trace of the ethanol is converted into acetic acid. No information is available to permit assessment of the relative contributions of these mechanisms under a given set of fermentation conditions (Rainbow, 1961).

The first step in the reaction, conversion of ethanol to acetaldehyde, is carried out by an enzyme (or enzymes), alcohol dehydrogenase. The second step, conversion of acetaldehyde to acetic acid, is accomplished by acetaldehyde dehydrogenase.

Nakayama (1959, 1960, 1961a,b) and Nakayama and De Ley (1965) have studied the enzyme systems involved in the acetic acid fermentation. The organism used in the study was isolated from an industrial vinegar plant. It did not utilize inorganic nitrogen and hence may not have been A. aceti.

The first paper (Nakayama, 1959) described a complete enzyme system associated with the cellular fragments and absent from the water-soluble fraction of the cells. This water-soluble extract contained a strong TPN-dependent, coenzyme A-independent aldehyde dehydrogenase.

A subsequent paper (Nakayama, 1960) described the purification and properties of this enzyme. The purified aldehyde dehydrogenase showed a broad substrate specificity in dehydrogenating acet-, propion-, n-butyl-, n-valer, n-capron-, n-oenanth-, and n-capryaldehydes. It was distinguished from the aldehyde dehydrogenase of yeast by its ability to dehydrogenate aldehydes having more than 10 carbon atoms in their chains and by its inability to dehydrogenate glyceraldehyde. A centrifugal supernatant from sonically disrupted cells effected the reduction of TPN by both ethanol and acetaldehyde. Increasing concentrations of semicarbazide prolonged the initial reduction of TPN, indicating that ethanol was dehydrogenated to form acetaldehyde which then reacted with the semicarbazide.

A second, coenzyme-independent aldehyde dehydrogenase, designated as acetaldehyde–ferricyanide reducing enzyme, was isolated and separated from the TPN-dependent aldehyde dehydrogenase (Nakayama, 1961a). The enzyme was concentrated about 30-fold from acetone-dried cells by extraction at pH 3.0, with a citrate–phosphate buffer. Subsequent processing included adsorption, elution, dialysis, and finally precipitation with ammonium sulfate.

The enzyme catalyzed electron transport from acetaldehyde to ferricyanide, methylene blue, or 2, 6-dichloroindophenol, but not to other coenzymes. It showed a broad specificity for aromatic and aliphatic aldehydes and was not inhibited by hydroxylamine.

An alcohol–cytochrome-533 reductase, a new type of alcohol dehydrogenase, also was isolated from acetone-dried cells (Nakayama, 1961b). This alcohol–cytochrome-533 reductase differed from the previously reported (Nakayama, 1960) NAD-linked alcohol dehydrogenase. The isolation technique was similar in many respects to that used for the coenzyme-independent aldehyde dehydrogenase.

The alcohol-cytochrome-533 reductase reduced ferricyanide or 2,6-dichloroindophenol, but not NAD or NADP in the presence of ethanol. It showed a broad specificity toward n-straight chain monohydroxy alcohols from ethanol to n-decyl alcohol, inclusive. Phenyl propyl and cinnamyl alcohols and 1,4-butylene glycol were oxidized, but not the methyl or isoalcohols that were tested.

Nakayama and De Ley (1965) found the alcohol–cytochrome-533 reductase in several strains of Acetobacter in the oxydans and mesoxydans groups. A similar enzyme, but having a higher optimum pH, was found in A. aceti. Evidence was obtained to show that the enzyme was located mainly in the cell wall and that it was a constitutive enzyme.

One of the attributes of A. aceti, A. rancens, and others is their ability to oxidize acetic acid to carbon dioxide and water, a phenomenon noted by Pasteur (1862a). The mechanism of this oxidation is not clear. One is tempted to ascribe it to the TCA cycle; however, Furukawa *et al.* (1973a,b) have shown that radioactivity supplied by acetic ^{14}C was incorporated into the cells of *A. aceti* and into succinic acid during the acetic acid fermentation of yeast-fermented apple juice. However, when yeast-fermented apple juice was supplemented with succinic ^{14}C and yeast-fermented orange juice was supplemented with citric ^{14}C there was little decrease in the amount of citric acid and little radioactivity in the acetic acid. The data indicated that acetic acid was oxidized to succinic but that no citric acid in the medium was oxidized to acetic acid.

Apparently Nakayama (1959) is the only investigator to advance explanations for the accumulation of high concentrations of acetic acid from the oxidation of ethanol by *Acetobacter*. He suggested that all of the bound coenzymes in the cells may be used for the conversion of ethanol to acetic acid so that the enzyme responsible for the oxidation of the latter cannot function; or possibly that the oxidation of ethanol to acetic acid and the further metabolism of acetic acid may be regulated, in an unknown manner, by the particle-bound enzymes, coenzymes, and cytochrome system.

X. Composition of Vinegar

While the acetic acid present in vinegar is responsible for its preservative and solvent properties, trace components have an important effect on the flavor and the improved piquancy of vinegar over that of a solution of acetic acid and water. The trace components are contributed by the raw materials, are produced as a result of the fermentations, or are derived from the interaction of compounds from these two sources. Thus ethanol and acetic acid react to form ethyl acetate, one of the flavor components of vinegar. During the acetic fermentation, trace components in the raw materials are subjected to the oxidative effect of the *Acetobacter*.

Some of the trace compounds found in highest concentrations in wine and in fermented cider and malt extract include acetaldehyde, acetal, ethyl acetate, and fusel alcohols. Synthetic ethanol contains only insignificant amounts of impurities and hence white distilled vinegar contains little or no fusel alcohols or their oxidation products.

A number of investigators have identified a wide variety of compounds in vinegars. A few (see discussion of studies of Jones and Greenshields below) have measured the concentration of trace components in vinegars.

De Ley (1959) found that *A. rancens, A. pasteurianum,* and *A. ascendens* formed acetoin from *dl*-lactate by means of a dual pathway. The lactate first was dehydrogenated to pyruvate. The latter was then decarboxylated in the presence of thiamine pyrophosphate (ThPP) to yield aldehyde-ThPP. Aldehyde-ThPP reacted with acetaldehyde to yield acetoin and ThPP, or the aldehyde-ThPP reacted with pyruvate to yield α-acetolactate and ThPP. The α-acetolactate was decarboxylated to acetoin and CO_2. No other microorganisms were known which possessed more than one mechanism for the formation of acetoin. Hromatka and Polesofsky (1962) showed that alcohols having 3,5 carbon atoms were oxidized by *Acetobacter* to the corresponding acids.

Suomalainen and Kangasperko (1963) found ethyl acetate but no other volatile flavorant in spirit vinegar. Only a small portion of the ethyl acetate (200 mg/l) was formed during the production of the vinegar. The major portion, up to 1500 mg/l, was formed during storage. Five phenolic acids were identified by Maekawa and Kodama (1964) as occurring in rice vinegar.

Aurand *et al.* (1966) identified 25 volatile compounds in four different types of vinegars, namely, cider, wine, white distilled (five different samples), and a special mixed tarragon-flavored vinegar. The compounds identified included 8 carbonyls, 9 esters, and 8 alcohols.

Jones and Greenshields (1969, 1970a,b, 1971) have studied the composition of malt vinegars. They identified 6 esters, 6 alcohols, 2 acids (besides acetic), acetoin, and acetaldehyde in one or more of several commercial vinegars. In the first study reported in 1970 they compared the compositions of commercial samples of charging wort with the composition of acetified wort immediately after production, after filtration, and after 6 months storage. All compounds found in the vinegar, except propionic and isobutyric acids, were found in the wort. *n*-Amyl acetate was found in samples of commercial vinegars but not in the experimental samples compared. There was no increase in the number of volatiles or in their concentrations during storage of the vinegar after acetification.

In the second 1970 study these authors found that each volatile compound occurred at approximately the same concentration in different acetified liquors regardless of the origin or method of production. They suggest that the process of acetification produces a pattern and concentration of volatile compounds which are in equilibrium with the bacterial cell's most favorable metabolic balance.

The fourth study compared the composition of malt vinegars produced by the Frings, a continuous, and a quick process. Previous observations were confirmed in that the concentrations of the individual volatile compounds do not vary markedly in finished vinegars despite the variations in strength and composition of the wort, the yeast strain, the acetifying culture, and the type of acetification process.

Kahn *et al.* (1972) identified 61 compounds occurring in one or more samples of six commercial vinegars. The latter included distilled and cider vinegars from both the submerged and trickling (quick vinegar) processes, as well as one sample each of malt and wine vinegars. Sixteen esters, 11 alcohols, 11 halogenated compounds, 8 hydrocarbons, 7 carbonyls, and 5 ethers and acetals were identified by a combination of gas chromatography and mass spectrometry. Another 21 compounds were shown to be present but could not be identified with certainty. The origins of many of the compounds, including 18 containing halogens found in vinegar, have not been determined. Possibly some of them originated from the waters used in vinegar production.

Twenty samples of wine vinegar were analyzed by Garcia *et al.* (1973). Most of the samples contained acetoin and butylene glycol but only 7 contained diacetyl. Zappavigna *et al.* (1974) found tyramine in all samples of wine vinegar they examined.

Yanazida *et al.* (1974) determined that vinegar mashes from rice, sake cake, malt, grapes, and apples contained alanine, aspartic acid, proline, and glutamic acid with smaller quantities of histidine, arginine, and methionine. The concentrations of these amino acids decreased during the acetic acid fermentation because of bacterial assimilation.

XI. Summary

This review traces the historical development of the technology, microbiology, and chemistry of vinegar from the early records of each to the present. Vinegar has played an important but little-emphasized role as a food adjunct in man's development of his civilization. Production methods and improvements developed slowly and empirically for centuries, and only in the last few years have they benefited from the application of the scientific method.

The amount of information presently available regarding the acetic fermentation is only a fraction of that which exists on its companion, the alcoholic fermentation. Much remains to be learned regarding the enzyme system of these remarkably efficient organisms. Another gap in our knowledge is that of their mode of survival. They are ubiquitous in nature; and this is difficult to understand, considering their behavior in laboratory culture. Frequent transfers are necessary to maintain the cultures in a viable condition, and a 1-minute interruption of the flow of air to a Frings Acetator results in the death of most of the cells.

Many unresolved differences concerning *Acetobacter* are recorded in the literature, an outstanding one being the disparate views of Shimwell and De Ley.

It is unlikely that in the near future there will be marked advances in methods of vinegar production and use. Vinegar will continue to be used primarily as a food preservative and enhancer. Preparation of special vinegars for food enhancement has received little attention and this phase needs more study.

No great latitude exists for the improvement of yields in the acetic fermentation. Yields of 94% of the theoretical, based on the weight of the alcohol, have been claimed by manufacturers. Similarly, the rate of acetification is not likely to be increased by a significant amount. However, it is possible that the use of more efficient strains of *Acetobacter* may achieve some decrease in cycle time and an increase in the final acidity of vinegar from a Frings Acetator.

A cheaper automated fermentor and a workable automated continuous process for vinegar are likely developments. While the tower fermentor of Greenshields and co-workers appears to be cheaper than the Frings Acetator, only inadequate comparative data exist on their relative productive capacities.

REFERENCES

Allgeier, R. J., Nickol, G. B., and Conner, H. A. (1974). *Food Prod. Dev.* **8**(5), 69–71; 8(6), 50–53 and 56.

Alwood, W. B. (1903). *U.S., Dep. Agric. Bur. Chem. Bull.* **71.**

Ameyama, M., and Kondo, K. (1966). *Agric. Biol. Chem* **30**(3), 203–211; *C.A.* **64**, 20239 (1966). Original not seen.

Ameyama, M., Fujisawa, H., and Kondo, K. (1965). *Nippon Nogei Kagaku Kaishi* **39**(11), 427–435; *C.A.* **64**, 7074 (1966). Original not seen.

Anonymous. (1670). *Phil. Trans. R. Soc. London* **5**, 2002–2004.

Antoniani, C., Cernuschl, E., Lanzani, G. A., and Pecile, A. (1958). *Clin. Ostet. Ginecol.* **60**, 225; cited by Asai (1968). Original not seen.

Asai, T. (1934). *J. Agric. Chem. Soc. Jpn.* **10**, 621, 731, 932, and 1124; cited by Asai (1968). Original not seen.

Asai, T. (1935). *J. Agric. Chem. Soc. Jpn.* **11**, 50, 331, 377, 499, 610, and 674; cited by Asai (1968). Original not seen.

Asai, T. (1968). "Acetic Acid Bacteria." Univ. Park Press, Baltimore, Maryland.

Asai, T., Iizuka, H., and Komagata, K. (1964). *J. Gen. Appl. Microbiol.* **10**, 95.

Aurand, L. W., Singleton, J. A., Bell, T. A., and Etchells, J. L. (1966). *J. Food Sci.* **31**(2), 172–177.

Beaman, H. R. (1967). *In* "Microbial Technology" (H. Peppler, ed.), Chapter 13, pp. 344–376. Van Nostrand-Reinhold, Princeton, New Jersey.

Beech, F. W. (1972). *J. Inst. Brew., London* **78**, 477–491.

Beijerinck, M. W. (1898). *Zentralbl. Bakteriol., Parasitenkd. Infektionskr., Abt. 2* **4**, 209; cited by Asai (1968). Original not seen.

Beijerinck, M. W. (1911). *Zentralbl. Bakteriol., Parasitenkd. Infektionskr., Abt. 2* **29**, 169; cited by Asai (1968). Original not seen.

Bitting, A. W. (1928). *Fruit Prod. J. Am. Vinegar Ind.* **8**(7), 25–27; 8(8), 16–18; 8(9), 13–15; 8(10), 18–20 and 27; 8(11), 16–18; 8(12), 18–20.

Bitting, A. W. (1929). *Fruit Prod. J. Am. Vinegar Ind.* **9**(1), 19–20 and 25; 9(2), 49–51 and 86; 9(3), 86.

Boerhaave, H. (1732). "Elementa Chemicae." Lugduni Batavorum **2**, 179–207; cited by Lafar (1898) and by Mitchell (1926). Original not seen.

Brannt, W. T. (1900). "Manufacture of Vinegar," 2nd ed. Baird, Philadelphia, Pennsylvania.

Brown, A. J. (1886a). *J. Chem. Soc., London* **49**, 172–187.

Brown, A. J. (1886b). *J. Chem. Soc., London* **49**, 432–439.

Brown, G. D., and Rainbow, C. (1956). *J. Gen. Microbiol.* **15**, 61–69.
Burgoon, D. W., Ciabattari, E. J., and Yeomans, C. (1960). U.S. Patent 2,966,345.
Cato. "Cato the Censor on Farming" (transl. by E. Brehaut), p. 107. Columbia Univ. Press, New York, 1933.
Cato and Varro. "Marcus Porcius Cato on Agriculture; Marcus Terentius Varro on Agriculture" (transl. by W. D. Hooper; rev. by H. B. Ash), pp. 73, 97, and 107. Harvard Univ. Press, Cambridge, Massachusetts, 1934.
Chemetron Corporation. (Undated). "Crystallization and Concentration. Subject: Vinegar," Tech. Bull., Votator Div., Chemetron Corp., Louisville, Kentucky.
Clark, J., and Goldblith, S. A. (1974). "Processing and Manufacturing of Food in Ancient Rome," contribution no. 2327. Department of Nutrition and Food Science, Massachusetts Instutute of Technology, Cambridge.
Cohee, R. F., and Steffen, G. (1959). *Food Eng.* **31**(3), 58–59.
Columella. "Lucius Junius Moderatus Columella on Agriculture and Trees" (transl. by E. S. Forster and E. H. Heffner), pp. 199, 207, and 223. Harvard Univ. Press, Cambridge, Massachusetts, 1955.
Cooksley, K. E., and Rainbow, C. (1962). *J. Gen. Microbiol.* **27**, 135–142.
Crawford, S. L. (1929). *Fruit Prod. J. Am. Vinegar Ind.* **9**(3), 80–83 and 85.
Cruess, W. V. (1948). "Commercial Fruit and Vegetable Products," 3rd ed. McGraw-Hill, New York.
Dave, B. A., and Vaughn, R. H. (1969). *Am. J. Enol. Viticult.* **20**(1), 56–65.
De Ley, J. (1959). *J. Gen. Microbiol.* **21**, 352–356.
De Ley, J. (1961). *J. Gen. Microbiol.* **24**, 31–50.
De Ley, J., and Schell, J. (1959). *Biochim. Biophys. Acta* **25**, 154.
Digbie, K. (1669). "The Closet of the Eminently Learned Sir Kenelme Digbie Kt." Printed by E. C. for H. Brome at the Star in Little Britain, London ("Mallinckrodt Food Classics," Vol. VI. Mallinckrodt Chemical Works, St. Louis, Missouri, 1967).
Dittmar, W. (1891). In "Encyclopaedia Britannica" (T. S. Baynes, ed.), Vol. 9, p. 98. R. S. Peale Company, Chicago, Illinois.
Dooley, J. R., and Lineberry, D. D. (1965). In "Symposium on New Developments in Bioengineering—Minneapolis," preprint, 12 p. Am. Inst. Chem. Eng., New York.
Duddington, C. L. (1961). "Microorganisms as Allies," Chapter 1. Macmillan, New York.
Dussance, H. (1871). "A General Treatise on the Manufacture of Vinegar." Baird, Philadelphia, Pennsylvania.
Ebner, H. (1967). *Zentralbl. Bakteriol., Parasitenkd. Infektionskr. Hyg., Abt. 1: Orig., Suppl.* **2**, 65–72.
Ebner, H., Pohl, K., and Enenkel, A. (1967). *Biotechnol. Bioeng.* **9**, 357–364.
Edwards, F. W., and Nanji, H. R. (1938). *Analyst* **63**, 410–420.
Enenkel, A., and Ebner, H. (1970). *160th Natl. Meet. Am. Chem. Soc., 1970.*
Faltings, V. (1952). *Angew. Chem.* **64**, 605.
Fetzer, W. R. (1930). *Food Ind.* **2**, 489.
Fowler, G. J., and Subramaniam, V. (1923). *J. Indian Inst. Sci.* **6**, 147–171.
Fowler, G. R., and Schwartz, M. H. (1973). In "Encyclopedia Britannica" (W. E. Preece, ed.), Vol. 6, p. 657. William Benton, Chicago, Illinois.
Frateur, J. (1950). *Cellule* **53**, Separate Vol. 3, 287.
Frateur, J., and Simonart, P. (1952). *Congr. Int. Ind. Agrar., 9th, 1952* Vol. II C, p. 15; cited by Suomalainen et al. (1965). Original not seen.
Frings, H. (1932). U.S. Patent 1,880,381.
Frings, H. (1937). U.S. Patent 2,094,592.

Furukawa, S., Takenaka, N., and Uedda, R. (1973a). *J. Ferment. Technol.* **51**(5), 321–326.

Furukawa, S., Takenaka, N., and Uedda, R. (1973b). *J. Ferment. Technol.* **51**(5), 327–334.

Gamova-Kayukova, N. I. (1950). *Mikrobiologiya* **29**(2), 137–146.

Garcia, O. R., Carballido-Esteved, A., and Castana-Torres, M. (1973). *Am. Bromatol.* **25**(2), 121–145; *C.A.* **80** 106,966 (1974). Original not seen.

Gilliland, R. B., and Lacey, J. P. (1966). *J. Inst. Brew., London* **72**, 291–303.

Gogúi, A. (1856). "Secrets de la cuisine française." Libraire de L. Hachette et Cie, Paris.

Greenshields, R. N. (1972). British Patent 1,263,059.

Greenshields, R. N., and Smith, E. L. (1971). *Chem. Eng. (London)* **249**, 182–190.

Greenshields, R. N., and Smith, E. L. (1974). *Process Biochem.* **9**(3), 11–13, 15, 17, and 28.

Haeseler, G. (1949). *Branntweinwirtschaft* **75**, 17.

Ham, J. (1824). British Patent 5012.

Hansen, A. E. (1935). *Food Ind.* **7**, 277.

Hansen, E. C. (1879a). *Medd. Carlsberg Lab.* **2**, cited by Brown (1886a). Original not seen.

Hansen, E. C. (1879b). *C. R. Trav. Lab. Carlsberg* **1**, No. II, 49, 96; cited by Asai (1968). Original not seen.

Hansen, E. C. (1894). *Res. Medd. Carlsberg Lab.* **3**, No. 3, 265; cited by Asai (1968). Original not seen.

Heid, J. L., and Joslyn, M. A. (1963). "Food Processing Operations." Avi Publ., Westport, Connecticut.

Heid, J. L., and Joslyn, M. A. (1967). "Fundamental of Food Processing Operations." Avi Publ., Westport, Connecticut.

Henneberg, W. (1897). *Zentralbl. Bakteriol., Parasitenkd. Infektionskr., Abt. 2* **3**, 933.

Henneberg, W. (1898). *Dtsch. Essigind.* No 14; "Handbuch der Garungsbakteriologie," 2nd ed., Vol. 2, p. 191. Parey, Berlin. Cited by Asai (1968). Original not seen.

Henneberg, W. (1926). "Handbuch der Garungsbakteriologie." Berlin. Cited by Shimwell (1954). Original not seen.

Herman, S., and Neuschul, P. (1931). *Biochem. Z.* **233**, 130.

Hildebrandt, F. M., Nickol, G. B., Dukowicz, M., and Conner, H. A. (1961). "Vinegar Newsletter," No. 33. U.S. Ind. Chem. Co., New York.

Hildebrandt, F. M., Nickol, G. B., and Conner, H. A. (1967). "Vinegar Newsletter," No. 50. U.S. Ind. Chem. Co., New York.

Hildebrandt, F. M. Nickol, G. B., and Kahn, J. H. (1969). "Vinegar Newsletter," No. 56. U.S. Ind. Chem. Co., New York.

Hippocrates. "Genuine Works of Hippocrates" (transl. by F. Adams), Vol. 1, p. 301. Lydenham Society, London, 1849.

Holy Bible. A. V. Num. 6:3; Ruth 2:14; Ps. 69:21; Prov. 10:26; Matt. 27:48; Mark 15:36; Luke 23:36; John 19:29.

Hoyer, D. P. (1899). *Dtsch. Essigind.* **3**, No. 1–25; "Bijdrage tot de Kennis von de Azijnbakteriën." Leidener Dissertation, University of Leiden, Waltman, Delft, 1898. Cited by Asai (1968). Original not seen.

Hromatka, O., and Ebner, H. (1949). *Enzymologia* **13**(6), 369–387.

Hromatka, O., and Ebner, H. (1950). *Enzymologia* **14**(2), 96–106.

Hromatka, O., and Ebner, H. (1951). *Enzymologia* **15**(2), 57–69.

Hromatka, O., and Ebner, H. (1955). U.S. Patent 2,707,683.

Hromatka, O., and Kastner, G. (1951). *Enzymologia* **15**(6); 337–350.

Hromatka, O., and Leutner, U. (1963a). *Branntweinwirtschaft* **103**(1), 1.

Hromatka, O., and Leutner, U. (1963b). *Branntweinwirtschaft* **103**(6), 107.

Hromatka, O., and Polesofsky, W. (1962). *Enzymologia* **24**(6), 341–359.

Hromatka, O., Ebner, H., and Csoklich, C. (1951). *Enzymologia* **15**(3), 134–153.

Huber, E. (1927a). *Dtsch. Essigind.* **31**(1), 12–15.

Huber, E. (1927b). *Dtsch. Essigind.* **31**(2), 28–30.

Imrie, F. K. E., and Greenshields, R. N. (1973). *Biotechnol. Bioeng. Symp.* **4.**

Janke, A. (1916). *Zentralbl. Bakteriol., Parasitenkd. Infektionskr., Abt. 2* **45**, 48.

Jarvis, D. C. (1959). "Folk Medicine," Appendix B. Holt, New York.

Johnson, G., Donnelly, B. J., and Johnson, D. K. (1968). *J. Food Sci.* **33**, 254–257.

Jones, D. D., and Greenshields, R. N. (1969). *J. Inst. Brew., London* **75**, 457–463.

Jones, D. D., and Greenshields, R. N. (1970a). *J. Inst. Brew., London* **76**, 55–60.

Jones, D. D., and Greenshields, R. N. (1970b). *J. Inst. Brew., London* **76**, 235–242.

Jones, D. D., and Greenshields, R. N. (1971). *J. Inst. Brew., London* **77**, 160–163.

Joslyn, M. A. (1970). *Kirk-Othmer Encycl. Chem. Technol., 2nd. Ed.* **21**, 254–269.

Kahn, J. H., Nickol, G. B., and Conner, H. A. (1972). *Agric. Food Chem.* **20**, 214–218.

Kaneko, T., Ohmori, S., and Masai, H. (1973). *J. Food Sci.* **38**, 350–353.

King, T. E., and Cheldelin, V. H. (1952a). *J. Biol. Chem.* **198**, 127–133.

King, T. E., and Cheldelin, V. H. (1952b). *J. Biol. Chem.* **198**, 135–141.

King, T. E., Kawasaki, E. H., and Cheldelin, V. H. (1956). *J. Bacteriol.* **72**, 418–421.

Kluyver, A. J., and de Leeuw, F. J. G. (1925). *Dtsch. Essigind.* **22**, 175.

Kondo, K., and Ameyama, M. (1958). *Bull. Agric. Chem. Soc. Jpn.* **22**, 369–379; *C.A.* **53**, 22229C (1959). Original not seen.

Kromer, R. (1972). "Dragoco Report 3/72," pp. 57–58. Dragoco, Inc., Totowa, New Jersey.

Kützing, F. T. (1837). *J. Prakt. Chem.* **II**, 385.

Lafar, F. (1895). *Zentralbl. Bakteriol., Parasitenkd. Infektionskr., Abt. 2* **1**, 129–150.

Lafar, F. (1898). "Technical Mycology," Vol. I. Griffin, London.

Lanzani, G. A., and Pecile, A. (1960). *Nature (London)* **185**, 175.

Lavoisier, A. L. (1793). "Traité élémentaire de chimie," 2nd ed., Vol. 1, p. 159; cited by Lafar (1898) and Vaughn (1954). Original not seen.

Liebig, J. (1839). *J. Prakt. Chem.* **I**, 35, 312.

Liebig, J. (1851). "Letters on Chemistry," p. 216. London. Cited by Mitchell (1926). Original not seen.

Liefson, E. (1954). *Antonie van Leeuwenhoek* **20**, 102–110.

Llaguno, C. (1964). Patronato "Juan de la Cierva" de Investigation Tecnica, **21**, Madrid.

Löw, I. (1901). *In* "The Jewish Encyclopedia" (I. Singer, ed.), Vol. 12, pp. 439–440. Funk & Wagnalls, New York.

Maekawa, K., and Kodama, M. (1964). *Agric. Biol. Chem.* **28**(7), 436–442.

Manteïfel, A. Ya. (1939). *Mikrobiologiya* **8**(3,4), 372–79.

Martinez, M. P. (1963). "Bibliografia de los vinagres." Ministero de Industrias, Dept. de Information Technica, Havana.

Masai, H., and Yamada, K. (1974). Japanese Patent 74108,295; *C.A.* **82**, 84490 (1975). Original not seen.

Masai, H., Ohmori, S., Kaneko, T., and Ebine, H. (1973). *Agric. Biol. Chem.* **37**(6), 1321–1325.

Mecca, F., Sapegno, A., and Spaggiari, P. G. (1969). *Chim. Ind. (Milan)* **51**, 985–989.

Mecca, F., Sapegno, A., and Spaggiari, P. G. (1970). *Chim. Ind. (Milan)* **52**, 880.

Meinschein, W. G., Rinaldi, G. G. L., Hayes, J. M., and Scholler, D. A. (1974a). *Biomed. Mass Spectrometry* **1**, 172–174.

Meinschein, W. G., Rinaldi, G. G. L., Hayes, J. M., and Scholler, D. A. (1974b). *Biomed. Mass Spectrometry* **1**, 412–414.

Meinschein, W. G., Rinaldi, G. G. L., Hayes, J. M., and Scholler, D. A. (1974c). *Biomed. Mass Spectrometry* **1**, 415–417.

Mitchell, C. A. (1926). "Vinegar: Its Manufacture and Examination," 2nd ed. Griffin, London.

Moldenke, H. N. and Moldenke, A. L. (1952). "Plants of the Bible." Ronald Press, New York.

Monties, B., and Barret, A. (1965). *Ann. Technol. Agric.* **14**(2), 167–172.

Mori, A., Kunno, N., and Terni, G. (1970). *J. Ferment. Technol.* **48**(4), 203–212.

Nakayama, T. (1959). *J. Biochem. (Tokyo)* **46**, 1217–1225.

Nakayama, T. (1960). *J. Biochem. (Tokyo)* **48**, 812–830.

Nakayama, T. (1961a). *J. Biochem. (Tokyo)* **49**, 158–163.

Nakayama, T. (1961b). *J. Biochem. (Tokyo)* **49**, 240–251.

Nakayama, T., and De Ley, J. (1965). *Antonie van Leeuwenhoek* **31**(2), 205–219.

Neuberg, C., and Nord, E. F. (1919). *Biochem. Z.* **96**, 158.

Neuberg, C., and Simon, E. (1928a). *Biochem. Z.* **197**, 259.

Neuberg, C., and Simon, E. (1928b). *Biochem. Z.* **199**, 232.

Neuberg, C., and Windisch, F. (1925). *Biochem. Z.* **166**, 454.

Nickol, G. B., Conner, H. A., Dukowicz, M., and Hildebrandt, F. M. (1964). "Vinegar Newsletter," No. 43. US Ind. Chem. Co., New York.

Ochs, I. L. (1950a). *Arch. Otolaryngol.* **52**, 935–941.

Ochs, I. L. (1950b). *J. Am. Med. Assoc.* **142**, 1361–1362.

Pasteur, L. (1862a). *C. R. Hebd. Seances Acad. Sci.* **54**, 265–270.

Pasteur, L. (1862b). *C. R. Hebd. Seances Acad. Sci.* **55**, 28–32.

Pasteur, L. (1864). *Ann. Chim. Phys.* [4] **57**, 58.

Pasteur, L. (1868). "Etudes sur le vinagre." Masson, Paris.

Persoon, C. H. (1822). *Mycol. Eur.* **42**(1), 96.

Platina (1475). *"De honesta voluptate."* Venice L. De Aguila ("Mallinckrodt Food Classics," Vol. V. Mallinckrodt Chemical Works, St .Louis, Missouri, 1967).

Pliny. "The Natural History of Pliny" (transl. by J. Bostock and H. A. Riley), Book XIV, Chapter 25, p. 267. Henry G. Bohn, London, 1856.

Pliny. "Pliny Natural History" (transl. by H. Rackham), pp. 73, 97, 107, 451, and 453. Harvard Univ. Press, Cambridge, Massachusetts, 1945.

Pliny. "The Natural History of C. Pliny Secundus" (transl. by P. Holland), Book 10, p. 106. McGraw-Hill, New York, 1964.

Prescott, S. C., and Dunn, C. G. (1949). "Industrial Fermentations," 2nd ed. McGraw-Hill, New York.

Rainbow, C. (1961). *Prog. Ind. Microbiol.* **3**, 43.

Rainbow, C. (1966). *Wallerstein Lab. Commun.* **29**, No. 98/99, 5–14.

Rainbow, C., and Mitson, G. W. (1953). *J. Gen. Microbiol.* **9**, 371–375.

Rao, M. R. R. (1955). "Pyruvate and Acetate Metabolism in *Acetobacter aceti* and *Acetobacter suboxydans*." Thesis, Urbana, Illinois. Cited by De Ley (1961). Original not seen.

Rao, M. R. R. (1957). *Annu. Rev. Microbiol.* **11**, 317–338.

Razumovskaya, Z. G., and Loitsyanskaya, M. S. (1956). *Mikrobiologiya* **25**, 727–774.

Rothenback, F. (1898). *Dtsch. Essigind.* 35, 207; cited by Asai (1968). Original not seen.

Rozier, F. (1786). "Dictionnaire universel d'agriculture," Vol. 4, p. 525. Cited by Vaughn (1954). Original not seen.

Schell, J., and De Ley, J. (1962). *Antonie van Leeuwenhoek* 28, 445–465.

Schierback, J. (1951). "The Manufacture of Vinegar." Technical Information Service Report No. 17. National Research Council, Ottawa, Canada.

Scopes, A. W. (1962). *J. Gen. Microbiol.* 28, 69–79.

Shaposhnikov, V. N., Dulman, T. M., and Bechtereva, M. N. (1940). *Symp. Centr. Sci. Res. Lab. Ferment. Ind., Publ. Food Ind.* p. 24; cited by Gamova-Kayukova (1950). Original not seen.

Shimwell, J. L. (1948). *Wallerstein Lab Commun.* 11, 27–36.

Shimwell, J. L. (1954). *J. Inst. Brew., London* 60, 136–141.

Shimwell, J. L. (1956). *J. Inst. Brew., London* 62, 339–343.

Shimwell, J. L. (1957). *J. Inst. Brew., London* 63, 45–56.

Shimwell, J. L. (1958). *Antonie van Leeuwenhoek* 24, 187–192.

Shimwell, J. L. (1959). *Antonie van Leeuwenhoek* 25, 49–67.

Shimwell, J. L., and Carr, J. G. (1959). *Antonie van Leeuwenhoek* 25, 353.

Shimwell, J. L., and Carr, J. G. (1960). *Antonie van Leeuwenhoek* 26, 169–187.

Simon, H., Rouschenback, P., and Frey, A. (1968). *Z. Lenbensm.-Unters. -Forsch.* 136(5), 279–284.

Slater, A. W. (1970). *Ind. Archaeol.* 7, 292–309.

Slater, L. E. (1953). *Food Eng.* 25, 88, 89, 135, 137, and 139.

Stainer, R. T. (1947). *J. Bacteriol.* 54, 191–194.

Stouthamer, A. H. (1959). *Antonie van Leeuwenhoek* 25, 241–264.

Suomalainen, H., and Kangasperko, J. (1963). *Z. Lebensm.-Unters. -Forsch.* 120(5), 353–356.

Suomalainen, H., Keranen, A. J. A., and Kangasperko, J. (1965). *J. Inst. Brew., London* 71, 41–45.

Tait, A. H., and Ford, C. A. (1876). U.S. Patent 181,999.

Tannahill, R. (1973). "Food in History." Stein & Day, New York.

Thompson (1931). *J. R. Asiat. Soc. Bengal, Sci.* [2], 1–25.

Titus Livius. "The History of Rome" (transl. by D. Spillan and C. Edmonds), Vol. 2, Book 21, Chapter 37, pp. 735–736. Harper, New York, 1895.

Tošić, J., and Walker, T. K. (1946). *J. Soc. Chem. Ind., London* 65, 180.

U.S. Industrial Chemicals Company (1969). "Ethyl Alcohol Handbook." New York.

Vaughn, R. H. (1942). *Wallerstein Lab. Commun.* 5, 5–25.

Vaughn, R. H. (1954). *In* "Industrial Fermentations" (L. A. Underkofer and R. J. Hickey, eds.), Vol. 1, Chapter 17, pp. 498–535. Chem. Publ. Co., New York.

Vinegar Institute, Atlanta, Georgia (1974). "Specification for White Distilled Vinegar." Vinegar Inst., Atlanta, Georgia.

Visser't-Hooft, F. (1925). "Biochemische onderzockingen over het geslacht Acetobacter." Dissertation, Tech. Univ. (Hoogeschool), Meinema, Delft. Cited by Asai (1968) and Vaughn (1954). Original not seen.

von Knieriem, W. and Mayer, A. (1873). *Landwirtsch. Vers. Stn.* 16, 305; cited by Asai (1968) and Lafar (1898). Original not seen.

Weiser, H. R., Mountney, G. J., and Gould, W. A. (1971). "Practical Food Microbiology and Technology." Avi Publ., Westport, Connecticut.

White, J. (1966). *Process Biochem.* 1(3), 139–165.

White, J. (1970). *Process Biochem.* 5(10), 54–56.

White, J. (1971). *Process Biochem.* 6(5), 21–25 and 50.

Wiame, J. M. (1949). *Bull. Tech. Vinaigr.* 4, 74.

Wiame, J. M., and Lambion, R. (1950). *Bull. Tech. Vinaigr.* 6, 146.

Wiame, J. M., and Lambion, R. (1951). *Bull. Tech. Vinaigr.* 7, 195.

Wiame, J. M., Harpigny, R., and Dothey, R. G. (1959). *J. Gen. Microbiol.* 20, 165–172.

Williams, P. J. le B., and Rainbow, C. (1964). *J. Gen. Microbiol.* 35, 237–247.

Wüstenfeld, H. (1930). "Lehrbuch der Essigfabrikation." Parey, Berlin.

Yamamato, T., Lanzani, G. A., and Cernuschl, E. (1959). *Clin. Ostet. Ginecol.* 61, 35; cited by Asai (1968). Original not seen.

Yanazida, F., Fukui, J., Kaneko, N., Yamamato, Y., and Koizumi, Y. (1974). *Nippon Jozo Kyokai Zasshi* 69(11), 759–764; *C.A.* 82, 123295a (1975). Original not seen.

Zappavigna, R., Brambati, E., and Cerutte, G. (1974). *Riv. Vitic. Enol.* 27(7), 285–294; *C.A.* 82, 29650a (1975). Original not seen.

Microbial Rennets

M. STERNBERG

Marschall Division Research, Miles Laboratories, Inc., Elkhart, Indiana

I.	Introduction	135
II.	Microbial Rennets from *Endothia parasitica, Mucor miehei,* and *Mucor pusillus* var. *Lindt*	136
	A. Production and Purification	136
	B. Characterization of Milk-Clotting Enzymes	139
	C. Analytical Methods	142
	D. Preparation of Cheese with Microbial Rennets	144
III.	Screening for New Microbial Rennets	146
	A. Bacteria	148
	B. Fungi	150
IV.	Conclusions	151
	References	153

I. Introduction

The replacement of calf rennet with microbial milk coagulants has progressed at a fast pace. The search for microbial rennets has assumed a sense of urgency since the review of the subject by Sardinas (1972), more so than in the last two decades.

Microbial rennets are proteolytic enzymes produced by microorganisms. They are able to induce the coagulation of milk in a way similar to that of animal rennets. The standardized water extract obtained from the fourth stomach or abomasum of milk-fed calves is called rennet, in contrast to the pure enzyme called chymosin or rennin (E.C. 3.4.4.3). Added to a properly fermented milk it induces the formation of a coagulum which by processing and aging eventually becomes one of hundreds of cheese varieties consumed in the world.

The need for rennet substitutes has existed for some time since most calves are now raised to become mature animals whose stomach does not therefore contain rennet. The total cow herd, to which calves are born, has also decreased while world cheese production has increased every year over the past several years. In an authoritative market survey, Christensen (1974) estimated the yearly world production of cheese at 12 billion pounds and increasing at a rate of 500 million pounds per year. Supplies of calf stomachs have decreased in the United States from a high of 30 million in 1960 to a low of 3 million in 1973.

Calf rennet has been used for so many years and over such wide geographic areas that cheese made with rennet sets the standard for taste, flavor, consistency, and texture, a standard which rennet substitute must meet.

Declining supplies and increasing costs of calf rennet have been aggravated by negative changes in the market of pepsin, an extender of the calf enzyme since the beginning of the century (Merker, 1919). Here too, the increased use in processed meats of pork stomachs, from which pepsin is extracted, and rising costs of petroleum solvents employed in processing operations have doubled the price and reduced the supply. In such an economic climate the replacement of calf rennet by microbial rennets had to accelerate.

Scientific and patent literature published since the comprehensive review of microbial rennets by Sardinas (1972) mostly covers subjects related to commercial microbial rennets. Much attention is given to the characterization, milk-clotting mechanism, and action specificity of the enzymes on synthetic substrates. Results of cheese trials made with commercial microbial rennets have also been given in a significant number of research articles. Interest in analytic methods to determine the milk-clotting ability of rennets and to distinguish between commercial, animal, and microbial varieties has increased. Published papers reflect less interest in screening of microorganisms for new calf rennet substitutes. Market penetration by the existing products limits the chances of competition from new microbial rennets due to reluctance on the part of cheese-makers to undertake costly large-scale experiments. Still, interest in new microbial rennets is alive, as witnessed by patents securing the rights on microbial strains that produce milk-clotting enzymes, and the occasional criticism leveled at the existing commercial microbial rennets.

In 1974 proteases from *Endothia parasitica*, *Mucor pusillus*, and *Mucor miehei*, all of which made their appearance as substitutes of calf rennet at the beginning of the decade, were already used for the manufacturing of 60% of the cheese made in the United States. Blends of rennet with porcine pepsin accounted for 25% and rennet extract for the remaining 15% (Christensen, 1974). Considering that the animal enzyme has had centuries to accustom us to the peculiar taste of every variety of cheese, the fast acceptance of microbial rennets as substitutes is a remarkable achievement of science and industry.

II. Microbial Rennets from *Endothia parasitica*, *Mucor miehei*, and *Mucor pusillus* var. *Lindt*

A. PRODUCTION AND PURIFICATION

Milk coagulants from strains of *Endothia parasitica*, *Mucor miehei*, and *Mucor pusillus* var. *Lindt* account for almost all of the commercial microbial rennets in the world. Table I summarizes the names of microorganisms, trade names, and suppliers of commercial microbial rennets, as obtained from advertisements and inquiry letters to manufacturers.

TABLE I
COMMERCIAL MICROBIAL RENNETS

Microorganism	Trade name	Supplier
Endothia parasitica	Suparen	Pfizer International
Endothia parasitica	Sure-Curd	Pfizer, Inc.
Mucor miehei	Fromase	Wallerstein Company
Mucor miehei	Hannilase	Chr. Hansen's Laboratory, Inc.
Mucor miehei	Marzyme	Miles Laboratories, Inc.
Mucor miehei	Rennilase	Novo Industri A/S
Mucor pusillus var. *Lindt*	Emporase	Dairyland Food Laboratories, Inc.
Mucor pusillus var. *Lindt*	Meito MR	Meito Sangyo
Mucor pusillus var. *Lindt*	Noury Rennet	Vitex

Production and recovery methods with the above organisms and related strains which are not used for commercial production have been described in patents and scientific papers (Sardinas, 1966; Arima and Iwasaki, 1964; Iwasaki *et al.*, 1967; Anstrup, 1968; Baxter Laboratories, 1968; Charles *et al.*, 1970; Somkuti and Babel, 1967).

Except for *M. pusillus,* all organisms are grown by submerged fermentations in media based on soybean meal, glucose, grain starch hydrolyzates, mineral salts, and occasionally whey or skim milk. *Mucor pusillus* grown in surface culture on bran (Krayushkina *et al.*, 1973; Pozsár-Hajnal *et al.* 1974b; Huang, 1970; Cserháti and Holló, 1972a) has resisted attempts at deep fermentation by secreting excessive amounts of proteolytic enzymes (Pozsár-Hajnal *et al.*, 1974a) and lipase (Somkuti and Babel, 1968). The medium used for deep fermentation consisted of 5% w/v wheat bran in a water suspension, inoculated with a 24-hour 5% vegetative cell suspension. After 2 days fermentation time, 860 clotting units were obtained as compared to half that much produced by surface growth (Pozsár-Hajnal *et al.*, 1974a). However, the proteolytic activity increased so much that the authors had to disqualify their preparation for cheese manufacture. Modification of the fermentation conditions failed to reduce the ratio of proteolytic vs. clotting activities. In a latter work, Pozsár-Hajnal and Hegedüs-Völgyesi (1975), using electrofocusing analysis, confirmed their suspicion that the milk-clotting enzyme produced by deep fermentation of *M. pusillus* was different from the rennet produced in a surface culture by the same organism.

All commercial microbial rennets are mixtures of enzymes containing, apart from a major milk-clotting component, other enzymes that may be damaging to the cheese process. Removal of the accompanying enzymes in a supplimentary purification step permits more variability in fermentation conditions and in the use of high-activity mutants; it may also be necessary for improving the results of cheesemaking.

Concern about lipase, an enzyme which may cause rancid flavor in cheese, has been expressed in patents covering the purification of *M. miehei* rennet. Schleich (1971) effected the removal of esterase activity by acidifying the preparation of *M. miehei* to a pH of 2–3.5 and by maintaining it at a temperature of 20–55°C until the esterase was decomposed. Precipitation of the lipase from *M. miehei* rennet took place by simple adjustment of the pH to 4.5–4.7 (Charles and Dolby, 1973). The precipitate containing the lipase impurity was separated by centrifuging. Hershberger and Sternberg (1974) purified microbial rennets from *M. miehei* and *M. pusillus* var. *Lindt* by reversible precipitation with polyanionic polymers, an extension of a general method for purifying proteins with polyacrylic acids (Sternberg and Hershberger, 1974). Treatment of crude microbial rennets with polyacrylic or polyethylenemaleic acids yielded a water-insoluble complex between the reagent and the rennets. After filtration, the complex was decomposed with calcium carbonate which resulted in the insoluble calcium salt of the reagent and the microbial rennets significantly purified with respect to lipase, ash, and nonenzymic protein. Somkuti (1974) removed the lipase from a crude microbial rennet preparation obtained from the culture filtrate of *M. pusillus* PCC 410 by the use of ammonium sulfate precipitation and gel filtration on Sephadex G-75 or G-100. He noticed that, although lipase may contribute to acceleration of cheese curing, it also bears the risk of poor organoleptic quality.

Excessive proteolytic activity in a microbial rennet preparation causes extensive breakdown of milk proteins, reduced yields of cheese, and bitter taste. Improvement of the ratio of clotting to proteolytic activity by purification methods is possible if the crude microbial rennet preparation contains several proteolytic enzymes having different clotting-to-proteolysis ratios (Sternberg, 1971; Edelsten and Siegaard Jensen, 1972). Treatment of *M. miehei* and *E. parasitica* rennets with silicates (Organon Laboratories, 1971) may result in reduction of proteolytic activity without affecting the clotting power. Morvai-Rácz (1974) fractionated a filtrate of a strain of *M. pusillus* var. *Lindt* with ammonium sulfate and ion exchange and achieved a 2.8 times better clotting-to-proteolysis ratio than in the initial preparation. However, the ratio he obtained after purification was only slightly higher than the one achieved with *M. pusillus* rennet produced by Meito Sangyo Ltd., either because the manufacturer purifies it or is using a strain and proprietary fermentation procedure that yields an improved preparation. Rotini and Sequi (1972) and Sequi *et al.* (1972) purified the crude rennet from *E. parasitica* by gel filtration and chromatography on CM-cellulose. After elution with a gradient acetate buffer they isolated two enzymes having different ratios of clotting to proteolytic activities. Electrophoresis of casein samples clotted with the crude rennet and the separated enzymes showed that the crude rennet had a more advanced protein fragmentation effect on the casein clot than one of the two purified fractions. They suggested using the purified rennet for practical application.

A patent issued to Meito Sangyo (1975) concerns inactivation of cellulase impurities present in *M. pusillus* rennet. Prevention of enzymic deterioration of cellulose tissues used for sheathing cheese, can be achieved by thermal treatment with solutions of NaCl. The milk-coagulating activity of the rennet is not affected by the treatment.

B. CHARACTERIZATION OF MILK-CLOTTING ENZYMES

Apparent practical similarities between calf and commercial microbial rennets have stimulated interest in determining resemblances at the molecular level of the purified enzymes. The study of bond specificity on oxidized B chain of insulin (Fig. 1) has shown that the purified enzymes from *E. parasitica*, *M. miehei*, and *M. pusillus* var. *Lindt* as well as chymosin do preferentially split bonds of aromatic and hydrophobic amino acid residues. The most susceptible bond to *E. parasitica* enzyme is Phe_{24}–Phe_{25}, while 3 of the 8 split bonds are common to chymosin (Williams *et al.*, 1972). Microbial rennets from *M. miehei* strain CBS 370.65 characterized by Ottesen and Rickert (1970) and strain NRRL 3420 characterized by Sternberg (1971) are closely related to chymosin with respect to a preference for splitting peptide bonds. All bonds hydrolyzed by the *M. miehei* enzymes, except for Tyr_{26}–Thr_{27}, are also hydrolyzed by chymosin. The enzymes of the two *M. miehei* strains have almost identical bond specificities except for the additional hydrolysis of Ala_{14}–Leu_{15} by *M. miehei* NRRL 3420 enzyme.

McCullough and Whitaker (1971) incubated the *M. pusillus* enzyme with oxidized B insulin for 20 hours at $37°C$, and found more bonds hydrolyzed than did Oka *et al.* (1973). Oka and his colleagues incubated the same enzyme for only 10 minutes at $40°C$. Results from both groups show more restricted specificity than chymosin and lack of attack of the Tyr_{26}–Thr_{27} bond, which is easily hydrolyzed by the enzymes from *M. miehei*.

Specificity of microbial rennets from *M. pusillus* var. *Lindt* (Oka *et al.*, 1973) *M. miehei* (Voynick and Fruton, 1971; Sternberg, 1972; Oka and Morihara, 1973a,b), and *E. parasitica* (Whitaker and Caldwell, 1973) on synthetic peptides showed preference for aromatic or bulky and hydrophobic amino acid residues on both sides of the peptide bond. Importance of secondary interactions and accessibility of the bond are stressed by the behavior of synthetic substrates containing Phe–Met bond. The Phe-Met bond located between the C-terminal phenylalanyl

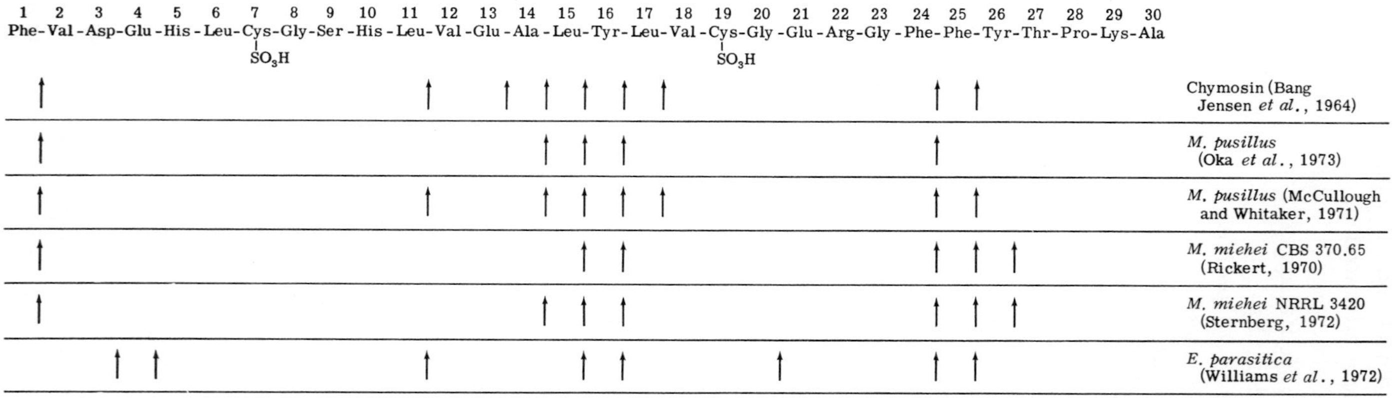

1 2 3 4 5 6 7 8 9 10 11 12 13 14 15 16 17 18 19 20 21 22 23 24 25 26 27 28 29 30

Phe-Val-Asp-Glu-His-Leu-Cys-Gly-Ser-His-Leu-Val-Glu-Ala-Leu-Tyr-Leu-Val-Cys-Gly-Glu-Arg-Gly-Phe-Phe-Tyr-Thr-Pro-Lys-Ala

SO_3H (at position 7) SO_3H (at position 19)

Chymosin (Bang Jensen et al., 1964)

M. pusillus (Oka et al., 1973)

M. pusillus (McCullough and Whitaker, 1971)

M. miehei CBS 370.65 (Rickert, 1970)

M. miehei NRRL 3420 (Sternberg, 1972)

E. parasitica (Williams et al., 1972)

FIG. 1. Bond specificity of chymosin and purified rennets from M. pusillus var. Lindt, M. miehei, and E. parasitica determined on oxidized B chain of insulin.

residue of para-κ-casein and the N-terminal methionyl residue of κ-caseino-macropeptide has been singled out (Jollès *et al.*, 1962; Delfour *et al.*, 1965) as the primary target of chymosin attack in the chain of events which eventually bring upon the coagulation of milk, yet neither chymosin (Hill, 1968; Voynick and Fruton, 1971) nor *M. Miehei* rennet (Sternberg, 1972) can hydrolyze the bond when present in dipeptides. However, inclusion of the Phe–Met bond in a larger peptide having a particular sequence (Raymond *et al.*, 1972) or anchoring it along a polylysine backbone (Sternberg, 1972) makes it accessible to enzymic hydrolysis.

Modification of functional groups with specific reagents helped establish the molecular determinants of enzymic activity. Nitration (Rickert and McBride-Warren, 1974b) and carbamylation (Rickert, 1972) of the rennet from *M. miehei* CBS 370.65 suggested that two tyrosyl, the terminal alanyl and a portion of the lysyl residues, could be modified without significant loss of proteolytic activity.

Dinitrofluorobenzene (Sternberg, 1972) gradually inactivated the *M. miehei* NRRL 3420 rennet, first by combining with the N terminals of alanyl and glycyl, then by substituting part of the tyrosyl and lysyl, and all of the hystidyl residues. The relation of hystidyl residues to the active center of microbial rennets from *M. pusillus* was shown by Yu *et al.* (1971c). More structural details followed publication of the X-ray precession photograph (Moews and Bunn, 1972).

Methyl esters of diazoacetylnorleucine and diazoacetylglycine are potent inhibitors of chymosin, *M. pusillus* (Takahashi *et al.*, 1972), and *M. miehei* microbial rennets (Sternberg, 1972). Stoichiometric incorporation of norleucine or glycine indicated that all these enzymes were similar to pepsin and a number of other microbial acid proteases because they possess an essential carboxyl group as part of the active site.

A carbohydrate moiety is present in the microbial rennet of *M. miehei*. The composition and type of linkage to the protein were elucidated by Rickert and McBride-Warren (1974a). Rickert and Elliott (1973), based on molecular weight studies, concluded that the three commercial enzymes from species of *Mucor* are distinctively different. Another *M. miehei* protease, separated by column electrofocusing from a train of *M. miehei* isolated in Cuba (Rickert and McBride-Warren, 1975), was characterized regarding its amino acid composition, molecular weight, helical content, total carbohydrate content, and approximate isoelectric point. Mobilities of the main milk-clotting enzymes from the Cuban and CBS 370.65 strains of *M. miehei*, determined by isoelectric focusing in polyacrylamide, were the same, yet differences were found in their circular dicroism, number of amino acid residues, and molecular weights. Therefore the two enzymes, though similar, are not identical.

Insight into the proteolysis of milk proteins and the milk-clotting mechanism was gained by incubating casein fractions or milk with purified enzymes. Tam and Whitaker (1972) studied the rate and extent of hydrolysis of caseins by pepsin, chymosin, and microbial rennets from *M. pusillus* var. *Lindt* and *E. parasitica*. All four enzymes had the fastest initial rates of hydrolysis at pH 6 on κ-casein followed by α-casein and β-casein. The extent of hydrolysis at the same pH was of the order of α-, κ-, and β-casein except for *E. parasitica* rennet. Comparison by disc-gel electrophoresis of macropeptides released by incubation of crystallized chymosin and purified rennets from *M. pusillus* and *E. parasitica* with acid casein and κ-casein (Kovács-Proszt and Sanner, 1973) showed some similarity with *M. miehei* but differences from *E. parasitica* rennet. Kinetic studies made by the same workers revealed that the calf stomach enzyme has nearly 2.5 times higher aggregation activity of κ-casein than the two microbial enzymes. However, maximum rates of aggregation at high enzyme concentrations are the same for all three enzymes, indicating that the mechanism of κ-casein aggregation and the product of the reaction leading to aggregation are the same (Sanner and Kovács-Proszt, 1973). Chymosin and purified *M. miehei* rennet release sialic acid and TCA-soluble nitrogen at identical rates from reconstituted milk. Amino acid composition of the κ-caseinoglycopeptides were similar and electrophoresis patterns of milk clots and fresh cheese were identical (Sternberg, 1972), indicating that at least the primary reaction of milk clotting (Alais, 1974) follows a similar path for purified calf and *M. miehei* rennets.

Reports of milk clotting by crude preparations of calf and microbial rennets tend to show more differences between the two groups of enzymes. Though from a practical standpoint such data are important, they are irrelevant to the understanding of the milk-clotting mechanism because of enzymic heterogeneity of the microbial and to some extent even the calf preparations.

At this point a few words should be said about toxicity. Based on animal toxicity data submitted by manufacturers, the preparations from *E. parasitica*, *M. pusillus* var. *Lindt*, and *M. miehei* have been approved in the United States by the Food and Drug Administration as milk-clotting enzymes for the production of cheese (Anonymous, 1974; Van Logten *et al.*, 1972; Mernier, 1973).

C. ANALYTICAL METHODS

The assay of clotting activity based on measuring the time necessary for coagulation of milk has survived all trials to replace it. In spite of deficiencies of reproducibility, substrate standardization, and end-point observation, this assay is used by industrial and research laboratories because of correlation to the actual performance of the enzyme in the cheese vat. In general, methods developed for calf rennet have been translated without change to determinations of microbial rennets and for expression of the clotting strength of the microbial preparation, as compared to a calf rennet standard.

Estimation of proteolytic activity instead of milk-clotting power has the advantage of direct measurement of an event leading to the clotting process. Determination of proteolytic breakdown of sodium caseinate in conditions similar to those of cheese-making (Clarke and Richards, 1973) followed by determination of perchloric acid-soluble peptides was suggested for assaying rennets. Practical use of proteolytic methods has been limited to identification of coagulants (Stavlund and Kiermeier, 1973). A spectrophotometric assay measures the proteolytic activity of chymosin on a synthetic hexapeptide reminiscent of the sequence from κ-casein which is involved in the milk-clotting mechanism (Raymond et al. 1972, 1973; Schattenkerk et al., 1971). An original method for determination of E. parasitica rennet, which is based on the activation of trypsinogen, was proposed by Whitaker (1972).

Identification of rennet substitutes, including microbial rennets, has been extensively reviewed by De Koning (1972a). The same author (De Koning, 1974a,b), on behalf of the International Dairy Federation, submitted a critical survey of procedures for the characterization of samples containing unknown rennets. A scheme directs investigators on how to identify and determine enzymes present in mixtures of calf rennet and one substitute from the group of pig pepsin, bovine pepsin, M. pusillus, E. parasitica, or M. miehei microbial rennets.

Gel-slab isoelectric focusing in a pH gradient of 3–10 has been used to identify calf and microbial rennets by their specific electrophoretic patterns and pI values (De Koning and Draaisma, 1973). Results obtained with electrophoresis in polyacrylamide (gelled in an alkaline buffer) were not very encouraging because of the multiplicity of observed bands and difficulty to assign a specific pattern for each rennet. Moreover, sodium chloride concentration influenced the relative mobilities and the number of protein zones, further confusing identification of components (Vámos-Vigyázó et al., 1973). Another gel-electrophoretic method, using an acidic buffer, succeeded in preserving the enzymic activities. Individual clotting activities were observed after electrophoresis by their ability to coagulate a milk overlay. Every sample gave only a few zones which could be identified by running in parallel controls of known enzymes. There were marked differences among the electrophoretic locations of rennets from E. parasitica, M. pusillus, and calf stomach. In contrast M. miehei rennet migrated in the same position as calf rennet and chymosin (Shovers et al., 1972).

Isoelectric focusing of surface and submerged cultures of *M. pusillus* var. *Lindt* gave substantial differences of the isoelectric points of the clotting enzymes from the two types of fermentation (Pozsár-Hajnal and Hegedüs-Völgyesi, 1975). Lipase could also be separated by the same method.

In a time-demanding procedure, mixtures of two or more enzymes from the group of calf rennet, pepsin, and rennets from *M. miehei* and *M. pusillus* were separated by DEAE cellulose chromatography in a piperazine buffer followed by selective denaturation and measurement of the milk-clotting activity (O'Leary and Fox, 1974). Animal rennet preparations were separated previously on DEAE cellulose with a linear gradient of NaCl (Sergeeva *et al.*, 1972).

We have not seen in the literature any further interest in perfecting existing methods for assaying lipolytic, amylolytic, and cellulolytic activities accompanying microbial rennet preparation.

D. PREPARATION OF CHEESE WITH MICROBIAL RENNETS

Consumer acceptance of cheese made with microbial rennets is an established fact. There is no perfect identity between cheese technology using calf rennet and technology using microbial coagulants, but industry has learned how to use the microbial rennets, first in mixtures with calf rennet, then as total replacers. A comparison of *M. miehei* enzyme with calf rennet was made in 12 pairs of vats of cheddar cheese at three cheese factories in Ontario (Dinesen *et al.*, 1975). Typical cheddar flavor of the cheese made with calf or microbial rennets was the same after 15 months of aging. Nonprotein nitrogen was higher in the whey of cheese made with microbial rennet than in the whey of calf rennet cheese, reflecting higher proteolytic activity of microbial rennet. Flavor of cheese made with calf rennet was superior to microbial rennet cheese at 3 weeks and at 3 months, but differences disappeared after aging for 6 months. The content of fat in whey is considered an indicator of cheese yield. Whey from 173 vats of cheese manufactured with calf rennet-pepsin had a mean whey fat of 0.32%, while 157 vats of cheese manufactured with *M. pusillus* rennet had mean whey fat of 0.39%, thus pointing out that the yields obtained with animal and microbial coagulants were similar (Nelson, 1975).

Comparative studies made with some of the purified microbial rennets and chymosin have indicated great similarities in milk-clotting mechanisms. However, commercial preparations contain a multiplicity of proteolytic and other enzymes which influence the quality of cheese. Cheese made with crystalline rennet from *M. pusillus* var. *Lindt* had less of a bitter taste than cheese made with crude preparation of the same microbial rennet (Yu *et al.*, 1971a). Microbial rennets produced by two manufacturers supposedly using the same species of microorganism may give different cheese results if the rennets have not been made with identical strains and fermentation procedures. The commercial rennets may also have different enzymic compositions.

Cheesemakers prefer to substitute calf rennet with microbial rennets on an equal clotting activity basis without making any changes in the cheese process. However, this is not always possible if positive results are to be obtained. Studies with commercial microbial rennet preparations that point out differences from calf rennet allow for corrections in the cheesemaking procedure which are necessary to obtain good cheese.

Dependence of coagulation time on the Ca^{2+} ion concentration was shown for all microbial rennets. Milk-clotting time with rennets from *E. parasitica, M. pusillus,* and *M. miehei* was less influenced by variations of calcium ion concentrations than calf rennet when $CaCl_2$ additions to the milk were lower than 0.008 mole. At calcium concentrations above 0.01 mole the *M. pusillus* and *M. miehei* rennets were slightly more sensitive than calf rennet (Houins *et al.*, 1973). Alais (1971) found the clotting activity of the rennet from *E. parasitica* less sensitive to calcium ion variations in milk than calf rennet, while the *Mucor* rennets were more sensitive. The most sensitive to calcium ion concentration in milk at pH 6.33 and 6.57 seems to be the rennet from *Mucor pusillus* (Alais and Lagrange, 1972). Brinkman and Duiven (1972) also found *M. pusillus* more sensitive to calcium ion concentration than calf rennet, with coagulating activity increasing when $CaCl_2$ content was increased. The clotting activity of *M. pusillus* rennet was reported to be independent of pH variations between pH 5.3 and pH 6, decreasing between pH 6 and 6.3. Rennets from *M. miehei* and *E. parasitica* were similar to calf rennet in that their clotting activity decreased along the pH interval 5.3–6.3; above pH 6.0 their activity decreased faster than that of *M. pusillus* (Houins *et al.*, 1973). Clotting time variations between pH 5.55 and 6.85 were found to be practically identical for calf and *M. miehei* rennet (Alais and Lagrange, 1972).

Proteolytic activity, beneficial up to a certain point, may cause problems of bitter taste, softening of texture, and loss of cheese yield. Compared to calf rennet, microbial rennets liberate more nonprotein nitrogen and have higher proteolytic activity (Houins *et al.*, 1973; Alais and Lagrange, 1972; Vanderpoorten and Weckx, 1972; Itoh and Thomasow, 1971; Rymaszewski *et al.*, 1973; Carini *et al.*, 1973). The bitter flavor of cheese made with microbial rennets (De Koning, 1972b) may be due to advanced proteolysis. Carini *et al.* (1974) prepared four varieties of Italian cheeses using commercial microbial rennets and calf rennet, and did not notice any bitterness of the ripe cheeses. Correlation between bitterness and starter cultures rather than rennet was reported (Lawrence et al., 1972).

Cserháti and Holló (1974b) analyzed peptides released from casein by rennets prepared in Hungary from E. parasitica and M. pusillus. They found that the peptides had a bitter but acceptable taste. Calf rennet also released bitter peptides by proteolytic breakdown of αs_1- and β-caseins (Pelissier et al., 1974). Electrophoretic methods were used to follow up patterns of proteolytic breakdown by microbial rennets of casein and casein fractions. Each coagulating enzyme gives a specific pattern after incubation with αs_1- and β-caseins. Microbial rennets were considerably more effective than calf rennet in breaking down β-casein, their action beginning earlier and proceeding farther (Vanderpoorten and Weckx, 1972; Weckx and Vanderpoorten, 1973; Itoh and Thomasow, 1971; Cserháti and Holló, 1974a).

A small amount of inhibitor of M. miehei rennet was found in raw milk, but pasteurization of the milk alleviated any practical problems (Alais et al., 1974). Endothia parasitica rennet stimulated acid production by starter cultures (Havlova, 1973). Hungarian workers found their E. parasitica culture less dependent on temperature for coagulating milk than M. pusillus rennet (Cserháti and Holló, 1974a).

Thermal stability of microbial rennets (Brinkman, 1972) is now considered to be less of an asset. Low levels of residual coagulants in whey break down the protein and restrict whey uses in the food industry (Harper and Lee, 1975). Heat treatment of the whey is slower to inactivate microbial rennets at particular pH values compared to calf rennet or mixtures of calf rennet and pepsin (Hyslop, 1975; Ernstrom, 1975). At pH values closer to neutral or slightly alkaline the Mucor rennets are not very stable and can be destroyed by heating for short times. This important difference between animal and microbial coagulants will have to be overcome by dairy technology or microbial rennet improvement.

Many cheese varieties have been prepared from milk with microbial rennets (Table II). The flavor, texture, and appearance in general compare favorably to controls of cheese made with animal rennet.

III. Screening for New Microbial Rennets

The market success of commercial microbial rennets seems to have reduced interest in searching for and evaluating microorganisms for production of new milk-clotting enzymes. Anyone wishing to penetrate and have a share of the market must have fermentation facilities, be ready to prove that his microbial rennet gives good cheese, and that it is superior to what is already being sold. Superiority would include lower cost to the cheesemaker, smaller ratios of defective vs. acceptable cheese

TABLE II
LITERATURE REFERENCES ON CHEESES PREPARED WITH
COMMERCIAL MICROBIAL RENNETS

Microbial rennet	Cheese	Reference
Endothia parasitica	Cheddar	Lawrence *et al.* (1972), Poznánski *et al.* (1974)
	Crescenza	Carini *et al.* (1973, 1974)
	Duch	Nadassky (1971)
	Gouda	Vanderpoorten and Weckx (1972)
	Herve	Weckx and Vanderpoorten (1973)
	Italico	Carini *et al.* (1974)
	Kortowski	Poznánski *et al.* (1974)
	Robiola	Carini *et al.* (1974)
	Taleggio	Carini *et al.* (1974)
	Tilsit	Poznánski *et al.* (1974)
Mucor miehei	Bel Paese	Doležálek *et al.* (1974)
	Camembert	Ramet and Alais (1972)
	Cheddar	Dinesen *et al.* (1975), Lawrence *et al.* (1972), Wigley (1974)
	Crescenza	Carini *et al.* (1973, 1974)
	Domiati	Hamdy (1972)
	Edam	Toma (1973)
	Emmental	Ramet and Alais (1973)
	Gouda	Vanderpoorten and Weckx (1972)
	Italico	Carini *et al.* (1974)
	Kachkaval	Šipka *et al.* (1973, 1974), Toma (1973)
	Saint Paulin	Aarnes (1971), Ramet and Alais (1973)
	Taleggio	Carini *et al.* (1974)
	Trapist	Šipka *et al.* (1973, 1974)
	White Brine	Šipka *et al.* (1973, 1974), Toma (1973)
Mucor pusillus	Cheddar	Lawrence *et al.* (1972), Poznánski *et al.* (1974), Wigley (1974)
	Crescenza	Carini *et al.* (1973, 1974)
	Edam	De Koning (1972b)
	Gouda	De Koning (1972b), Vanderpoorten and Weckx (1972)
	Grana	Alberini and Nizzola (1974)
	Italico	Carini *et al.* (1974)
	Kortowski	Poznánski *et al.* (1974)
	Robiola	Carini *et al.* (1974)
	Taleggio	Carini *et al.* (1974)
	Tilsit	Poznánski *et al.* (1974)

batches, applicability in the manufacturing of a large variety of cheeses, and an enzyme preparation that is stable during storage yet capable of rapid inactivation when residual in the cheese whey. For all or some of the enumerated reasons the three major commercial rennets have not been challenged. The number of existing commercial microbial rennets may even decrease if the experience of cheesemakers favors one source against another.

Microorganisms other than *E. parasitica, M. miehei,* and *M. pusillus* which have been mentioned as producers of milk-clotting enzymes subsequent to the review by Sardinas (1972) are listed in Table III.

A. BACTERIA

Production of enzymes using bacterial strains is generally attractive because of high titers of active compounds and shorter fermentation time than required for fungi. In spite of these incentives, efforts to develop a commercial fermentation of microbial rennets using bacteria have not been successful because of the invariable strong and nonspecific proteolytic action of bacterial milk-clotting enzymes with consequent losses of fat and nitrogen in the whey, reduced yield, and poor quality of the aged cheese. Occasional claims of positive cheese trial results are mentioned in the literature, raising hope that an adequate bacterial strain may eventually be found.

Bacillus polymyxa rennet was supplied in limited amounts for experimental purposes by the Japanese company Godo Shusei under the name Milcozyme (Imai *et al.,* 1972). Compared to calf rennet it has a broad temperature optimum between 48 and 60°C and higher proteolytic activity. A much higher optimal temperature, namely, 70°C, was found by Reps *et al.* (1975). The *B. polymyxa* rennet strongly hydrolyzed all casein fractions and gelatin (Itoh and Thomasow, 1971). Whey from cheese made with *polymyxa* rennet contains more nonprotein nitrogen than the control of cheese prepared with calf rennet (Fogarty and Griffin, 1973). Cheddar, Tilsit, and Kortowski cheese manufactured with *B. polymyxa* rennet ripened more intensively than those produced with calf rennet, but the cheese yield was much lower and had a brittle character due to intensive casein degradation (Poznánski *et al.,* 1974; Reps *et al.,* 1974a). Other differences from calf rennet were: inactivation temperature, variability of the product of coagulation time and enzyme quantity, and lower curd tension (Reps *et al.,* 1974b).

Bacillus subtilis, favored in the past as a potential source of bacterial rennets (Sardinas, 1972), continues to draw attention in several literature reports. The extensive proteolytic activity of milk-clotting enzymes from *B. subtilis* prevented significant commercial production. Experiments of cheddar cheese manufacture with rennet from a mutant of *B. subtilis* gave a softer curd with more protein and higher fat loss in the whey than the control. Yields of cheese were also 10% lower, apparently due to proteolytic breakdown of the curd (Puhan and Irvine, 1973).

TABLE III
MICROORGANISMS REPORTED TO PRODUCE MICROBIAL RENNETS[a]

Bacteria	References
Bacillus cereus	Babbar *et al.* (1974), Brandl (1971), Islam and Blanshard (1973), Krishnamurthy *et al.* (1973)
Bacillus megatherium	Babbar *et al.* (1974), Krishnamurthy *et al.* (1973)
Bacillus mesentericus	Babbar *et al.* (1974), Dimitroff and Mashev (1970), Dimitroff and Prodanski (1973), Goranova and Stefanova-Kondratenko (1975)
Bacillus polymyxa	Brandl (1971), Fogarty and Griffin (1973), Imai *et al.* (1972), Imai and Irie (1972), Itoh and Thomasow (1971), Poznánski *et al.* (1974), Reps *et al.* (1974a,b, 1975)
Bacillus sp.	Cserháti and Holló (1972b, 1974b)
Bacillus subtilis	Babbar *et al.* (1974), Brandl (1971), Dutta *et al.* (1971), Havlova and Dolezalek (1973), Krishnamurthy *et al.* (1973), Puhan and Irvine (1973), Rao and Mathur (1974)
Streptococcus liquefaciens	Krishnamurthy *et al.* (1973)
Streptomyces erythreus	Abdel-Fattah *et al.* (1974)
Fungi	
Absidia ramosa	Sannabhadti and Srinivasan (1974)
Agaricales sp.	Fedorova and Shivrina (1974)
Aphyllophorales	Fedorova and Shivrina (1974)
Aspergillus fischeri	Abdel-Fattah and El-Hawwary (1974)
Aspergillus flavus	Abdel-Fattah and El-Hawwary (1974)
Aspergillus giganteus	Abdel-Fattah and El-Hawwary (1974)
Aspergillus hennenbergii	Abdel-Fattah and El-Hawwary (1974)
Aspergillus nidulans	Krishnamurthy *et al.* (1973)
Aspergillus niger	Abdel-Fattah and Mabrouk (1971)
Aspergillus versicolor	Abdel-Fattah and El-Hawwary (1974)
Byssochlamys fulva	Chu *et al.* (1973), Jedrychowski *et al.* (1974), Rymaszewski *et al.* (1973)
Candida lipolytica	Yoshino *et al.* (1972)
Chlamydomucor oryzae	Ellis *et al.* (1974)
Cunninghamella echinulata	Abdel-Fattah and El-Hawwary (1974)
Dothiorella ribis	Yu *et al.* (1971b)
Epicoccum purpurascens	Abdel-Fattah and El-Hawwary (1974)
Irpex lacteus	Kawai (1971)
Mucor hiemalis	Abdel-Fattah and El-Hawwary (1974)
Mucor lamprosporus	Wiken and Bakker (1974)
Mucor renninus	Zvyagintsev *et al.* (1972)
Penicillium chrysogenum	Abdel-Fattah and El-Hawwary (1974)
Penicillium citrinum	Abdel-Fattah *et al.* (1972), Abdel-Fattah and El-Hawwary (1972, 1973a,b)
Penicillium expansum	Abdel-Fattah and El-Hawwary (1974)
Penicillium martensii	Abdel-Fattah and El-Hawwary (1974)
Physarum polycephalum	Farr (1974), Farr *et al.* (1974)
Rhizopus arrhizus	Ellis *et al.* (1974)
Rhizopus chinensis	Ellis *et al.* (1974)
Rhizopus chungkuoensis	Ellis *et al.* (1974)
Rhizopus hangchao	Ellis *et al.* (1974)
Rhizopus microsporus	Ellis *et al.* (1974)
Rhizopus niveus	Ellis *et al.* (1974)
Rhizopus oligosporus	Ellis *et al.* (1974)
Rhizopus oryzae	Ellis *et al.* (1974)
Rhizopus rhizopodiformis	Ellis *et al.* (1974)
Rhizopus stolonifer	Ellis *et al.* (1974)
Trametes ostreiformis	Kobayashi *et al.* (1975)

[a] Reported since Sardinas' review (1972) and excluding *E. parasitica*, *M. pusillus*, and *M. miehei*.

Opposite results were obtained by Krishnamurthy *et al.* (1973), who claimed that rennets from strains of *B. subtilis, B. cereus,* and *B. megaterium* gave the same yield and cheese quality as calf rennet. A *B. subtilis* rennet, called Milkozym by the investigators, has been developed in Czechoslovakia where it is used on a limited scale for cheese production (Havlova and Dolezalek, 1973).

Bacillus mesentericus rennet isolated and characterized from the fermentation of strain 11-11 (Dimitroff and Mashev, 1970) was tried with satisfactory results in the manufacturing of Kachkaval, a cheese made from sheep and cow's milk (Dimitroff and Prodanski, 1973). An electrophoretic study of casein hydrolysis by the rennet from strain E-76 of *B. mesentericus* showed degradation of αs_1- and β-casein fractions; however, the pattern was similar to calf rennet when the concentrations of both enzymes were in the limits commonly used in cheesemaking (Goranova and Stefanova-Kondratenko, 1975).

A milk-clotting enzyme was identified in a culture filtrate of *Streptomyces erythreus*. Screening of 20 other species of *Streptomyces* indicated that 17 had some milk-clotting activity. The investigators, also noticing peptonization of the curd, thus concluded that only *S. erythreus* culture produced a rennetlike clot (Abdel-Fattah *et al.*, 1974).

B. FUNGI

The commercial success of fungal rennets has encouraged further exploration of this large subphylum for enzymes similar to rennets.

1. Agaricales

An investigation of 79 strains belonging to 51 species of higher basidial fungi revealed that culture filtrates of strains of *Flammulina velutipes* had the highest milk-clotting activity. No assessments were made of the proteolytic activities (Fedorova and Shivrina, 1974).

2. Rhizopus

A total of 347 strains belonging to 10 species of *Rhizopus* and 9 strains of *Chlamydomucor oryzae* were examined, and milk-clotting, antibiotic, and amylolytic activities were recorded. Milk-clotting and amylolytic activities were widespread throughout species of *Rhizopus* growing above 37°C. There was a wide range of proteolytic, amylolytic, and antibiotic activity among strains of a single species (Ellis *et al.*, 1974).

3. Physarum polycephalum

A patent covering the production of a milk-clotting enzyme with this Myxomycete has been issued to Nestle in Switzerland (Farr, 1974). The major protease from the culture, filtrate, containing three proteolytic enzymes, was isolated and characterized. It specifically cleaves the Phe–Met bond in κ-casein; however, the peptide map of a hydrolyzate of the B chain of oxidized insulin was different from the map obtained in the same conditions with chymosin (Farr et al., 1974).

4. Trametes ostreiformis

The culture filtrates of *Trametes ostreiformis* contain a microbial rennet having an estimated molecular weight of 35,000. The enzyme specifically coagulated κ- and β-casein. Results of Gouda cheesemaking showed that the 5-month-old cheese had a slightly bitter taste and the same fat, water, and protein content as the control made with calf rennet. The bitter taste may be due to excessive proteolytic activity (Kobayashi et al., 1975).

5. Byssochlamys fulva

Strains of this organism have drawn attention as potential producers of microbial rennets since the screening work of Knight (1966). An enzyme called *Byssochlamyopepdidase* A was partially purified from the culture of *B. fulva* IMI 83277 by ammonium sulfate precipitation, DEAE-cellulose chromatography, and gel filtration. It had an optimal pH around 2.9 and a temperature optimum of 60°C. Milk-clotting activity was minimal at pH 6.2 or above. The texture and appearance of the milk curd as well as nonprotein nitrogen liberated in the whey were similar to those of rennin (Chu et al., 1973). No differences were found in the production processes of Camembert, Edam, and Kortowski cheese manufactured with a *B. fulva* rennet as compared to cheese made with calf rennet, although certain chemical differences were observed (Jedrychowski et al., 1974). The *B. fulva* rennet hydrolyzed casein fraction in decreasing order; αs-, β-, and κ-casein (Rymaszewski et al., 1973).

6. Candida lipolytica

Candida lipolytica produces an exocellular rennet, isolated by ammonium sulfate precipitation and gradient chromatography on CM-Sephadex. The curd obtained by coagulating milk with *C. lipolytica* rennet was softer than calf rennet curd (Yoshino et al., 1972).

IV. Conclusions

Microbial rennets have successfully challenged coagulants from animal sources, moving by power of their low cost into the void made by calf rennet shortages. In spite of gratifying practical performance there is still a need for these rennets to meet in every respect the properties of calf rennet.

Commercial microbial rennets have higher heat stability than calf rennet. While residual calf rennet in cheese whey is inactivated during pasteurizing, microbial rennets are resistant, displaying proteolytic action that damages the whey for food applications. Efforts will increase to develop heat-sensitive microbial rennets unless inactivation of the enzymes is achieved through acceptable changes of whey-pasteurizing technology.

Microbial rennets are used for manufacturing fewer cheese varieties than calf rennet. The replacement of calf rennet by microbial rennets for manufacturing a particular type of cheese is a time-consuming process of small- and large-scale experiments that involve financial risks and sometimes approval of government regulatory agencies. Expansion of the application range will define the limits of animal rennet replacement and the need for other microbial rennets.

There is a greater chance of variability of individual batches of microbial rennets than of calf rennet due to spontaneous microbial mutations or changes in fermentation conditions. Commercial preparations of microbial rennets are complex mixtures containing not only several proteases with various degrees of milk-clotting and proteolytic activities, but also other enzymes, such as lipases, cellulases, and β-galactosidases. Conceivably the quality of cheese may be related to more than the major clotting enzyme. The defining of individual contributions of enzymes in microbial rennet preparations for the making of good cheese has yet to be done. Purified microbial rennets of controlled composition hopefully will, in the next developmental stage, assure the reproducibility of cheese quality. The blending of purified microbial enzymes may even give preparations that are superior to calf rennet for fast ripening and flavor development.

The potential of immobilized microbial rennets has not yet been explored. Except for one report (Cheryan et al., 1975) all publications on milk clotting with immobilized enzymes refer to animal rennets. The basic problem regarding the continuous clotting of milk still needs to be solved, but investigators seem to prefer working with calf rennet or pepsin commercially available in a crystallized form (Cheryan et al., 1974; Hicks et al., 1974; Bourdeau et al., 1974; Whelan et al., 1973; Arima et al., 1974; Brown and Swaisgood, 1975). However, once continuous clotting processes become important for actual cheese production, the better stability of microbial rennets compared to animal enzymes will make them attractive for use in immobilized form.

Substitution of calf rennet by microbial rennets is already an acknowledged fact. The successful duplication of elusive cheese qualities by the use of microbial enzymes points out that, given the right economic circumstances, microbial fermentations can replace agricultural and husbandry products with equivalent materials. In our world of diminishing food supplies, microbial rennets are an accomplishment in the right direction.

REFERENCES

Aarnes, A. G. (1971). *Diary Res. Inst. Norway* 158, 1–21

Abdel-Fattah, A. F., and El-Hawwary, N. M. (1972). *J. Gen. Appl. Microbiol.* 18, 341–348.

Abdel-Fattah, A. F., and El-Hawwary, N. M. (1973a). *Acta Biol. Acad. Sci. Hung.* 24, 95–101.

Abdel-Fattah, A. F., and El-Hawwary, N. M. (1973b). *Z. Allg. Mikrobiol.* 13, 373–379.

Abdel-Fattah, A. F., and El-Hawwary, N. M. (1974). *J. Gen. Microbiol.* 84, 327–331.

Abdel-Fattah, A. F., and Mabrouk, S. S. (1971). *J. Gen. Appl. Microbiol.* 17, 509–512.

Abdel-Fattah, A. F., Mabrouk, S. S., and El-Hawwary, N. M. (1972). *J. Gen. Microbiol.* 70, 151–155.

Abdel-Fattah, A. F., El-Hawwary, N. M., and Amr, A. S. (1974). *Acta Microbiol. Pol.* 6, 27–32.

Alais, C. (1971). *Tech. Lait.* 26, 63–65.

Alais, C. (1974). *Chimia* 28, 597–604.

Alais, C., and Lagrange, A. (1972). *Lait* 52, 407–427.

Alais, C., Ducroo, P., and Delecourt, R. (1974). *Lait* 54, 517–527.

Alberini, B., and Nizzola, I. (1974). *Mondo Latte* 28, 537–580.

Anonymous, (1974). "Code of Federal Regulations," Title 21, Sect. 121.1199.

Anstrup, K. (1968). British Patent 1,108,287.

Arima, K., and Iwasaki, S. (1964). U.S. Patent 3,151,039.

Arima, S., Shimazaki, K., Yamazumi, T., and Kanamaru, Y. (1974). *Rakuno Kagaku No Kenkyu* 23, A83–A87.

Babbar, I., Srinivasan, A., Chakravorty, S. C., Krishnaiyengar, M. K., Dudani, A. T., and Iya, K. K. (1974), Indian Patent 96,317.

Bang-Jensen, V., Foltman, B., and Rombatus, W. (1964). *C. R. Trav. Lab. Carlsberg* 34, 326–345.

Baxter Laboratories, (1968). French Patent 1,556,473.

Bourdeau, J. P., Seris, J. L., Porning, R., and Fossati, P. (1974). German Offen. 2,357,113.

Brandl, E. (1971). *Oesterr. Milchwirtsch.* 26, 69–71.

Brinkman, D. and Duiven, M. (1972). *Ind. Aliment. Agric.* 89, 1755–1758.

Brown, R. J., and Swaisgood, H. W. (1975). *J. Dairy Sci.* 58, 796.

Carini, S., Lodi, R., and Todesco, R. (1973). *Latte* 1, 13–22.

Carini, S., Todesco, R., and Delforno, G. (1974). *Latte* 2, 780–782, 788.

Charles, R. L., and Dolby, S. C. (1973). German Offen. 2,232,996.

Charles, R. L., Gertzmann, D. P., and Melachouris, N. (1970). U.S. Patent 3,549,390.

Cheryan, M., Van Wyk, P. J., Olson, N. F., and Richardson, T. (1974). *J. Dairy Sci.* 58, 477–481.

Cheryan, M., Van Wyk, P. J., Olson, N. F., and Richardson, T. (1975). *Biotechnol. Bioeng.* 17, 585–598.

Christensen, V. W. (1974). *Natl. Meet., Am. Inst. Chem. Eng.* June, 1974.

Chu, F. S., Nei, P. Y. W., and Leung, P. S. C. (1973). *Appl. Microbiol.* **25**, 163–168.

Clarke, N. H., and Richards, E. L. (1973). *N. Z. J. Dairy Sci. Technol.* **8**, 152–155.

Cserháti, T., and Holló, J. (1972a). *Nahrung* **16**, 431–440.

Cserháti, T., and Holló, J. (1972b). *Gordian* **72**, 405–407.

Cserháti, T., and Holló, J. (1974a). *Gordian* **74**, 257–260.

Cserháti, T., and Holló, J. (1974b). *Nahrung* **18**, 625–633.

De Koning, P. J., (1972a). *Annu. Bull., Int. Dairy Fed.* Part 4, pp. 1–15.

De Koning, P. J. (1972b). *Annu. Technol. Agr.* **21**, 357–366.

De Koning, P. J. (1974a). *Annu. Bull., Int. Dairy Fed.* **80**, 1–10.

De Koning, P. J. (1974b). *Mondo Latte* **28**, 647–650.

De Koning, P. J., and Draaisma, J. T. M. (1973). *Ned. Melk-Zuiveltijdschr.* **27**, 368–378.

Delfour, A., Jollès, J., Alais, C., and Jollès, P. (1965). *Biochem. Biophys. Res. Commun.* **19**, 452–455.

Dimitroff, D., and Mashev, N. (1970). *Nauchn. Tr. Vissh. Inst. Khranit. Vkusova Prom., Plovdiv.* **17**, 125–133.

Dimitroff, D., and Prodanski, P. (1973). *Milchwissenschaft* **28**, 568–571.

Dinesen, N., Emmons, D. B., Beckett, D., Reiser, B., Lammond, E., and Irvine, D. M. (1975). *J. Dairy Sci.* **58**, 795.

Doležálek, J., Hladik, J., Studenovsky, J., and Brezina, P. (1974). *Int. Dairy Congr. Pro., 19th,* Vol. 1E, p. 690.

Dutta, S. M., Kuila, R. K., Srinivasan, R. A., Babbar, I. J., and Dudani, A. T. (1971). *Milchwissenschaft* **26**, 683–685.

Edelsten, D., and Siegaard Jensen, J. (1972). *Arsskr., K. Vet.-Landbohoejsk., Copenhagen* pp. 62–70.

Ellis, J. J., Wang, H. L., and Hesseltine, C. W. (1974). *Mycologia* **66**, 593–599.

Ernstrom, C. A. (1975). *70th Annu. Meet., Am. Dairy Sci. Assoc.,* Manhattan, Kansas.

Farr, R. D. (1974). U.S. Patent 3,852, 478.

Farr, R. D., Horisberger, M., and Jollès, P. (1974). *Biochim. Biophys. Acta* **334**, 410–416.

Fedorova, L. N., and Shivrina, A. N. (1974). *Mikol. Fitopatol.* **8**, 22–25.

Fogarty, W. M., and Griffin, P. J. (1973). *Ir. J. Agric. Res.* **12**, 97–102.

Goranova, L., and Stefanova-Kondratenko, M. (1975). *Lait* **541**, 58–67.

Hamdy, A. (1972). *Indian J. Dairy Sci.* **25**, 73–76.

Harper, W. J., and Lee, C. R. (1975). *J. Food Sci.* **40**, 282–284.

Havlova, J., and Doležálek, J. (1973). *Prum. Potravin* **24**, 51–53.

Hershberger, D. F., and Sternberg, M. (1974). German Offen. 2,419,232.

Hicks, C. L., Ferrier, L. K., Olson, N. F., and Richardson, T. (1974). *J. Dairy Sci.* **58**, 19–24.

Hill, R. D. (1968). *Biochem. Biophys. Res. Commun.* **33**, 659–663.

Houins, G., Deroanne, C., and Coppens, R. (1973). *Lait* **53**, 610–624.

Huang, C. M. (1970). Ph.D. Thesis, Purdue University, Lafayette, Indiana.

Hyslop, D. B., Swanson, A. M., and Lund, D. B. (1975). *J. Dairy Sci.* **58**, 795.

Imai, T., and Yoshio, I. (1972). German. Patent 1,767, 816.

Imai, T., Irie, Y., and Kanazawa, Y. (1972). U.S. Patent 3,661,594.

Islam, M. A., and Blanshard, J. M. V. (1973). *J. Dairy Res.* **40**, 427–440.

Itoh, T., and Thomasow, J. (1971). *Milchwissenschaft* **26**, 671–675.

Iwasaki, S., Tamura, G., and Arima, K. (1967). *Agric. Biol. Chem.* **31**, 546–551.

Jedrychowski, L., Reps, A., Poznánski, S., and Rymaszewski, J. (1974). *Int. Dairy Congr. Proc., 19th*, Vol. 1E, pp. 707–708.

Jollès, P., Alais, C., and Jollès, J. (1962). *Arch. Biochem. Biophys.* **98**, 56–57.

Kawai, M. (1971). *Agric. Biol. Chem.* **35**, 1517–1525.

Knight, S. G. (1966). *Can. J. Microbiol.* **12**, 420–422.

Kobayashi, F., Yabuki, M., Hoshino, K., and Sakamoto, M. (1975). *Nippon Nogei Kagaku Kaishi* **49**, 81–92.

Kovács-Proszt, G., and Sanner, T. (1973). *J. Dairy Res.* **40**, 263–272.

Krayushkina, E. A., Zherebtsov, N. A., and Zvyagintsev, V. I. (1973). *Prikl. Biokhim. Mikrobiol.* **9**, 883–885.

Krishnamurthy, T. S., Sannabhadti, S. S., and Srinivasan, R. A. (1973). *J. Food Sci. Technol.* **10**, 118–122.

Lawrence, R. C., Creamer, L. K., Gilles, J., and Martley, F. G. (1972). *N. Z. J. Dairy Sci. Technol.* **7**, 32–37.

McCullough, J. M., and Whitaker, J. R. (1971). *J. Dairy Sci.* **54**, 1575–1578.

Meito Sangyo. (1975). Netherlands Patent 7409–947.

Merker, H. M. (1919). *J. Dairy Sci.* **2**, 482–486.

Mernier, M. (1973). *Ann. Falsif. Expert. Chim.* **66**, 110–122.

Moews, P. C., and Bunn, C. W. (1972). *J. Mol. Biol.* **68**, 389–390.

Morvai-Rácz, M. (1974). *Acta Aliment Acad. Sci. Hung.* **3**, 37–48.

Nadassky, S. (1971). *Veda Vyzk. Potravin. Prum.* **22**, 143–151.

Nelson, J. H. (1975). *J. Dairy Sci.* **11**, 1739–1750.

Oka, T., and Morihara, K. (1973a). *Arch. Biochem. Biophys.* **156**, 543–551.

Oka, T., and Morihara, K. (1973b). *Arch. Biochem. Biophys.* **156**, 552–559.

Oka, T., Ishino, K., Tsuzuki, H., Morihara, K., and Arima, K. (1973). *Agric. Biol. Chem.* **37**, 1177–1184.

O'Leary, P. A., and Fox, P. F. (1974). *J. Dairy Res.* **41**, 381–387.

Organon Laboratories. (1971). Netherlands Patent 1,249,636.

Ottesen, M., and Rickert, W. (1970). *C. R. Trav. Lab. Carlsberg* **37**, 301–325.

Pelissier, J. P., Mercier, J. C., and Ribadeau Dumas, B. (1974). *Ann. Biol. Anim., Biochim., Biophys.* **14**, 343–362.

Poznánski, S., Reps, A., Kowalewska, J., Rymaszewski, J., and Jedrychowski, L. (1974). *Milchwissenschaft* **29**, 742–746.

Pozsár-Hajnal, K., and Hegedüs-Völgyesi, E. (1975). *Acta Aliment. Acad. Sci. Hung.* **4**, 63–79.

Pozsár-Hajnal, K., Vámos-Vigyázó, L., and Hegedüs-Völgyesi, E. (1974a). *Acta Aliment. Acad. Sci. Hung.* **3**, 73–82.

Pozsár-Hajnal, K., Vámos-Vigyázó, L., and Hegedüs-Völgyesi, E. (1974b). *Acta Aliment. Acad. Sci. Hung.* **3**, 83–92.

Puhan, Z., and Irvine, D. M. (1973). *J. Dairy Sci.* **56**, 317–322.

Ramet, J. P., and Alais, C. (1972). *Lait* **52**, 654–663.

Ramet, J. P., and Alais, C. (1973). *Lait* **53**, 154–162.

Rao, L. K., and Mathur, D. K. (1974). *Int. Dairy Congr. Proc., 19th*, Vol. 1E, p. 684.

Raymond, M. N., Garnier, J., Bricas, E., Cilianu, S., Blasnic, M., Chaix, A., and Lefrancier, P. (1972). *Biochimie* **54**, 145–154.

Raymond, M. N., Bricas, E., Salesse, R., Garnier, J., Garnot, P., and Ribadeau Dumas, B. (1973). *J. Dairy Sci.* **56**, 419–422.

Reps, A., Poznánski, S., Rymaszewski, J., Jarosz, M., and Jedrychowski, L. (1974a). *Rocz. Inst. Przem. Mlecz.* **16**, 27–39.

Reps, A., Poznánski, S., Kaszycka, A., and Wasilewski, R. (1974b). *Rocz. Inst. Przem. Mlecz.* **16**, 41–54.

Reps, A., Poznánski, S., and Jakubowski, J. (1975). *Milchwissenschaft* **30**, 65–68.

Rickert, W. S. (1970). *C. R. Trav. Lab. Carlsberg* **38**, 1–17.

Rickert, W. S. (1972). *Biochim. Biophys. Acta* **271**, 93–101.

Rickert, W. S., and Elliott, J. R. (1973). *Can. J. Biochem.* **51**, 1638–1646.

Rickert, W. S., and McBride-Warren, P. A. (1974a). *Biochim. Biophys. Acta* **336**, 437–444.

Rickert, W. S., and McBride-Warren, P. A. (1974b). *Biochim. Biophys. Acta* **371**, 368–378.

Rickert, W. S., and McBride-Warren, P. A. (1975). *Can. J. Biochem.* **53**, 269–274.

Rotini, O. T., and Sequi, P. (1972). *Ann. Technol. Agric.* **21**, 367–383.

Rymaszewski, J., Poznánski, S., Reps, A., and Ichilczyk, J. (1973). *Milchwissenschaft* **28**, 779–784.

Sannbhadti, S. S., and Srinivasan, R. A. (1974). *Int. Dairy Congr. Proc., 19th*, Vol. 1E, pp. 360–361.

Sanner, T., and Kovács-Proszt, G. (1973). *Biochim. Biophys. Acta* **303**, 68–76.

Sardinas, J. L. (1966). U.S. Patent 3,275,453.

Sardinas, J. L. (1972). *Adv. Appl. Microbiol.* **15**, 39–73.

Schattenkerk, C., Holtkamp, I., Hessing, J. G. M., Kerling, K. E. T., and Havinga, E. (1971). *Recl. Trav. Chim. Pays-Bas* **90**, 1320–1322.

Schleich, H. (1971). U.S. Patent 3,616,233.

Sequi, P., Petruzzelli, G., and Rotini, O. T. (1972). *Agrochimica* **16**, 293–299.

Sergeeva, E. G., Dolgikh, T. V., and Zvyagintsev, V. I. (1972). *Prikl. Biokhim. Mikrobiol.* **8**, 354–357.

Shovers, J., Fossum, G., and Neal, A. (1972). *J. Dairy Sci.* **55**, 1532–1534.

Šipka, M., Stojanović, L., Petković, L., Ignjatović, S., and Mladenović, S. (1973). *Mljekarstvo* **23**, 3–11.

Šipka, M., Stojpanović, L., Petcović, L., Ignjatović, S., and Mladenović, S. (1974). *Milchwissenschaft* **29**, 656–660.

Somkuti, G. A. (1974). *J. Dairy Sci.* **57**, 898–899.

Somkuti, G. A., and Babel, F. J. (1967). *Appl. Microbiol.* **15**, 1309–1312.

Somkuti, G. A., and Babel, F. J. (1968). *Appl. Microbiol.* **16**, 617–619.

Stavlund, K., and Kiermeier, F. (1973). *Z. Lebensm.-Unters. -Forsch.* **152**, 138–144.

Sternberg, M. (1971). *J. Dairy Sci.* **54**, 159–167.

Sternberg, M. (1972). *Biochim. Biophys. Acta* **285**, 383–392.

Sternberg, M., and Hershberger, D. (1974). *Biochim. Biophys. Acta* **342**, 195–206.

Takahashi, K., Mizobe, F., and Chang, W. J. (1972). *J. Biochem. (Tokyo)* **71**, 161–164.

Tam, J. J., and Whitaker, J. R. (1972). *J. Dairy Sci.* **55**, 1523–31.

Toma, C. (1973). *Ind. Aliment. (Bucharest)* **24**, 613–615.

Vámos-Vigyázó, L., Metwalli, O. M., Morvai-Rácz, M., and Hegedüs-Völgyesi, E. (1973). *Acta Aliment Acad. Sci. Hung.* **2**, 81–97.

Vanderpoorten, R., and Weckx, M. (1972). *Ned. Melk- Zuiveltijdschr.* **26**, 47–59.

Van Logten, M. J., den Tonkelaar, E. M., Kroes, R., and Van Esch, G. J. (1972). *Food Cosmet. Toxicol.* **10**, 649–654.

Voynick, I. M., and Fruton, J. S. (1971). *Proc. Natl. Acad. Sci. U.S.A.* **68**, 257–259.

Weckx, M., and Vanderpoorten, R. (1973). *Milchwissenschaft* **28**, 332–338.

Whelan, K., Capozza, R. C., and Schmitt, E. E. (1973). South African Patent 7,204,363.

Whitaker, J. R. (1972). *J. Dairy Sci.* **55**, 719–25.

Whitaker, J. R., and Caldwell, P. V. (1973). *Arch. Biochem. Biophys.* **159**, 188–200.

Wigley, R. C. (1974). *Int. Dairy Congr. Proc., 19th,* Vol. 1E, p. 687.

Wiken, T. O., and Bakker, G. (1974). U.S. Patent 3,857,969.

Williams, D. C., Whitaker, J. R., and Caldwell, P. V. (1972). *Arch. Biochem. Biophys.* **149**, 52–61.

Yoshino, U., Chang, J. E., Miwa, K., and Yamauchi, K. (1972). *Nippon Nogei Kagaku Kaishi* **46**, 675–8.

Yu, J. H., Kim, Y. S., and Hong, Y. M. (1971a). *Hanguk Sikp'um Kwahahoe Chi,* **3**, 6–14.

Yu, J. H., Kim, Y. S., Hong, Y. M., and Arima, K. (1971b). *Hanguk Sikp'um Kwahakhoe Chi* **3**, 89–93.

Yu, J., Tamura, G., and Arima, K. (1971c). *Agric. Biol. Chem.* **35**, 1194–1199.

Zvyagintsev, V. I., Krasheninin, P. F., Sergeeva, E. G., Buzov, I. P., Mosichev, M. S., and Rubtsova, N. A. (1972). *Prikl. Biokhim. Mikrobiol.* **8**, 913–917.

Biosynthesis of Cephalosporins

TOSHIHIKO KANZAKI[1] AND YUKIO FUJISAWA

Microbiological Research Laboratories, Central Research Division,
Takeda Chemical Industries, Ltd.,
Jusohonmachi, Yodogawa-ku, Osaka, Japan

I.	Introduction	159
II.	Occurrence of Cephalosporins and Related Metabolites	160
III.	Genetic Studies and Strain Improvement	169
IV.	Fermentation	172
	A. Cephalosporin C Production by *Cephalosporium acremonium*	172
	B. Cephalosporin Production by Streptomycetes and Fungi	176
V.	Biosynthesis	176
	A. Amino Acids and Acetic Acid as Precursors and Their Metabolism	177
	B. Peptides as Intermediates	186
	C. Formation of the Dihydrothiazine Ring System	189
	D. Cephalosporin C Synthesis after Ring Closure	190
VI.	Enzymes Related to Cephalosporin Metabolism	196
	A. Acetylhydrolase	196
	B. Arylamidase	196
VII.	Conclusions	196
	References	198

I. Introduction

It is nearly three decades since the cephalosporin-producing fungus, *Cephalosporium acremonium*, was isolated by Brotzu (1948) in Sardinia. Seven years later, the first cephalosporin compound, cephalosporin C, was discovered by Newton and Abraham (1955) as a contaminant of a crude preparation of penicillin N (Newton and Abraham, 1954) which was a product of the Brotzu's strain. They isolated cephalosporin and found it to be stable to acid and to penicillinase, and relatively active against gram-negative bacteria though its activity was about only one-tenth that of penicillin N. These characteristics attracted the attention of investigators because the appearance of pathogens resistant to penicillin, mainly owing to the presence of penicillinase, had become a serious clinical problem. In addition, the discovery of the method for preparing 7-aminocephalosporanic acid (Morin *et al.*, 1962) stimulated the extensive development of cephalosporin antibiotics similar to the surprising progress in chemotherapy induced by the discovery of 6-aminopenicillanic

---

[1] Present address: Corporate Planning Division, Takeda Chemical Industries, Ltd., Nihonbashi, Chuo-ku, Tokyo, Japan.

acid. The significant development of cephalosporin antibiotics as human medicine during the last several years reminds us of the history of penicillin.

Total chemical syntheses of penicillin and cephalosporin have been established (Sheehan and Henery-Logan, 1959; Woodward et al., 1966; Woodward, 1966). However, these methods have not been competitive with microbial synthesis. The development of semisynthetic cephalosporins has simultaneously stimulated the advancement of cephalosporin fermentation since the supply of the starting material, cephalosporin C, of semisynthetic cephalosporins is still dependent on microbial fermentation. Nevertheless, cephalosporin producers had been substantially limited to only the Brotzu's strain until recently and almost all work on cephalosporin fermentation had been based on the Brotzu's strain or its descendants. In addition, all of the commercial cephalosporins for parenteral use still seem to be exclusively dependent on the line of the Brotzu's strain. Lately, cephalosporin producers, however, have been found to be widely distributed in microorganisms such as eukaryotic *Paecilomyces* (Higgens et al., 1974; Kitano et al., 1974a; Pisano and Vellozzi, 1974) and prokaryotic *Streptomyces* (Nagarajan et al., 1971; Stapley et al., 1972).

Although both penicillin and cephalosporin have old histories, the most important aspects of the mechanism of the biosynthetic pathways have remained uncertain because of the difficulty in studying the metabolism, which is the general problem in studying the metabolism of secondary metabolites. The discovery of the tripeptide and the tetrapeptides in the mycelia of *C. acremonium* by Loder and Abraham (1971a,b) has paved the way for the rapid progress of substantial work on the biogenesis of cephalosporin. The introduction of a variety of antibiotic negative mutants and ^{13}C isomers into the biosynthetic studies is also yielding fruitful results. The finding of cephalosporins in the *Streptomyces* species is expected to bring forth further advancement in elucidating the features of the biosynthetic pathway of cephalosporin.

II.   Occurrence of Cephalosporins and Related Metabolites

The occurrence of compounds with β-lactam rings is now not unusual and not restricted to antibiotics. As is well known, penicillin was found in several *Penicillium* molds (Sanders, 1949) including *P. notatum* (Fleming, 1929) and *P. chrysogenum* (Raper et al., 1944) in the early stage. With advancement of research, fungi other than penicillia were demonstrated to produce penicillins. These include eukaryotic species of *Aspergillus* (Dulaney, 1947), *Malbranchea* (Rode et al., 1947), *Cephalosporium* (Burton and Abraham, 1951; Roberts, 1952), *Emericellopsis*

(Grosklags and Swift, 1957), *Paecilomyces* (Fleischman and Pisano, 1961), *Epidermophyton* (Uri *et al.*, 1963), *Trichophyton* (Uri *et al.*, 1963), *Anixiopsis* (Kitano *et al.*, 1974a), *Arachnomyces* (Kitano *et al.*, 1974a), *Diheterospora* (Higgens *et al.*, 1974), *Scopulariopsis* (Higgens *et al.*, 1974) and *Spiroidium* (Kitano *et al.*, 1974a), and prokaryotic microorganisms of *Streptomyces* (Miller *et al.*, 1962; Nagarajan *et al.*, 1971).

The discovery of a single β-lactam antibotic, nocardicin A, is worthy of note (Aoki *et al.*, 1975). Both bleomycin and phleomycin were also found to involve the β-lactam ring in their molecules (Takita *et al.*, 1972). For nonantibiotic compounds, the β-lactam ring is demonstrated in products of a plant, *Pachysandra terminalis* (Kikuchi and Uyeo, 1967) and of *Pseudomonas tabaci* (Stewart, 1971).

Before 1971, descriptions of cephalosporin-producing strains were restricted within an extremely narrow limit, i.e., one strain of *Cephalosporium* (Newton and Abraham, 1955) and a few species of its sexual stage, *Emericellopsis* (Elander *et al.*, 1961). Recently, cephalosporin-producing microorganisms have been found to be widely distributed in nature and to be very similar to the above penicillin producers. Those involve molds of *Anixiopsis* (Kitano *et al.*, 1974a), *Arachnomyces* (Kitano *et al.*, 1974a), *Diheterospora* (Higgens *et al.*, 1974), *Paecilomyces* (Higgens *et al.*, 1974; Kitano *et al.*, 1974b; Pisano and Vellozzi, 1974), *Scopulariopsis* (Higgens *et al.*, 1974) and *Spiroidium* (Kitano *et al.*, 1974a), and several species of *Streptomyces* (Nagarajan *et al.*, 1971; Stapley *et al.*, 1972; Higgens *et al.*, 1974; Hasegawa *et al.*, 1975). The successive findings of cephalosporin producers thus seem to be partly due to the advancement of the analytical methods for cephalosporins. One is the method using a specific mutant of *Pseudomonas*, highly sensitive to cephalosporins, whereby Kitano *et al.* (1974a, 1975a) succeeded in detecting trace amounts.

Cephalosporins and various metabolites related to the biosynthesis of β-lactam antibiotics in cephalosporin-producing strains are summarized in Table I.

Cephalosporin C is produced by the strains indicated in Table I. Higgens *et al.* (1974) state that the majority of the deacetoxycephalosporin C-producing strains accumulate cephalosporin C and deacetylcephalosporin C. Indeed, in almost every strain, the production of cephalosporin C is accompanied with that of deacetylcephalosporin C and deacetoxycephalosporin C. From these facts, it may be generalized that those strains which produce one of the three cephalosporin compounds concomitantly produce two other cephalosporin antibiotics.

The mechanism of synthesis of deacetylcephalosporin C in microorganisms has not yet been determined except for the case of the mutants

TABLE I

Metabolites Related to β-Lactam Antibiotics from Cephalosporin-Producing Strains

Compound	R_1	R_2	R_3	Microorganism	References
Cephalosporin C	HOOC·CH(CH$_2$)$_3$—CO 　　| 　　NH$_2$	H	OCOCH$_3$	*Cephalosporium acremonium* *Cephalosporium polyaleurum* *Arachnomyces minimus* *Anixiopsis peruviana* *Spiroidium fuscum* *Paecilomyces persicinus* *Paecilomyces carneus*	Newton and Abraham, 1955 T. Yamano, T. Kanzaki; H. Suide, and K. Tubaki, unpublished data, 1973, Tubaki, 1973 Kitano *et al.*, 1974a Kitano *et al.*, 1974a Kitano *et al.*, 1974a Pisano and Vellozzi, 1974, Kitano *et al.*, 1974b Higgens *et al.*, 1974, Kitano *et al.*, 1974b
Deacetylcephalosporin C	HOOC·CH(CH$_2$)$_3$—CO 　　| 　　NH$_2$	H	OH	Mutants of *C. acremonium* *C. polyaleurum*	Fujisawa *et al.*, 1973, 1975a T. Kanzaki, Y. Fujisawa, H. Shirafuji, and T. Fukita, unpublished data, 1974

Compound			Organism	Reference
Deacetoxy-cephalosporin C	$HOOC \cdot CH(CH_2)_3—CO$ $\quad\quad\;\; NH_2$	H	H	*Arachnomyces minimus*	Kitano *et al.*, 1974a
				Anixiopsis peruviana	Kitano *et al.*, 1974a
				Spiroidium fuscum	Kitano *et al.*, 1974a
				P. persicinus	Pisano and Vellozzi, 1974, Kitano *et al.*, 1974b
				P. carneus	Higgens *et al.*, 1974, Kitano *et al.*, 1974b
				Mutants of *C. acremonium*	Kanzaki *et al.*, 1974a, Queener and Capone, 1974, Liersch *et al.*, 1974, Queener *et al.*, 1974, Fujisawa *et al.*, 1975c
				C. polyaleurum	T. Kanzaki, Y. Fujisawa, H. Shirafuji, and T. Fukita, unpublished data, 1974
				C. chrysogenum[a]	Higgens *et al.*, 1974
				Cephalosporium sp.	Higgens *et al.*, 1974
				Emericellopsis sp.	Higgens *et al.*, 1974
				Diheterospora chlamydosporia	Higgens *et al.*, 1974
				Scopulariopsis sp.	Higgens *et al.*, 1974
				Streptomyces lipmanii	Higgens *et al.*, 1974
				S. clavuligerus	Higgens *et al.*, 1974
				Arachnomyces minimus	Kitano *et al.*, 1974a
				Anixiopsis peruviana	Kitano *et al.*, 1974a
				Spiroidium fuscum	Kitano *et al.*, 1974a
				P. persicinus	Pisano and Vellozzi, 1974, Kitano *et al.*, 1974b

(Continued)

TABLE I (*Continued*)

Compound	R₁	R₂	R₃	Microorganism	References
A16886A	HOOC·CH(CH₂)₃—CO, NH₂	H	OCONH₂	*P. carneus*	Higgens *et al.*, 1974, Kitano *et al.*, 1974b
A16884A	HOOC·CH(CH₂)₃—CO, NH₂	OCH₃	OCOCH₃	*S. clavuligerus*	Nagarajan *et al.*, 1971, Higgens and Kastner, 1971
A16886B or Cephamycin C	HOOC·CH(CH₂)₃—CO, NH₂	OCH₃	OCONH₂	*S. lipmanii*	Nagarajan *et al.*, 1971, Higgens and Kastner, 1971
Cephamycin B	HOOC·CH(CH₂)₃—CO, NH₂	OCH₃	OCO—C=CH(OCH₃)—C₆H₄—OH	*S. clavuligerus*	Nagarajan *et al.*, 1971, Higgens and Kastner, 1971
				S. lactamdurans	Stapley *et al.*, 1972
				S. griseus	Stapley *et al.*, 1972
				S. chartreusis	Stapley *et al.*, 1972
				S. cinnamonensis	Stapley *et al.*, 1972
				S. fimbriatus	Stapley *et al.*, 1972
				S. halstedii	Stapley *et al.*, 1972
				S. rochei	Stapley *et al.*, 1972
				S. viridochromogenes	Stapley *et al.*, 1972
				S. heteromorphus	Hasegawa *et al.*, 1975
				S. panayensis	Hasegawa *et al.*, 1975
Cephamycin A	HOOC·CH(CH₂)₃—CO, NH₂	OCH₃	OCO—C=CH(OCH₃)—C₆H₄—OSO₃H	Same as cephamycin B producers	Stapley *et al.*, 1972, Hasegawa *et al.*, 1975
C-2801X	HOOC·CH(CH₂)₃—CO, NH₂	OCH₃	OCO—C=CH(OCH₃)—C₆H₃(OH)₂	*S. heteromorphus*	Fukase *et al.*, 1976, Hasegawa *et al.*, 1975
				S. panayensis	Fukase *et al.*, 1976, Hasegawa *et al.*, 1975

F-1	HOOC·CH(CH₂)₃—CO $\mid$ NH₂	H	SCH₃	Mutant of *C. acremonium*	Kanzaki *et al.*, 1974b
F-2	HOOC·CH(CH₂)₃—CO $\mid$ NH₂	H	S₂O₃H	Mutant of *C. acremonium*	Y. Fujisawa, K. Kitano, K. Katamoto, K. Nara, and T. Kanzaki, unpublished data, 1975
C-43-219	HOOC·CH(CH₂)₃—COa $\mid$ NH₂	H	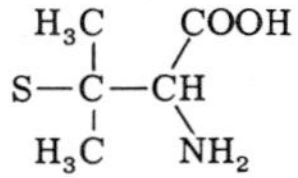 H₃C COOH $\mid$ $\mid$ S—C—CH $\mid$ $\mid$ H₃C NH₂	Mutant of *C. acremonium*	Kitano *et al.*, 1975b
N-Acetyl-deacet-oxyceph-alosporin C	HOOC·CH(CH₂)₃—CO $\mid$ CH₃CO—NH	H	H	Mutants of *C. acremonium*	Traxler *et al.*, 1975
C-1778a	HOOC·(CH₂)₃—CO	H	H	*C. chrysogenum*	Kitano *et al.*, 1975b
C-1778b	HOOC·(CH₂)₃—CO	H	OH	*C. chrysogenum*	Kitano *et al.*, 1975b
C-1778c	HOOC·(CH₂)₃—CO	H	OCOCH₃	*C. chrysogenum*	Kitano *et al.*, 1975b
C-2	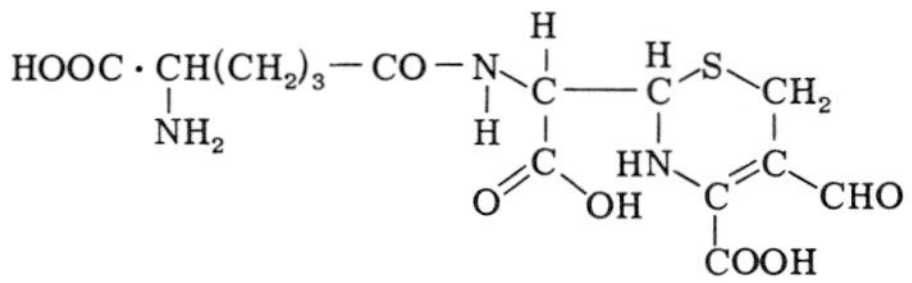 HOOC·CH(CH₂)₃—CO—N—... (see structure below)			Mutants of *C. acremonium*	Fujisawa and Kanzaki, 1975a

(Continued)

TABLE I (Continued)

Compound		Microorganism	References
Penicillin N	$HOOC \cdot CH(CH_2)_3-CO-N$ ... structure	All strains that produce cephalosporins	Abraham et al., 1954, Newton and Abraham, 1954
6-Aminopenicillanic acid	structure	Mutants of *C. acremonium*	Lemke and Nash, 1972
Tetrapeptide P1	α-Aminoadipic acid, cysteine, β-hydroxyvaline, and glycine	*C. acremonium*	Loder and Abraham, 1971a
Tetrapeptide P2	α-Aminoadipic acid, cysteine, valine, and glycine	*C. acremonium*	Loder and Abraham, 1971a
Tripeptide P3	δ-(L-α-Aminoadipyl)-L-cysteinyl-D-valine	*C. acremonium*	Loder and Abraham, 1971a
Peptide S-1	Dimer of δ-(L-α-aminoadipyl)-L-cysteinyl-D-valine	Mutants of *C. acremonium*	Kanzaki et al., 1974a; Shirafuji et al., 1975
Peptide S-2	Disulfide of δ-(L-α-aminoadipyl)-L-cysteinyl-D-valine and methanethiol	Mutants of *C. acremonium*	Kanzaki et al., 1974a; Shirafuji et al., 1975

[a] The majority of the deacetoxycephalosporin C-producing strains, *C. chrysogenum*, *Cephalosporium* sp., *Emericellopsis* sp., *Diheterospora chlamydosporia*, *Scopulariopsis* sp., *S. lipmanii*, and *S. clavuligerus*, produce cephalosporin C and deacetylcephalosporin C.

of *C. acremonium*. In the case of the latter, deacetylcephalosporin C is synthesized *de novo* by some of the cephalosporin C negative mutants (Fujisawa *et al.*, 1973, 1975a). In contrast, it may be accumulated through enzymic or chemical hydrolysis of cephalosporin C in other deacetyl-cephalosporin C-producing microorganisms, since acetylhydrolase has been demonstrated in the culture broth of a *C. acremonium* mutant (Fujisawa *et al.*, 1973, 1975b; Kanzaki *et al.*, 1974a; Liersch *et al.*, 1974) and the *Streptomyces* species (Brannon *et al.*, 1972), and a trace amount of deacetylcephalosporin C in the fermentation broths of *C. acremonium* was reported to be formed by chemical hydrolysis (Huber *et al.*, 1968).

It is known that deacetoxycephalosporin C is formed through *de novo* synthesis in the mutants of *C. acremonium* defective in cephalosporin C- or deacetylcephalosporin C-productivity (Kanzaki *et al.*, 1974a; Liersch *et al.*, 1974; Queener *et al.*, 1974; Fujisawa *et al.*, 1975c). These strains have been reported to produce slight amounts of deacetylcephalo-sporin C (Liersch *et al.*, 1974). Details are not known about the biosynthesis of deacetoxycephalosporin C in other microorganisms.

The finding of cephalosporins from streptomycetes is an epoch-making event in the following respects. One is the fact that prokaryotes can also produce cephalosporins and such organisms are widely distributed in nature. The other is the demonstration of the occurrence of a new family of cephem compounds, 7-methoxycephem derivatives. The new cephalosporins are noteworthy for the fact that the introduction of the methoxyl group to the cephem nucleus increased the activity against gram-negative bacteria, especially against the *Proteus* species (A. K. Miller *et al.*, 1972).

Compound F-1 is detected in culture broths of potent cephalosporin C-producing strains (Kanzaki *et al.*, 1974a,b). It may be produced through chemical or biochemical reaction between cephalosporin C and methanethiol originating from added methionine by the action of methionase.

A mutant of *C. acremonium*, No. 1011, produced a cephem compound F-2 in the culture broth (Y. Fujisawa, K. Kitano, K. Katamoto, K. Nara, and T. Kanzaki, unpublished data, 1975). The compound F-2 was determined to be a Bunte salt, in which the *O*-acetyl group of cephalosporin C was replaced by thiosulfate nonbiologically (Demain *et al.*, 1963b). Therefore, compound F-2 may be produced from thiosulfate, which is formed from methionine added to the medium and cephalosporin C.

A *C. acremonium* mutant No. C-43-219 accumulated a small amount of a new cephem compound, C-43-219 (Kitano *et al.*, 1975b). The compound was identified as 7-(D-5-amino-5-carboxyvaleramido)-3-(1,1-dime-thyl-2-amino-2-carboxyethylthiomethyl)-3-cephem-4-carboxylic acid, which may be formed from cephalosporin C and penicillamine.

A deacetoxycephalosporin C-producing mutant of *C. acremonium*, CP71, blocked in the synthesis of cephalosporin C, was found to accumulate a new cephem compound (Traxler *et al.*, 1975). The new compound was identified as *N*-acetyldeacetoxycephalosporin C. The titers of this compound in the three independently obtained deacetoxycephalosporin C-producing mutants varied from 0.05 to 0.2 mg/ml.

Kitano *et al.* (1975b) isolated and identified a family of cephem compounds with the glutaryl group instead of the α-aminoadipyl residue. These compounds are accumulated in a trace amount in the culture broths of *C. chrysogenum*. Hydrophilic β-lactam compounds have been found to involve only the α-aminoadipyl group or its *N*-acetyl derivative as a side chain. The finding of this family of cephalosporins is important, but the mechanism of its biosynthetic pathways remains to be determined.

A new cephalosporoate (C-2) was found to be accumulated in the culture broths of *C. acremonium* mutants Nos. 20, 29, 36, and 40, which were defective in acetyl CoA:deacetylcephalosporin C acetyltransferase and therefore could not accumulate cephalosporin C but could accumulate deacetylcephalosporin C (Fujisawa and Kanzaki, 1975a). The compound C-2 was isolated from the culture filtrate of the deacetylcephalosporin C-producing mutant No. 40 and identified as D-5-amino-5-carboxyvaleramido(5-formyl-4-carboxy-2H,3H,6H-tetrahydro-1,3-thiazinyl)glycine by its chemical and physical properties. It is supposed that compound C-2 is formed through hydrolysis of the β-lactam ring of 7-(5-amino-5-carboxyvaleramido)-3-formyl-3-cephem-4-carboxylic acid which may be synthesized by oxidation of deacetylcephalosporin C.

Penicillins fall into two types: hydrophobic and hydrophilic. From the viewpoint of antibiotic production, there are interesting differences between strains producing them. Organisms that produce a hydrophilic penicillin, penicillin N, never produce hydrophobic penicillins and fail to respond to the addition of side-chain precursors, and vice versa. A hydrophilic penicillin, isopenicillin N, possessing the L-α-aminoadipyl group at C-7 as the side chain, is observed in culture broths of hydrophobic penicillin producers. However, the amount is not only very small, but the major portion is contained in mycelia. Furthermore, biochemical approaches also suggest it to be an intermediate of the synthesis of hydrophobic penicillin. Thus, it is considered not to be a final product. Isopenicillin N has not yet been detected in the cultures of penicillin N producers.

Microorganisms that produce penicillin N are very similar to those producing cephalosporins. From the point of view of antibiotic production, on the other hand, fungi and streptomycetes that produce cephalosporins always produce penicillin N, but not all penicillin N-producing strains produce cephalosporins. In other words, only some of the penicillin

N producers can produce cephalosporins. Interestingly enough, this relationship also holds true in mutants of the *Cephalosporium* species (Lemke and Nash, 1972; Shirafuji *et al.*, 1975). Derivation of antibiotic-negative mutants from *C. acremonium* gives rise to mutants defective in both penicillin N and cephalosporin C, and penicillin N-positive and cephalosporin C-negative ones, but never yielded penicillin N-negative and cephalosporin C-positive strains. There must be a definite biochemical relationship behind these phenomena.

Cephalosporin-producing organisms produce only 6-aminopenicillanic acid as penam compounds besides penicillin N. When *C. acremonium* was cultivated in complex medium, 6-aminopenicillanic acid was produced (Lemke and Nash, 1972), but it was not detected in cultures of synthetic medium (Smith *et al.*, 1967). The function of 6-aminopenicillanic acid in the metabolism of β-lactam antibiotics in cephalosporin-producing organisms remains to be determined, although it is believed to be a shunt metabolite.

The so-called "tripeptide theory" where a common precursor of β-lactam antibiotics is a tripeptide, α-aminoadipylcysteinylvaline, has been supported by most of β-lactam antibiotic investigators. The tripeptide was first isolated from mycelia of *P. chrysogenum* by Arnstein and Morris (1960). Loder and Abraham (1971a) later isolated three peptides named P3, P2, and P1 from mycelia of *C. acremonium*. The peptides P3, P2, and P1 were identified as δ-(L-α-aminoadipyl)-L-cysteinyl-D-valine, a tetrapeptide consisting of α-aminoadipic acid, cysteine, valine and glycine, and a tetrapeptide consisting of α-aminoadipic acid, cysteine, β-hydroxyvaline, and glycine, respectively.

Independent of this work, Kanzaki *et al.* (1974a) obtained several kinds of antibiotic-negative mutants of *C. acremonium*. They noticed some mutants which accumulated two peptides extracellularly among both penicillin N and cephalosporin C-negative strains. The peptides, S-1 and S-2, were determined to be the dimer of δ-(L-α-aminoadipyl)-L-cysteinyl-D-valine and the disulfide of δ-(L-α-aminoadipyl)-L-cysteinyl-D-valine and methanethiol, respectively. Methanethiol may be derived from methionine added to the culture medium. The amounts of S-1 and S-2 were about 500 and 50 μg/ml, respectively.

III.   Genetic Studies and Strain Improvement

Until recently, cephalosporin-producing strains had been substantially limited to only the single strain of *C. acremonium*, Brotzu's strain, and the object of investigation had been restricted within this strain and its descendants. Studies on genetics and strain improvement of cephalosporin-producing microorganisms had also been undertaken mainly with

C. acremonium. Few genetic investigations had been made, probably because its cytological characteristics are unfavorable for analyses. The great importance of cephalosporins as chemotherapeutics also had not been recognized until recently. Few reports concerning *Streptomyces* species have appeared, since cephalosporin-producing streptomycetes have only recently been discovered.

Cephalosporium molds belong to an imperfect fungi family (*Moniliaceae,* similar to *Penicillium*) so that the genetic studies on the *Cephalosporium* species should depend on parasexual methods. *Streptomyces* is a genus of prokaryote, but studies have mostly centered on simple mutagenesis and heterokaryosis.

Cephalosporium acremonium has four cell types: hyphae, germlings, arthrospores, and conidia (Nash and Huber, 1971). Cytological inspection of these cells disclosed that they were mostly uninucleate and rarely multinucleate (Nüesch *et al.,* 1970). This condition restricts hetero-karyosis, subsequent formation of diploids, and recombination. However, Nüesch *et al.* (1970) successfully obtained heterokaryons and proto-trophic recombinants though the improvement of antibiotic potential was not satisfactory.

Fantini (1962) investigated the genetics of *Emericellopsis* species, *E. salmosynnemata* and *E. terricola* var. *glabra,* which produced penicillin N and represent the sexual stage of *Cephalosporium.* The process of heterokaryosis was found to be limited in frequency and extent, and sexual meiotic recombination rarely occurred. These observations coin-cide well with those of *Cephalosporium* fungus. Some of the hetero-karyons synthesized from auxotrophs, inferior in antibiotic production to the original wild strain, showed higher yields than the parents. How-ever, the yields were the same as or lower than those of the original. This is also true of diploids. In penicillin production by *Penicillium* molds, similar results have been obtained. Strains selected through parasexual methods, even when superior to the parents, have never been employed for industrial purposes because of their instability (Pathak and Elander, 1971). To date, it seems that there has been no unanimity of opinion with respect to the usefulness of improved strains through parasexual methods.

T. Kanzaki, Y. Fujisawa, H. Shirafuji, and T. Fukita (unpublished data, 1974) produced a number of heterokaryons and a few hetero-zygous diploidlike strains by crossing the auxotrophs of *C. acremonium.* However, the heterokaryons were mostly unbalanced, unlike those of Fantini (1962), and were not satisfactory for antibiotic production. One of the diploidal strains was moderately improved in cephalosporin C production as compared with the original wild-type strain.

The existence of mycophage has been recognized in *Cephalosporium* species (Day and Ellis, 1971), as in *Penicillium* fungi. Lemke and Brannon (1972) surmise that the high frequency of sectoring of *Penicillium* molds is due to the formation of aneuploidy and that there is probably a close relationship between mitotic nonconformity and the presence of mycophage. Such observations, however, have not been reported with the *Cephalosporium* species. Further investigations are necessary on the role of mycophage.

Methionine has been recognized to be a more expedient source of sulfur for the synthesis of cephalosporin C than sulfate. Specific mutants of *C. acremonium*, altered for sulfur metabolism and for their potential to synthesize cephalosporin C from sulfate, were derived (Niss and Nash, 1973). One of the mutants was M8650-*sp*-1, corresponding to the *cys*-3 mutant of *Neurospora crassa* in which the locus exerts coordinate control over the synthesis of sulfate permease as well as arylsulfatase. This mutant was facilitated for sulfate transport and repressed for arylsulfatase, and utilizes sulfate as effectively as methionine in providing sulfur for cephalosporin C. In this connnection, the parent strain, M8650, is considered to be a derepressed mutant for arylsulfatase synthesis. The sulfatase repression in M8650-*sp*-1 may be related to the accumulation of sulfide that regulates sulfatase synthesis, since sulfide is believed to be a corepressor of sulfatase in fungi (Metzenberg and Parson, 1966). Generally speaking, the mutants of *C. acremonium* are increasingly derepressed for arylsulfatase and concomitantly exhibit increased potentials for synthesis of antibiotics from methionine (Dennen and Carver, 1969).

Another mutant, IS-5, with enhanced potential to use sulfate for cephalosporin C production, had potency levels more than twofold that of the parent N-16, with sulfate as the sulfur source (Komatsu *et al.*, 1975). Cephalosporin C production by this mutant was sensitive to methionine, as opposed to the parent. In addition to the mutant IS-5, all of several mutants with an increased potential to produce higher levels of cephalosporin C from sulfate were methionine sensitive. Therefore, the increase in productivity from sulfate and that of methionine sensitivity may be metabolically related and caused by the same mutational event.

Practical approaches to improvement of antibiotic potential of *C. acremonium* have been attempted by some workers. Fasani *et al.* (1974) found phenethyl alcohol to be a good mutagen for improvement of antibiotic yields. Elander *et al.* (1974) examined various mutagens with respect to increasing cephalosporin C productivity and demonstrated that UV was the best mutagen.

With the *Streptomyces* species, mutants defective in the control of amino acids belonging to aspartic acid families were isolated (Godfrey, 1973). A potent strain, *leu⁻* LA473, which exhibited threefold productivity over the parent level, among strains lacking control in the *ilv–leu* region. The genetic investigation indicated the possibility that valine biosynthesis might be rate-limiting in those mutants.

Godfrey (1974) also developed a method whereby mutation could be directed toward preselected markers including some affecting the antibiotic production and demonstrated that N-methyl-N'-nitro-N-nitrosoguanidine could preferentially mutate the DNA replicating region by the use of *S. lipmanii.*

IV. Fermentation

A. CEPHALOSPORIN C PRODUCTION BY *Cephalosporium acremonium*

Cephalosporin C fermentation has been studied mainly with *C. acremonium* CMI 49,137, derived from the Brotzu's strain, and its descendants. An interesting feature is that methionine markedly stimulates cephalosporin C production by the fungus (Demain and Newkirk, 1962; Ott *et al.,* 1962). It was reported earlier that synthesis of penicillin N by *C. acremonium* was stimulated prominently by D-methionine (Miller *et al.,* 1956; Kavanagh *et al.,* 1958). Demain and Newkirk (1962) and Ott *et al.* (1962) confirmed that the yields of cephalosporin C as well as penicillin N were enhanced by the addition of D-methionine to a complex medium. An essential role of methionine is considered to be sulfur supply for β-lactam antibiotics (Caltrider and Niss, 1966), but *C. acremonium* requires excessive amounts of methionine for the maximum production of antibiotics (Demain and Newkirk, 1962). This is probably due to the existence of intracellular D-amino acid oxidase, which is reactive with D-methionine and D-amino acids related to methionine (Nüesch *et al.,* 1970; Benz *et al.,* 1971). The superiority of D-methionine over L-methionine for antibiotic biosynthesis may be due to its slower uptake and availability. Of a large number of sulfur compounds, only DL-methionine-DL-sulfoxide and s-methyl-L-cysteine were substituted for methionine to a considerable extent (Demain and Newkirk, 1962). These compounds are believed to act after conversion to methionine.

There is evidence to support the idea that methionine has an obligatory role in stimulation of antibiotic synthesis other than sulfur supply in *C. acremonium.* The demonstration that norleucine was capable of completely replacing methionine in a synthetic medium with inorganic sulfur eliminated the possibility that methionine was the exclusive sulfur source for the synthesis of β-lactam antibiotics (Demain *et al.,* 1963a). The methionine-replacing activity of norleucine, however, was not observed in a complex medium (Caltrider and Niss, 1966). Recently, the report of Drew and Demain (1973) showed that the inactivity of norleucine in complex medium was due to the lack of uptake of this amino acid by mycelia grown in such a medium. In chemically defined media supplemented with a suboptimal level of methionine, norleucine stimulated the antibiotic production up to the same level elicited by optimal methionine. At the optimal methionine level, the further increase in cephalosporin C production by norleucine was not observed. These data strongly suggest that these two amino acids act through a common mechanism, and the stimulation by methionine may be based on a regulatory mechanism. However, methionine was found to neither repress nor inhibit cysteine desulfhydrase or any other enzyme related to cysteine metabolism. The findings denied the earlier idea that methionine represses the synthesis of a cysteine-degrading enzyme. Furthermore, methionine was shown to have a greater stimulatory effect on cephalosporin C production when present during the growth rather than the production phase in a resting cell system that completely dissociated growth from formation of cephalosporin C (Demain, 1963b). These findings suggest that methionine has multiple effects on the cephalosporin C fermentation.

The significance of internal methionine pools in cephalosporin C formation was investigated by the aid of several auxotrophic mutants of *C. acremonium* CW 19, which were genetically blocked at different points in sulfur metabolism (Drew and Demain, 1974, 1975a). The mutant S-1 (*met⁻*), blocked in conversion of cystathionine to homocysteine, failed to produce the antibiotic in spite of normal growth and the presence of sufficient sulfate, while the parent produced moderate levels of cephalosporin C from sulfate as the sole source of sulfur (Drew and Demain, 1975b). The finding indicates that methionine has an obligatory role in stimulation of cephalosporin C production in addition to the sulfur source. The double mutant 11-8 (*met⁻*), which cannot convert sulfate to cysteine or cysteine to methionine, did not produce the antibiotic from cysteine even though enough methionine was added to support normal growth (Drew and Demain, 1975c). Addition of norleucine also stimulated the antibiotic production from cysteine in the double mutant.

An interesting experiment showing methionine's role in the stimulation of cephalosporin C formation was carried out with the mutant H (*cys⁻*), which cannot convert methionine to cysteine and sulfate to cysteine (Drew and Demain, 1975d). The antibiotic production by this cysteine auxotroph was stimulated by DL-alanyl-DL-methionine, which acts as a source of internal methionine in the presence of cysteine (Fig. 1). These facts show that methionine has a role in stimulating cephalosporin C synthesis other than that of being a sulfur donor, but the exact nature of the nonprecursor role is still unknown.

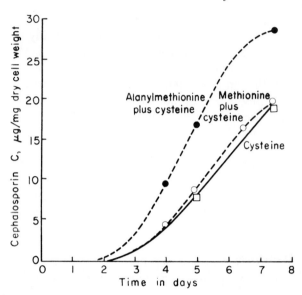

FIG. 1. Alanylmethionine stimulation of cephalosporin C production by mutant H (Drew and Demain, 1975d).

A noteworthy observation is that the maximum rate of cephalosporin C production coincides with the conversion of slender hyphal filaments to swollen hyphal fragments, arthrospores (Smith *et al.*, 1967). As compared with inorganic sulfate-grown cells, methionine-grown cells had a different morphology (Caltrider *et al.*, 1968). The cells formed under submerged culture consisted of hyphae, arthrospores, conidia, and germlings. Each cell type was isolated by density-gradient centrifugation, and its synthetic capacity for cephalosporin C and penicillin N was determined. Arthrospores synthesized 40% more of the antibiotics than any other morphological types (Nash and Huber, 1971). The data reveal that the antibiotic synthesis is proportional to the extent of differentiation into arthrospores. Isolation of an antibiotic-negative mutant, however, clearly eliminated an obligatory relationship between morphogenesis and antibiotic synthesis because the mutant normally produced arthrospores. Recently, it was reported that norleucine, which underwent methionine-replacing activity in synthetic medium, also induced fragmentation of mycelia (Drew and Demain, 1973, 1974).

Sulfatase is thought to catalyze the release of endogenously stored sulfur to sulfate in fungi (Hussey *et al.*, 1965). The sulfatase of *C. acremonium* was regulated by exogenous sulfur compounds, i.e., it was repressed in $2 \times 10^{-2} M$ sulfate, and derepressed in $5 \times 10^{-4} M$ sulfate (Dennen and Carver, 1969). Methionine, cysteine, and homocysteine caused sulfatase derepresssion. The sulfatase activity was proportional to the antibiotic synthesis. This relationship suggests that mutants with increased cephalosporin C productivity from methionine have decreasing ability to utilize this amino acid for other sulfur requirements, and are derepressed for sulfatase synthesis.

Methionine uptake rate is considered to play an important role in the biosynthesis of cephalosporin C. Some seleno-methionine-resistant mutants, which were impaired for the transport system of methionine, produced little antibiotic. The L-methionine uptake system was repressed by thiosulfate and cysteine and inhibited by cysteine and homocysteine (Nüesch et al., 1972). These data give an explanation for the fact that a mutant blocked for the condensation of inorganic sulfur with an organic compound, probably serine, shows a marked increase in cephalosporin C productivity.

The excessive amount of methionine necessary for the optimal production of cephalosporin C can partly be explained by the fact that C. acremonium possesses intracellular D-amino acid oxidase (Nüesch et al., 1970; Benz et al., 1971). The greater part of the added methionine was deaminated to form 2-keto-4-methylthiobutyric acid by the D-amino acid oxidase. The enzyme was strictly stereospecific for D-amino acid as substrate, and attacked D-methionine and a small number of D-amino acids related to methionine. The L-amino acid transaminase of C. acremonium converted 2-keto-4-methylthiobutyric acid to L-methionine which was then transferred to cysteine via the reverse transsulfuration pathway (Nüesch et al., 1972; Liersch et al., 1973).

The stimulation of cephalosporin C production with ε-N-acetyl-L-lysine and ε-aminocaproic acid was observed. However, it remains unknown whether the effect of these compounds is specific or mediated on the antibiotic synthesis. Lysine, α-aminoadipic acid, and α-ketoadipic acid were inactive (Demain and Newkirk, 1962; Demain et al., 1963a). Therefore, it does not appear that the supply of α-aminoadipic acid is a rate-limiting step in the synthesis of antibiotics by C. acremonium.

The cephalosporin C fermentation by the fungus in synthetic medium could be divided into two stages: a growth phase and an antibiotic production phase (Demain, 1963b). The mycelia showed varying degrees of heterogeneity during fermentation (Smith et al., 1967). These studies suggest a relationship between morphological changes and ability to synthesize antibodies. The production of cephalosporin C occurs efficiently under aerated and agitated conditions after rapid cell growth (Auden et al., 1969; Feren and Squires, 1969). Moderate decreases in the efficiency of aeration resulted in a fall in the production of cephalosporin C and a rise in that of penicillin N (Smith et al., 1967).

A relationship exists between cephalosporin C production and fatty acid composition of *C. acremonium* (Huber and Redstone, 1967). The production was paralleled by an increase in the amount of linoleic acid in the cells grown in the presence of methyl oleate.

B. CEPHALOSPORIN PRODUCTION BY STREPTOMYCETES AND FUNGI

A number of *Streptomyces* species have been found to produce new cephalosporins. Stapley *et al.* (1972) investigated in detail the production of cephamycins A and B by *S. griseus* MA-2837 and of cephamycin C by *S. lactamdurans* MA-2908. The yield of cephamycins A and B by *S. griseus* was increased almost 20-fold, from 9 to 170 μg/ml, by variation of the medium formula, by lowering the temperature of incubation, and by selection of a superior natural isolate. Cephamycins A and B could also be produced in a chemically defined fermentation medium. An improved isolate, *S. griseus* MA-4125, accumulated 120 μg/ml of cephalosporins. *Streptomyces lactamdurans* MA-2908 produced from 20 to 120 μg/ml of cephamycin C in complex media. The leucine mutant LA473, derived from *S. lipmanii*, was able to produce 290 μg/ml of 7-methoxycephalosporin C (A16884A) in a synthetic medium (Godfrey, 1973). The yield of cephalosporins produced by streptomycetes does not seem to reach the level of cephalosporin C accumulated by *C. acremonium* at present since cephalosporins have only recently been isolated from the *Streptomyces* species.

A few cases exist where production of antibiotics has been investigated in *n*-paraffin medium. In the course of studies on the production of antibiotics by *n*-paraffin-utilizing microorganisms, Kitano *et al.* (1974b) found that *Paecilomyces carneus* and *Paecilomyces persicinus* produced 255 and 36 μg/ml of cephalosporins from *n*-paraffin, respectively.

V. Biosynthesis

The so-called "tripeptide theory" has dominated the theme of β-lactam biosynthesis since the isolation of δ-(α-aminoadipyl)cysteinyl-valine from *P. chrysogenum* (Arnstein and Morris, 1960). Loder and Abraham (1971a) isolated three sulfur-containing peptides from mycelia of *C. acremonium*. The main peptide, P3, was shown to be δ-(L-α-amino-adipyl)-L-cysteinyl-D-valine. The existence of a D-valine residue in peptide P3 aroused interest in the mechanism of β-lactam ring formation. The tripeptide was synthesized from δ-(L-α-aminoadipyl)-L-cysteine and L-valine in a broken-cell system (Loder and Abraham, 1971b; Abraham, 1974). Further investigations have demonstrated that tritium-labeled isomers of δ-(L-α-aminoadipyl)-L-cysteinylvaline were incorporated into penicillin N by whole extract, although cephalosporin C formation has not been reported (Fawcett *et al.*, 1974). However, these facts seem to support the tripeptide theory.

Recent studies with ^{13}C-labeled precursors have confirmed earlier data obtained from incorporation experiments of ^{14}C compounds into cephalosporin C (Neuss *et al.*, 1971a,b). Labeling studies with ^{13}C-labeled compounds have given information on problems related to the origin of the cephalosporin ring system (Neuss *et al.*, 1973; Kluender *et al.*, 1973, 1974).

Recently, terminal steps in the biosynthesis of cephalosporin C were elucidated with the aid of cephalosporin C-negative mutants and by the use of cell-free systems. Some of the mutants were found to exclusively accumulate deacetylcephalosporin C in culture broths (Fujisawa *et al.*, 1973, 1975a), while others were shown to mainly accumulate deacetoxycephalosporin C in the broths (Kanzaki *et al.*, 1974a; Queener and Capone, 1974; Liersch *et al.*, 1974; Queener *et al.*, 1974; Fujisawa *et al.*, 1975c). In addition, studies with cell-free systems support a biosynthetic scheme that deacetoxycephalosporin C is converted to deacetylcephalosporin C, followed by acetylation to form cephalosporin C (Fujisawa *et al.*, 1973, 1975c; Liersch *et al.*, 1974; Fujisawa and Kanzaki, 1975b).

The occurrence of new cephalosporins from the *Streptomyces* species was reported for the first time by Nagarajan *et al.* (1971), Hamill *et al.* (1971), Gorman *et al.* (1971), and Stapley *et al.* (1971). The biochemical versatility of streptomycetes was recognized again by this discovery. Structural elucidation and labeling studies of cephalosporins from the *Streptomyces* species indicate that α-aminoadipic acid, cysteine, and valine, similar to fungal cephalosporin C, are precursors of the antibiotics (Hamill *et al.*, 1971; Inamine and Birnbaum, 1972; Miller *et al.*, 1972; Whitney *et al.*, 1972).

A. Amino Acids and Acetic Acid as Precursors and Their Metabolism

The structure of cephalosporin C can be divided into four residues of D-α-aminoadipic acid (A), L-cysteine (B), a C_5 fragment (C-2, C-3, C-4, C-16, C-17) which corresponds to α,β-dehydrovaline (C), and an acetoxy group (D), as shown by the broken lines in Fig. 2 (Abraham and Newton, 1961). The experiments were done to ascertain whether possible exogenous precursors play a part in the biosynthesis of cephalosporin C. The results obtained with these compounds will now be described.

1. Acetic Acid Metabolism

The first attempt to elucidate the biosynthesis of cephalosporin C was made by Trown *et al.* (1962) using radioactive precursors. Cephalosporin

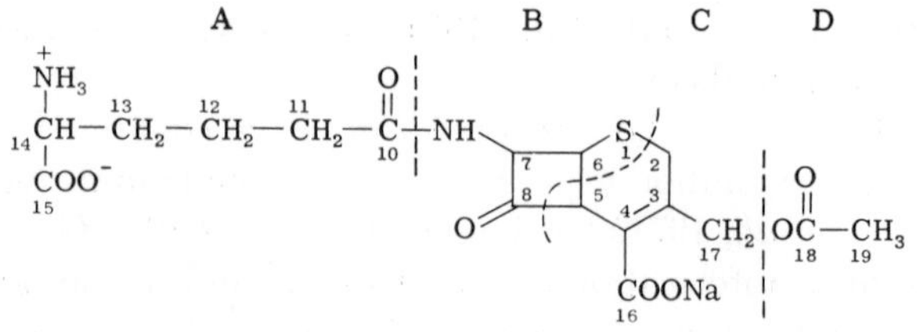

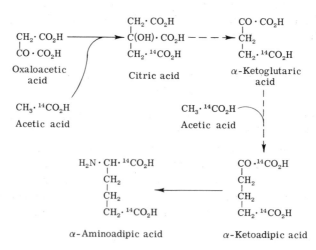

FIG. 2. Structure of cephalosporin C and its numbering.

C, produced by *C. acremonium* in the presence of acetate-1-^{14}C, was radioactive. About 90% of its radioactivity was incorporated into the α-aminoadipyl and acetoxy groups of the antibiotic. The radioactivity of C-1 of the D-α-aminoadipic acid from labeled cephalosporin C was about the same as that of the labeled carbon of the acetoxy group. This is consistent with the view that C-1 of the α-aminoadipic acid was derived from acetate.

Cephalosporin C, which was formed in a complex medium supplemented with α-ketoglutarate-5-^{14}C, contained most of its radioactivity in C-6 of the α-aminoadipyl side chain (Trown *et al.*, 1963a). These results support a hypothesis that the carbon skeleton of α-aminoadipic acid is synthesized from acetyl CoA and α-ketoglutarate in *C. acremonium* (Fig. 3). The hypothesis was confirmed further by the recent studies with ^{13}C-labeled compounds (Neuss *et al.*, 1971a,b). Cephalosporin C labeled in the presence of sodium acetate-2-^{13}C contained the label in the D-α-aminoadipyl side chain, C-11, C-12, C-13, and C-14, and in C-19

FIG. 3. Incorporation of acetate into α-aminoadipyl moiety of cephalosporin C (Trown *et al.*, 1963a).

of the acetoxy group. The radioactivity from sodium acetate-1-^{13}C was found in C-10, C-15, and C-18 (Fig. 2).

2. Lysine Metabolism

Cephalosporin C and penicillin N synthesized by *C. acremonium* have the α-aminoadipyl side chain. The amino acid is considered to be an intermediate for the biosynthesis of lysine in higher fungi. Lysine isolated from the mycelia of *C. acremonium,* which were grown in the presence of DL-α-aminoadipic acid-2-^{14}C, was radioactive. The ^{14}C of cephalosporin C, formed in the presence of DL-α-aminoadipic acid-2-^{14}C or L-α-aminoadipic acid-6-^{14}C, was present almost exclusively in the α-aminoadipyl group (Trown *et al.,* 1963b; Warren *et al.,* 1967a). Therefore, α-aminoadipic acid acts as a side-chain precursor of cephalosporin C.

The washed mycelia took up DL-α-aminoadipic acid-1-^{14}C rapidly and the radioactivity appeared in saccharopine, lysine, and protein. An experiment with L-lysine-U-^{14}C showed that no significant reversal of the lysine pathway existed, although a trace amount of the radioactivity was incorporated into glutamic acid and saccharopine (Abraham *et al.,* 1964). Therefore, the endogenous formation of α-aminoadipic acid in *C. acremonium* probably takes place through the pathway postulated for the synthesis of lysine in higher fungi or yeasts (Fig. 4).

The behavior of different optical isomers of α-aminoadipic acid was studied in *C. acremonium.* L-α-Aminoadipic acid-6-^{14}C was taken up much more rapidly than the D-isomer or α-ketoadipic acid-6-^{14}C by washed mycelia or protoplasts (Warren *et al.,* 1967a; Fawcett *et al.,* 1973). The radioactivity of L-α-aminoadipic acid-6-^{14}C appeared in intracellular δ-aminovaleric acid, saccharopine, lysine, protein, compounds which behaved like penicillin N, cephalosporin C, and deacetylcephalosporin C, respectively, on paper chromatography and electrophoresis, and also a peptide, which included α-aminoadipic acid, cysteine, and valine. The radioactivity was also observed in the D-α-aminoadipyl side chains of extracellular penicillin N and cephalosporin C. On the other hand, ^{14}C from D-α-aminoadipic acid-6-^{14}C was found in penicillin N and cephalosporin C, but the specific radioactivity was diluted relatively high (Warren *et al.,* 1967a). Accordingly, L-α-aminoadipic acid appears to be more available for antibiotic synthesis than the D-isomer.

Considering the phylogenetic position of streptomycetes, the existence of α-aminoadipic acid in the β-lactam antibiotics produced by the organisms is of interest. The origin of this amino acid was investigated using 7-methoxycephalosporin A16886B-producing *S. clavuligerus* (Whitney *et al.,* 1972). Unlike cephalosporin C, the radioactivity from DL-lysine-1-^{14}C was incorporated into the α-aminoadipic acid side chain in

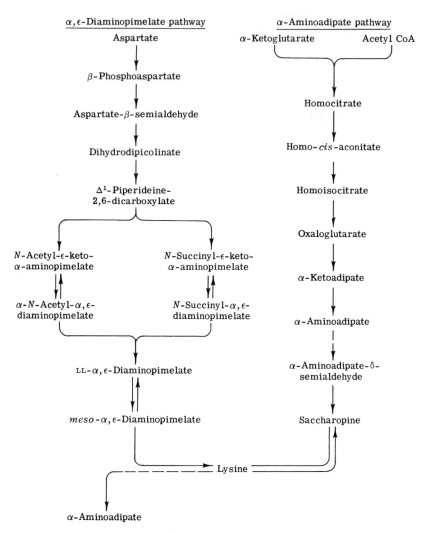

FIG. 4. Biosynthesis and metabolism of lysine.

A16886B. Similar results were obtained using S. *lipmanii,* which produced 7-methoxycephalosporin A16884A (Kirkpatrick *et al.,* 1973). The lysine mutant, LA423, was shown to be defective in diaminopimelate decarboxylase and to accumulate diaminopimelate under derepressed conditions. L-Aspartic acid-U-^{14}C was incorporated into diaminopimelate by whole cells of LA423 derepressed for lysine biosynthesis. Incorporation of L-lysine-U-^{14}C into the antibiotic was also shown using whole cells of LA423. These results support the hypothesis that α-aminoadipic acid is a catabolic product of lysine in this organism. It is believed, therefore, that the antibiotic-producing streptomycetes synthesize lysine by decarboxylation of α,ϵ-diaminopimelate and are capable of catabolizing lysine to α-aminoadipic acid; i.e., the *Streptomyces* species have the same lysine metabolism as bacteria (Fig. 4).

3. Valine Metabolism

Studies of the incorporation of radioactivity from DL-valine-1-^{14}C with *C. acremonium* have shown that the α,β-dehydro-γ-hydroxyvaline fragment of cephalosporin C is derived from valine (Trown *et al.*, 1963b). The role of valine in the biosynthesis of penicillin N and cephalosporin C by the fungus was then investigated using suspensions of washed mycelia. L-Valine-1-^{14}C was taken up rapidly but D-valine-1-^{14}C and α-ketoisovalerate-1-^{14}C were taken up relatively slowly (Warren *et al.*, 1967b). Recently, the uptake rate of L-valine by the protoplasts of *C. acremonium* was shown to be much larger than that of the D-isomer (Fawcett *et al.*, 1973). The addition of D-valine to suspensions of washed mycelia inhibited the production of penicillin N, but not that of cephalosporin C. This might support the view that L-valine is the precursor of the D-penicillamine fragment of the penicillin molecule in *C. acremonium*. The structure of cephalosporin C suggests that γ-hydroxyvaline might take part in its biosynthesis. However, the addition of DL-γ-hydroxyvaline decreased the yield of cephalosporin C, but not that of penicillin N. The inhibition was restored by L-valine but not by the D-isomer (Abraham *et al.*, 1964; Warren *et al.*, 1967b). Therefore, the participation of free γ-hydroxyvaline in the biosynthesis seems to be denied, although the definite role of the compound is not clear because of the use of a mixture of its optical isomers.

Extracellular penicillin N and cephalosporin C were rapidly labeled with L-valine-1-^{14}C. D-Valine-1-^{14}C appeared to be incorporated into the antibiotics after the conversion of D-valine into the L-isomer. An intracellular peptide which contained α-aminoadipic acid, cysteine, and valine was labeled more rapidly than extracellular penicillin N by ^{14}C from L-valine-1-^{14}C. The incorporation of L-valine-1-^{14}C into penicillin N occurred more rapidly than that into cephalosporin C (Warren *et al.*, 1967b). This apparent sequence of labeling by valine suggests the conversion of the peptide to penicillin N, followed by the formation of cephalosporin C from penicillin N. However, the effect of D-valine and γ-hydroxyvaline on antibiotic production as described above does not readily support the hypothesis that cephalosporin C is formed from penicillin N.

Radioactive cephalosporin C produced by *C. acremonium* in the presence of DL-valine-1-^{14}C contained 90% of its ^{14}C in the C_5 fragment shown in Fig. 2 (Trown *et al.*, 1963b). The finding supports the hypothesis that the carbon skeleton of valine is incorporated intact into this fragment. The addition of L-valine-1-^{14}C to the resting cells of *C. acremonium* resulted in the formation of radioactive cephalosporin C (Demain, 1963c). Furthermore, these results were confirmed by the use of ^{13}C-labeled valine (Neuss *et al.*, 1971b). The experiments with DL-valine-1-^{13}C showed the localization of the label specifically in the C-16 carboxyl of cephalosporin C. When DL-valine-2-^{13}C was used, the label in the molecule was found exclusively in C-4 (Fig. 2). Further labeling studies with (2S,3S)-[4-^{13}C]valine clearly showed that the ^{13}C label was located in the C-17 exocyclic methylene carbon of cephalosporin C. In the case of penicillin N, the ^{13}C valine was incorporated with retention

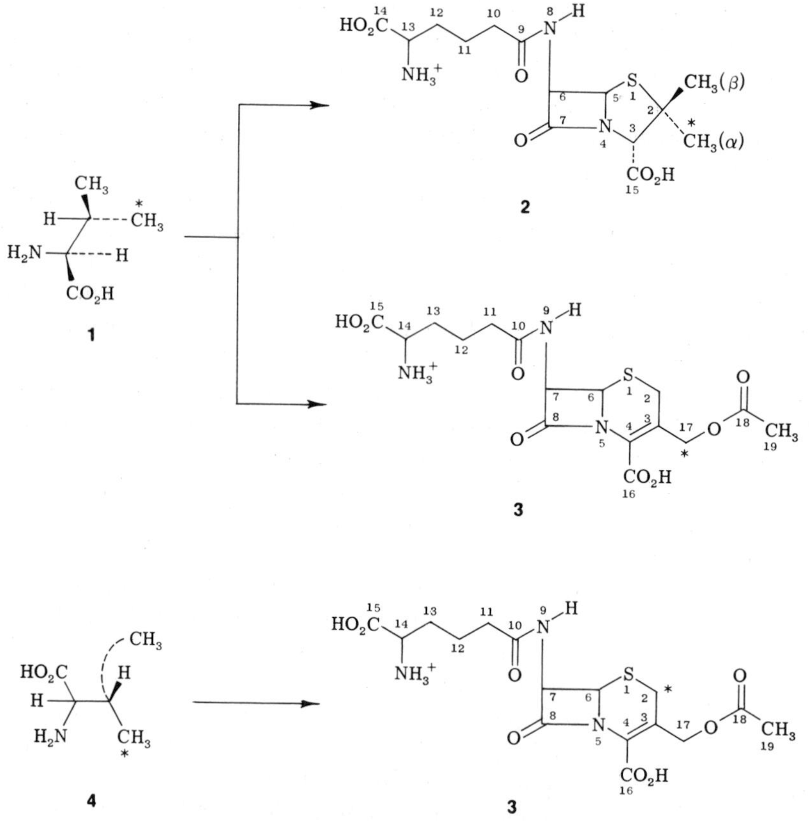

FIG. 5. Incorporation of (2S,3S)-[4-^{13}C]valine and (2RS,3S)-[4-^{13}C]valine into β-lactam antibodies (Neuss *et al.*, 1973; Kluender *et al.*, 1973). Reprinted with permission from the *Journal of the American Chemical Society*. Copyright by the American Chemical Society. (1) (2S,3S)-[4-^{13}C]Valine; (2) penicillin N; (3) cephalosporin C; and (4) (2RS,3S)-[4-^{13}C]valine.

of configuration, as shown in Fig. 5 (Kluender et al., 1973). When cephalosporin C was pulse labeled with (2RS,3S)-[4-¹³C]valine, the ¹³C label was incorporated exclusively into C-2 (Neuss et al., 1973). These results clearly show the asymmetric incorporation of ¹³C chiral label in valine into β-lactam antibiotics and demonstrate the value of valine with a chiral isotopic label as precursor in the biosynthesis of the antibodies. Studies of the fate of the diastereotopic deuterium-labeled methyls of L-valine, in the course of their incorporation into penicillin N and cephalosporin C, were carried out (Kluender et al., 1974). After labeling with (2S,3S)-[4,4,4-²H₃]valine and (2S,3R)-[4,4,4-²H₃]valine, the resulting penicillin N and cephalosporin C were converted into their respective N-acyl methyl ester derivatives, and subjected to mass spectrometric analyses. The α-methyl of the penicillin N derivative derived from (2S,3S)-[4,4,4-²H₃]valine contained three deuteriums. With cephalosporin C originating from (2S,3S)-[4,4,4-²H₃]valine, two deuterium atoms were retained at the exocyclic C-17 position. All three deuteriums in (2S,3R)-[4,4,4-²H₃]valine are also incorporated intact into the β-methyl group of the penicillin N derivative. The cephalosporin C derivative derived from (2S,3R)-[4,4,4-²H₃]valine contained two deuteriums in the endocyclic C-2 position of the cephem nucleus (Fig. 6). These data deny the possibility that the formation of the 3-cephem nucleus proceeds via a 2-cephem intermediate.

Recently, Huang et al. (1975) presented the results of biosynthetic studies with (2S,3S)-methylvaline-¹⁵N,3-*methyl*-d₃ and experimental evidence for the nonparticipation of α,β-dehydrovalinyl intermediates in the biosynthesis of δ-(L-α-aminoadipyl)-L-cysteinyl-D-valine.

4. Sulfur Metabolism

Experiments were undertaken to confirm whether cystine is incorporated into the β-lactam-dihydrothiazine ring of cephalosporin C (Trown et al., 1963b), as into the β-lactam-thiazolidine ring of benzylpenicillin (Arnstein, 1958). The radioactivity from a mixture of DL-cystine-3-¹⁴C and *meso*-cystine-3-¹⁴C, added to a complex medium, was incorporated into cephalosporin C. The majority of the total radioactivity in the antibiotic was found in the C_2 fragment (C-6, C-7) of the β-lactam ring. However, significant portions were also observed in the C_5 fragment of the dihydrothiazine ring and the α-aminoadipyl group (Fig. 2). This suggests the degradation of cysteine to pyruvate, which is metabolized through a currently accepted pathway for the biosynthesis of valine, since some cysteine desulfhydrase activity is detected in the extracts of *C. acremonium* (Wixom and Howell, 1965). The incorporation of some ¹⁴C of pyruvate-3-¹⁴C into the acetoxy and α-aminoadipyl groups of cephalosporin C may be explained by its oxidative decarboxylation to acetate-2-¹⁴C.

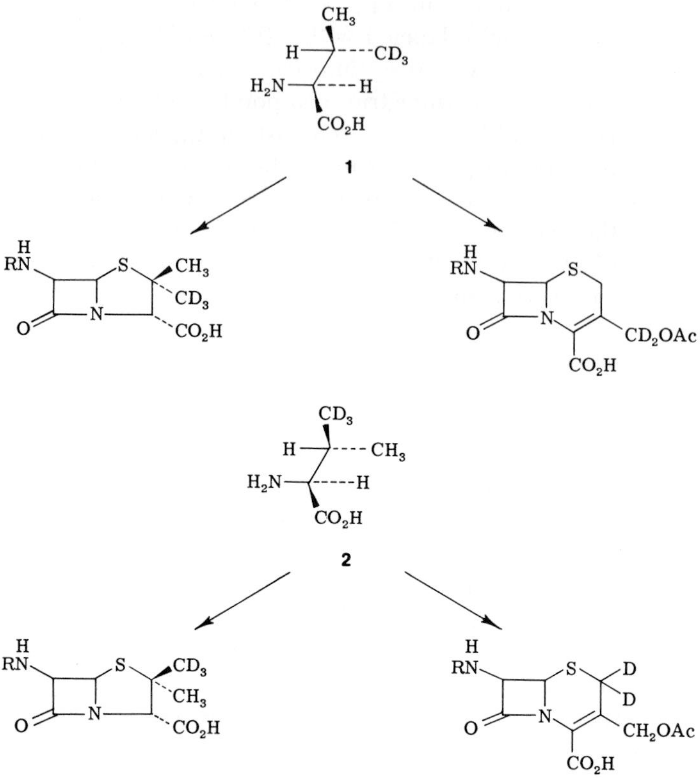

FIG. 6. Incorporation of (2S,3S)-[4,4,4-^{2}H$_3$]valine and (2S,3R)-[4,4,4-^{2}H$_3$]valine into β-lactam antibiotics (Kluender *et al.*, 1974). Reprinted with permission from the *Journal of the American Chemical Society.* Copyright by the American Chemical Society. (1) (2S,3S)-[4,4,4-^{2}H$_3$]Valine, and (2) (2S,3R)-[4,4,4-^{2}H$_3$]valine.

The β-lactam antibiotics possess a sulfur atom, derived from cysteine, in the molecule. The amino acid can be formed via either the sulfate reduction pathway or the reverse transsulfuration pathway (Fig. 7). The sulfur of penicillin produced by *P. chrysogenum* is derived mainly via the sulfate reduction pathway from sulfate (Stevens *et al.*, 1953; Segel and Johnson, 1963), while the atom for the β-lactam antibiotics formed by *C. acremonium* is derived preferentially from methionine via the reverse transsulfuration pathway (Caltrider and Niss, 1966).

Methionine stimulates the production of penicillin N and cephalosporin C by *C. acremonium* (Demain and Newkirk, 1962; Ott *et al.*, 1962). The amino acid is thought to play multiple roles in the synthesis of the antibiotics by the fungus, as stated above. Caltrider and Niss (1966) demonstrated that the sulfur atom in the β-lactam antibiotics produced by *C. acremonium* was derived from methionine on the basis of experiments

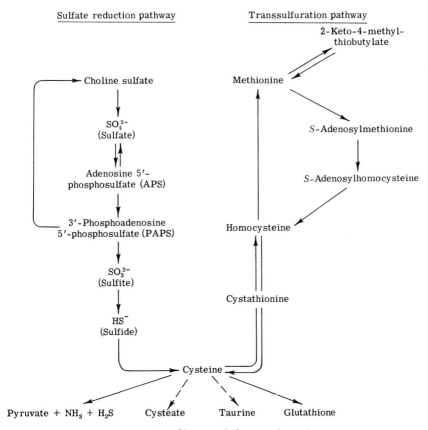

FIG. 7. Sulfur metabolism in fungi.

with L-methionine-^{35}S. Labeled homocysteine and cystathionine, the intermediates of the reverse transsulfuration pathway, were found in the soluble pool of cells grown in the presence of L-methionine-^{35}S. The incorporation of DL-serine-3-^{14}C into the antibiotics was stimulated by methionine. L-Cysteine had a sparing effect on the incorporation of L-methionine-^{35}S and DL-serine-3-^{14}C into the antibiotics. These findings are consistent with the hypothesis that the reverse transsulfuration pathway is operative in the transfer of sulfur between methionine and cysteine in *C. acremonium* (Fig. 7). The existence of the transsulfuration pathway in this fungus was confirmed using various mutants blocked for the pathway of biosynthesis of the sulfur-containing amino acids (Nüesch *et al.*, 1970).

The biosynthesis of 7-methoxycephalosporin A16886B by *S. clavuligerus* was investigated using labeled amino acids (Kirkpatrick *et al.*, 1973). The findings that labels from DL-serine-3-^{14}C and DL-methionine-^{35}S are incorporated into A16886B indicate that the transsulfuration route to cysteine may be operative in *S. clavuligerus*. Label from methionine-^{14}C-*methyl* was incorporated into the 7-methoxy group. The relative percentage of incorporation of DL-cystine-3,3'-^{14}C and L-cystine-3,3'-^{3}H suggests the improbability of the C-7 methoxy group arising from an intermediate containing a double bond between C-6 and C-7.

B. Peptides as Intermediates

A possible intermediate peptide in the biosynthesis of penicillin N and cephalosporin C has been considered to be the tripeptide, δ-(α-aminoadipyl)cysteinylvaline, which had been obtained in the form of its sulfonic acid from *P. chrysogenum* by Arnstein and Morris (1960). In experiments with valine-1-^{14}C, Abraham *et al.* (1964) detected a trace amount of radioactivity in a compound, which appeared to be the tripeptide, in mycelia of *C. acremonium*. They then obtained a peptide complex in very small amounts from the organism, which yielded α-aminoadipic acid, cysteic acid, and valine on hydrolysis. This complex may be labeled by incubating L-valine-1-^{14}C or DL-α-aminoadipic acid-6-^{14}C with washed mycelia under agitation. Further investigation by Loder and Abraham (1971a) showed that the peptides were precipitated as cuprous mercaptides together with glutathione and could be resolved into three peptides, P1, P2, and P3. Peptide P3 was shown to be δ-(L-α-aminoadipyl)-L-cysteinyl-D-valine, which was different from the hitherto proposed configuration of the constitutive amino acid. Peptide P2 contained α-aminoadipic acid, cysteine, valine, and glycine, and peptide P1 consisted of α-aminoadipic acid, cysteine, β-hydroxyvaline, and glycine. If peptide P3 is a precursor of penicillin N or cephalosporin C, the isomerization of the α-aminoadipyl residue must occur after the tripeptide formation. The existence of D-valine in peptide P3 reveals that the conversion of L-valine to the D-isomer takes place during or before the formation of the tripeptide. The presence of β-hydroxyvaline in peptide P1 has caused discussion as to whether β-hydroxyvaline is involved in the biosynthesis of the thiazolidine ring.

The biosynthesis of the peptides was investigated using both intact mycelia and extracts of *C. acremonium*. Dipeptide, δ-(α-aminoadipyl)cysteine, did not enter the mycelia intact but was hydrolyzed to its constituent amino acids (Loder *et al.*, 1969). When DL-valine-1-^{14}C was added to the mycelia, more ^{14}C was incorporated into peptide P3 than into peptide P1 or peptide P2. In a broken-cell system, labeled peptides P2 and P3 were formed from δ-(L-α-aminoadipyl)-L-cysteine and DL-valine-1-^{14}C, but not peptide P1. Further experiments indicated that peptide P3 was labeled from L-valine and not from the D-isomer. No synthesis of labeled peptide P1 was observed when glycine-1-^{14}C was added to a complete system of ultrasonic extract together with δ-(L-α-aminoadipyl)-L-cysteine and DL-β-hydroxyvaline, as shown in Table II (Loder and Abraham, 1971b; Abraham, 1974). These indicate that the biosynthesis of the α-aminoadipyl peptides starts from the N-terminal amino acid. The biosynthesis of these peptides was catalyzed by the 20,000 g particulate fraction of the broken-cell system, despite the fact that the synthesis of glutathione, which is very similar to the tripeptide in its structure, is catalyzed by the supernatant fraction. The presence of glycine in peptides P1 and P2 suggests the possibility that these tetrapeptides are derived from intermediates on the biosynthetic pathway of β-lactam antibiotics.

Attempts to elucidate the role of the peptide intermediates in the synthesis of β-lactam antibiotics were made using the β-lactam-negative

TABLE II

BIOSYNTHESIS OF CYSTEINE-CONTAINING PEPTIDES IN
BROKEN-CELL SYSTEMS OF *C. acremonium*[a]

| | | | \multicolumn{4}{c}{Radioactivity (nCi) found in} | | | |
Expt.	Extract	Addition	GSH	Aad-Cys-Val complex	Peptide P2	Peptide P3
I	Complete system	[^{14}C]Val + L-Aad-Cys[b]		3.4	1.1	1.1
		[^{14}C]Val + L-Aad + L-Cys		<0.1		
		[^{14}C]Gly + L-Glu-Cys	21.8			
		[^{14}C]Gly + L-Glu + L-Cys	16.6			
II	Complete system	[^{14}C]Val + L-Aad-Cys		1.4	0.7	0.7
		[^{14}C]Gly + L-Glu-Cys	10.0			
	Supernatant	[^{14}C]Val + L-Aad-Cys		<0.1		
		[^{14}C]Gly + L-Glu-Cys	12.0			
	Particulate fraction	[^{14}C]Val + L-Aad-Cys		2.2	0.7	0.8
		[^{14}C]Gly + L-Glu-Cys	3.8			
III	Complete system	[^{14}C]Aad + L-Cys-L-Val		<0.1		
		[^{14}C]Aad + L-Cys-D-Val		<0.1		
		[^{14}C]Aad + L-Cys + L-Val		<0.1		
IV	Complete system	[^{14}C]Val + L-Aad-Cys		4.1	1.5	1.5
		[^{14}C]Val + D-Aad-Cys		<0.1		
		[^{14}C]Val + L-Aad + L-Cys		<0.1		

[a] From Loder and Abraham (1971b).
[b] Abbreviations: Aad-Cys, δ-(α-aminoadipyl)-L-cysteine; Cys-Val, L-cysteinyl-valine; Glu-Cys, γ-(L-glutamyl)-L-cysteine; Aad, α-aminoadipic acid; Aad-Cys-Val complex, a mixture of peptides, including P2 and P3, before resolution.

188 TOSHIHIKO KANZAKI AND YUKIO FUJISAWA

mutants. Lemke and Nash (1972) divided mutant strains of *C. acremonium* into four phenotypic classes according to their potential to synthesize antibiotics: (A) strains with increased potential to synthesize antibiotics, (B) strains unable to synthesize antibiotics, (C) strains able to synthesize only penicillin N, and (D) strains able to synthesize penicillin N and 6-aminopenicillanic acid. Mutants of the second class, β-lactam-negative mutants, were characterized ever further. One characterization involved prototrophic- and consistently antibiotic-negative mutants. These mutants may be further subdivided based on their capacity to synthesize peptides implicated as intermediates in the synthesis of β-lactam antibiotics. This supports the hypothesis that the peptides are intermediates in antibiotic synthesis. The other characterization involved a strain which was auxotrophic for lysine and synthesized a trace amount of antibiotic as well as intermediate peptides when grown in the presence of α-aminoadipic acid and lysine, but not in the presence of lysine alone. Conditional cosynthesis of the antibiotic and the peptides by this lysine-requiring mutant provided experimental evidence that the peptides are related to the biosynthesis of antibiotics by *C. acremonium.* Recent investigation showed that antibiotic synthesis was detected with presumptive heterokaryons formed between certain pairs of mutants of the fungus, which were defective in the synthesis of intermediate peptides (Nash *et al.*, 1974).

Recently, Fawcett *et al.* (1974) reported that tritium-labeled isomers of δ-(L-α-aminoadipyl)-L-cysteinylvaline were incorporated into penicillin N with whole extract of *C. acremonium,* though the detection of cephalosporin C was unsuccessful (Table III). This is the first experiment in which the tripeptide was manifested to be the biosynthetic precursor of β-lactam antibiotic.

TABLE III

INCORPORATION OF TRITIUM FROM LABELED ISOMERS OF
δ-(L-α-AMINOADIPYL)-L-CYSTEINYLVALINE INTO PENICILLIN N[a]

Expt.	Enzyme system	Tritium-labeled tripeptide[b]	Radioactivity (nCi) in penicillaminic acid[e]
1	Whole extract	LLL(γ)[c]	0 (0%)
1	Whole extract	LLD(γ)	8 (0.17%)
2	Whole extract	LLD(γ)	12 (0.25%)
2	Whole extract	LLD(α)[d]	16 (0.33%)

[a] From Fawcett *et al.* (1974).
[b] Added radioactivity: 4.8 μCi.
[c] Specific activity of valine-γ-labeled peptide: 20 μCi/μmole.
[d] Specific activity of valine-α-labeled peptide: 10 μCi/μmole.
[e] The radioactivity in penicillin N has been taken to be that found in the penicillaminic acid spot on paper after oxidation.

Penicillin N-negative mutants were obtained from *C. acremonium* with a method using *Sarcina lutea* PCI 1001 as an indicator (Kanzaki *et al.*, 1974a; Shirafuji *et al.*, 1975). All mutants synthesized neither penicillin N nor cephalosporin C, but some of the mutants, N-2, N-36, and N-79, accumulated the dimer of δ-(L-α-aminoadipyl)-L-cysteinyl-D-valine (about 500 μg/ml), and the disulfide of the tripeptide and methanethiol (about 50 μg/ml) in culture broths. One of the deacetoxycephalosporin C-negative mutants, No. 20–5, also accumulated the above tripeptide derivatives in the culture broth (Y. Fujisawa, K. Kitano, K. Katamoto, K. Nara, and T. Kanzaki, unpublished data, 1975). These data suggest that genetic mutation blocked for the conversion of the peptides to antibiotics resulted in the accumulation of the implicated intermediate, δ-(L-α-aminoadipyl)-L-cysteinyl-D-valine, followed by the conversion to the tripeptide derivatives.

C. FORMATION OF THE DIHYDROTHIAZINE RING SYSTEM

As stated above, the biochemical demonstration that the 3-cephem structure of cephalosporin C is formed via the tripeptide seems near at hand because of the isolation and recent progress in the studies of the tripeptide and its derivatives from *C. acremonium* and *P. chrysogenum*. It is generally believed that the essential difference between penicillin and cephalosporin biosynthesis exists in the oxidation step of the valinyl moiety. However, at what stage the oxidative reaction occurs remains unknown. A few hypothetical mechanisms for the origin of the dihydrothiazine ring have been proposed by investigators, although there are few biochemical studies on the formation of the ring system.

The first possibility is that 3-cephem and penam antibiotics are formed from a common α,β-dehydrovaline derivative (I) of the tripeptide, as shown in Fig. 8 (Demain, 1963a, 1966). The dihydrothiazine ring of cephalosporin would be formed by dehydrogenation between one of the valine methyl groups and the thiol group, followed by the formation of deacetoxycephalosporin (Abraham and Newton, 1961). The hypothesis seems to be supported by the experiments with ^{13}C-valine described previously (Neuss *et al.*, 1973).

The second hypothesis is that penicillin N and cephalosporin C, respectively, diverge at an early stage of the biosynthetic process and are synthesized from different peptides (Abraham and Newton, 1965). Monocyclic, β-lactam intermediates (II and III) were proposed as hypothetical compounds (Fig. 8). Cyclization of these intermediates would result in the formation of the dihydrothiazine ring system.

The third possibility is explained by the ring expansion theory whereby 3-cephem antibiotic is synthesized by ring expansion of penam com-

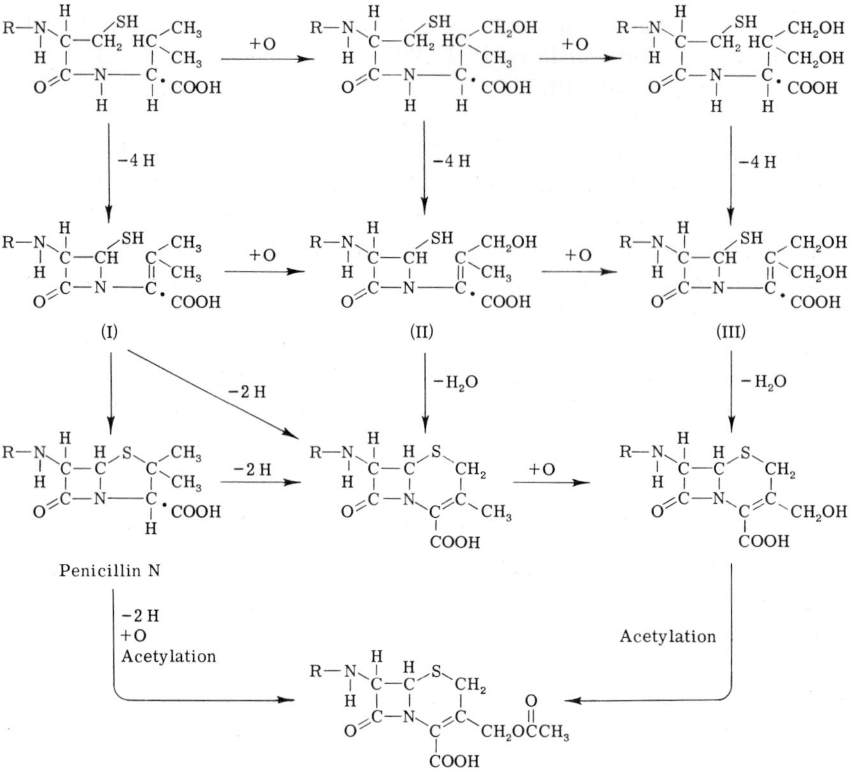

Fig. 8. Hypothetical scheme for the formation of β-lactam-dihydrothiazine ring system (Demain, 1963a; Abraham et al., 1965). R: $HOOC \cdot CH(CH_2)_3 \overset{O}{\overset{\|}{C}}O-$ with NH_2 on the CH.

pound (Abraham and Newton, 1965, 1967; Abraham et al., 1965). The theory was supported chemically by Morin et al. (1963) and Wolfe et al. (1963). It is also known that some mutants of C. acremonium produce only penicillin N and that all of the cephalosporin-producing microorganisms form penicillin N. These facts might support the idea that penicillin N is a precursor of cephalosporin C. However, at present there is no direct evidence for this hypothesis under biological conditions.

D. CEPHALOSPORIN C SYNTHESIS AFTER RING CLOSURE

Recently, an attempt was made to elucidate the cephalosporin C biosynthetic pathway in C. acremonium using specific mutants which were blocked for the synthesis of cephalosporin C. Fujisawa *et al.* (1973, 1975a) derived a number of cephalosporin C-negative mutants from a cephalosporin C-potent producer No. 52-54 by a method using *Alcaligenes faecalis* ATCC 8750 as an indicator, and showed that four of the mutants, Nos. 20, 29, 36, and 40, exclusively accumulated deacetylcephalosporin C (850 μg/ml), with a D-α-aminoadipyl side chain at C-7, in the culture broth. The finding that deacetylcephalosporin C was formed at the early period of fermentation while not even a small amount of cephalosporin C was detected throughout the process (Fig. 9) was unexpected, because deacetylcephalosporin C in broth had been believed to be formed by chemical hydrolysis of preformed cephalosporin C (Huber *et al.*, 1968). The examination of intra- and extracellular cephalosporin C acetyl-hydrolase made it more likely that deacetylcephalosporin C was produced through *de novo* synthesis in these mutants. In fact, there was no difference in levels of the acetylhydrolase between the parent and the mutants (Fujisawa *et al.*, 1973; 1975b). These results suggest that deacetyl-cephalosporin C is a precursor of cephalosporin C and is synthesized by enzymic acetylation of deacetylcephalosporin C. The existence of acetyl CoA:deacetylcephalosporin C acetyltransferase was demonstrated for the first time in the cell-free extracts of the cephalosporin C-producing strain, but was not detected in mutants Nos. 20, 29, 36, and 40 (Fujisawa *et al.*, 1973). The acetyltransferase activity was also recovered in a revertant

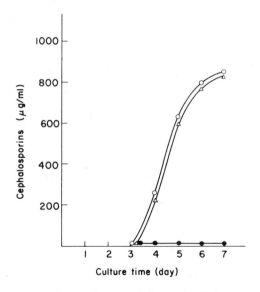

FIG. 9. Time course of the production of deacetylcephalosporin C by the cephalosporin C-negative mutant No. 40 (Fujisawa *et al.*, 1973). ○——○, Total cephalosporins; △——△, deacetylcephalosporin C; ●——●, cephalosporin C.

strain No. 40-B-35 (Table IV). One can conclude, therefore, that the deacetylcephalosporin C accumulation by these mutants is due to the lack of acetyltransferase. The enzyme existed in the 20,200 g supernatant fraction of the sonicates prepared from the parental mycelia. The optimum pH for its activity was 7.0 to 7.5. The enzyme essentially required Mg^{2+} as a cofactor (Fujisawa and Kanzaki, 1975b). The enzyme activity was found to be inhibited by cephalosporin C, deacetoxycephalosporin C, and penicillin N to various degrees (Liersch et al., 1974). These data strongly support the idea that deacetylcephalosporin C is an intermediate of cephalosporin C biosynthesis, and isomerization of the L-α-aminoadipyl moiety takes place before deacetylcephalosporin C synthesis, which is different from the hitherto proposed biosynthetic scheme of cephalosporin C.

The accumulation of deacetoxycephalosporin C, possessing an α-aminoadipyl side chain with D configuration, by C. acremonium mutants was reported independently by three groups (Kanzaki et al., 1974a; Queener and Capone, 1974; Liersch et al., 1974). Deacetylcephalosporin C-negative mutants were obtained from the deacetylcephalosporin C-producing strain No. 40 by a technique using Bacillus brevis IFO 12334, which is about 10 times more sensitive to deacetylcephalosporin C than to deacetoxycephalosporin C, as an indicator (Fujisawa et al., 1975c). One of the mutants, No. 40-20, accumulated deacetoxycephalosporin C (150 $\mu g/ml$) and a trace amount of deacetylcephalosporin C (Fig. 10). These results suggest that deacetylcephalosporin C is synthesized by enzymic

TABLE IV

ACETYL CoA: DEACETYLCEPHALOSPORIN C ACETYLTRANSFERASE
ACTIVITY IN CELL-FREE EXTRACTS OF VARIOUS STRAINS[a]

Strain	Main product	^{14}C-CPC formed (cpm)	Rate of incorporation (%)
ATCC 14553	CPC[c]	1200	0.75
No. 81	CPC → DCPC[b]	656	0.41
No. 52-54	CPC	5360	3.35
No. 20	DCPC	0	0
No. 29	DCPC	45	0.02
No. 36	DCPC	0	0
No. 40	DCPC	21	0.01
No. 40-B-35	CPC	2200	1.37

[a] From Fujisawa and Kanzaki (1975b).
[b] DCPC is produced from CPC by the action of CPC acetylhydrolase.
[c] Abbreviations: CPC, cephalosporin C; DCPC, deacetylcephalosporin C.

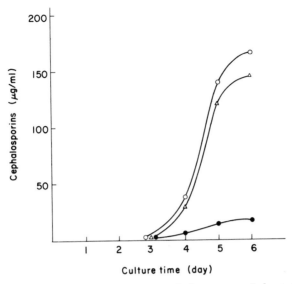

Fig. 10. Time course of the production of deacetoxycephalosporin C by the deacetylcephalosporin C-negative mutant No. 40-20 (Fujisawa *et al.*, 1975c). O——O, Total cephalosporins; △——△, deacetoxycephalosporin C; ●——●, deacetylcephalosporin C.

hydroxylation of deacetoxycephalosporin C. In fact, the existence of labile deacetoxycephalosporin C hydroxylase activity was observed in cell-free extracts of the parent but was not detected in the mutant (Liersch *et al.*, 1974; Fujisawa *et al.*, 1975c, 1976). These results are consistent with the hypothesis that deacetylcephalosporin C is formed from deacetoxycephalosporin C by the action of the hydroxylase. Based on these facts, a new biosynthetic scheme for cephalosporin C was proposed, as shown in Fig. 11 (Kanzaki *et al.*, 1974a; Fujisawa *et al.*, 1975c). Queener and Capone (1974) isolated cephalosporin C-nonproducing strains after exposing a potent cephalosporin C producer M8650-13 to ultraviolet light. They found that one of the blocked mutants, MH-63, accumulated deacetoxycephalosporin C (1500 μg/ml) and deacetylcephalosporin C (400 μg/ml) in the culture broth. They proposed a hypothetical branched pathway for biosynthesis of penicillin N and cephalosporin C, as shown in Fig. 12 (Queener *et al.*, 1974). Liersch *et al.* (1974) reported the derivation of deacetoxycephalosporin C-producing mutant and the formation of cephalosporin C from deacetoxycephalosporin C, which suggested the existence of deacetoxycephalosporin C hydroxylase.

Another type of deacetylcephalosporin C-producing mutant was obtained from cephalosporin C-negative mutants of *C. acremonium* (Fujisawa *et al.*, 1973, 1975b). The mutant No. 81 accumulated cephalosporin

α-Aminoadipic acid   Cysteine   Valine

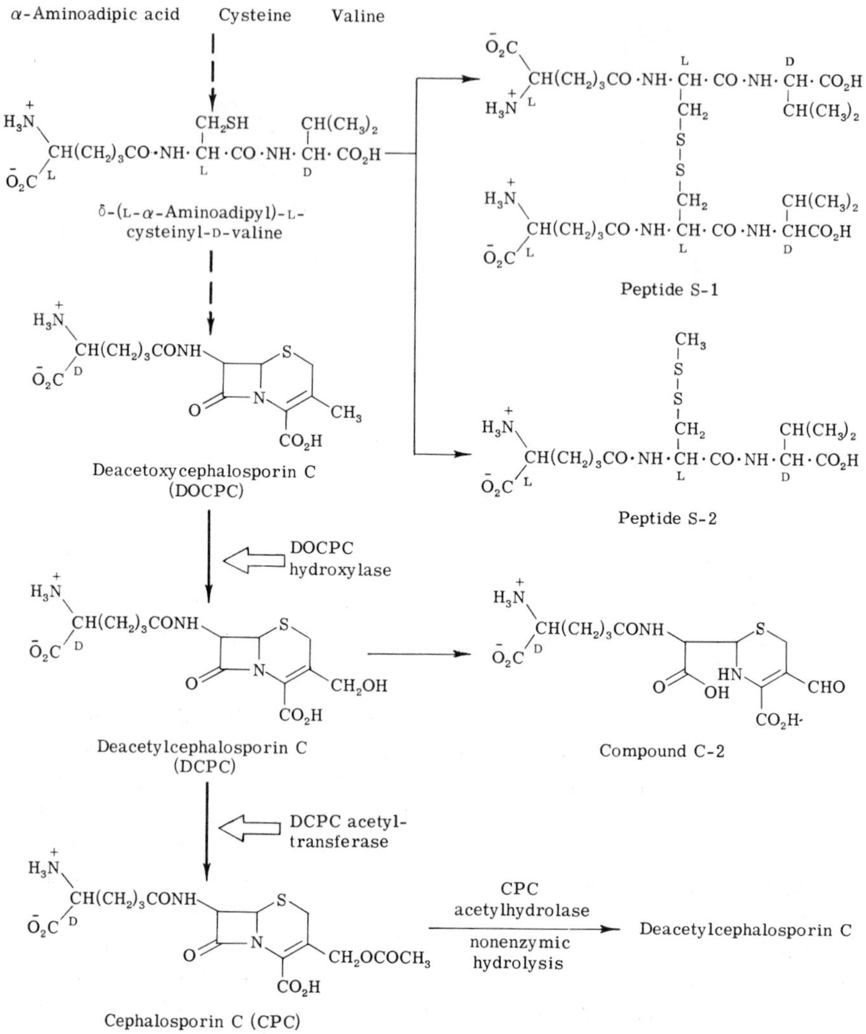

$\bar{O}_2C$
  $CH(CH_2)_3CO \cdot NH \cdot \overset{L}{CH} \cdot CO \cdot NH \cdot \overset{D}{CH} \cdot CO_2H$
$H_3\overset{+}{N}$ ᴸ CH_2 $CH(CH_3)_2$
               S
               S
            CH_2 $CH(CH_3)_2$
$H_3\overset{+}{N}$
  $CH(CH_2)_3CO \cdot NH \cdot \overset{L}{CH} \cdot CO \cdot NH \cdot \overset{D}{CHCO_2H}$
$\bar{O}_2C$ ᴸ

Peptide S-1

$H_3\overset{+}{N}$ CH_2SH $CH(CH_3)_2$
  $CH(CH_2)_3CO \cdot NH \cdot \overset{L}{CH} \cdot CO \cdot NH \cdot \overset{D}{CH} \cdot CO_2H$
$\bar{O}_2C$ ᴸ

δ-(L-α-Aminoadipyl)-L-
cysteinyl-D-valine

                  CH_3
                   S
                   S
             CH_2 $CH(CH_3)_2$
$H_3\overset{+}{N}$
  $CH(CH_2)_3CO \cdot NH \cdot \overset{L}{CH} \cdot CO \cdot NH \cdot \overset{D}{CH} \cdot CO_2H$
$\bar{O}_2C$ ᴸ

Peptide S-2

$H_3\overset{+}{N}$
  $CH(CH_2)_3CONH$
$\bar{O}_2C$ ᴰ

Deacetoxycephalosporin C
(DOCPC)

DOCPC
hydroxylase

$H_3\overset{+}{N}$
  $CH(CH_2)_3CONH$
$\bar{O}_2C$ ᴰ
        CH_2OH
        CO_2H

Deacetylcephalosporin C
(DCPC)

$H_3\overset{+}{N}$
  $CH(CH_2)_3CONH$
$\bar{O}_2C$ ᴰ
        HN CHO
        OH $CO_2H \cdot$

Compound C-2

DCPC acetyl-
transferase

$H_3\overset{+}{N}$
  $CH(CH_2)_3CONH$
$\bar{O}_2C$ ᴰ
        CH_2OCOCH_3
        CO_2H

Cephalosporin C (CPC)

CPC
acetylhydrolase
nonenzymic
hydrolysis → Deacetylcephalosporin C

FIG. 11. Hypothetical scheme of cephalosporin C biosynthesis and its metabolism in *C. acremonium* (Kanzaki *et al.*, 1974a; Fujisawa *et al.*, 1975c).

C in the early stage of fermentation, and almost all of the cephalosporin C was then finally converted to deacetylcephalosporin C. These findings lead to a conclusion that the accumulation of deacetylcephalosporin C by the mutant No. 81 is performed through the hydrolysis of cephalosporin C, which has been produced by the strain, by extracellular cephalosporin C acetylhydrolase.

Compound C-2 is mainly found in the culture broths of deacetylcephalosporin C-producing mutants, which were defective in acetyl

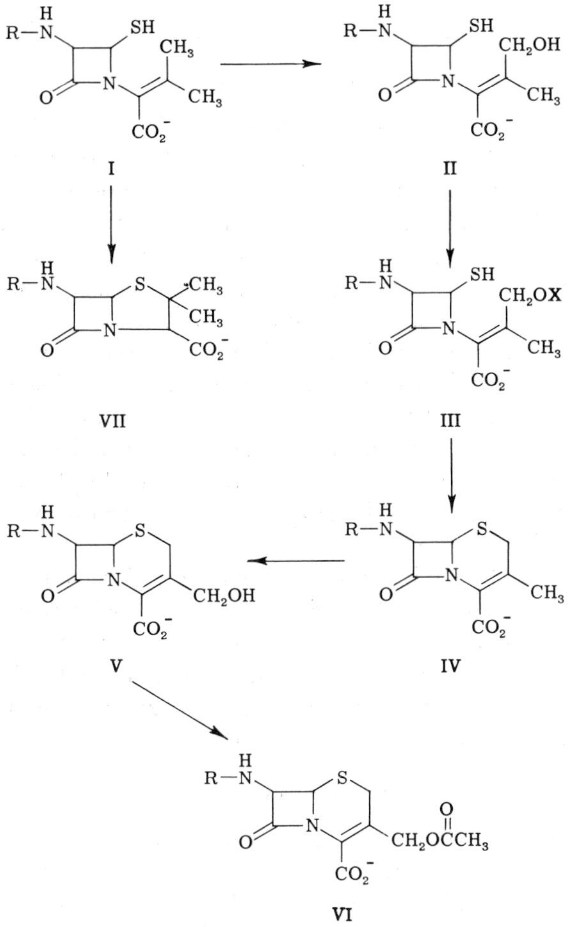

FIG. 12. Hypothetical branched pathway of biosynthesis of penicillin N and cephalosporin C (Queener *et al.*, 1974). (IV) Deacetoxycephalosporin C, (V) deacetylcephalosporin C, (VI) cephalosporin C, (VII) penicillin N.

CoA: deacetylcephalosporin C acetyltransferase, though one can observe its occurrence in cephalosporin C culture broths in a trace amount (Fujisawa and Kanzaki, 1975a). Considering this fact and the mode of biochemical reaction, compound C-2 appears to be synthesized through the corresponding 3-formyl cephalosporin, which is susceptible to chemical hydrolysis in a neutral medium, to form compound C-2. Therefore, the formyl cephalosporin may be a by-product from deacetylcephalosporin C rather than an intermediate of cephalosporin C biosynthesis.

VI. Enzymes Related to Cephalosporin Metabolism

A. Acetylhydrolase

A trace amount of deacetylcephalosporin C is usually detected in the culture broth of *C. acremonium*. Kinetic studies showed that the formation of deacetylcephalosporin C was due to the nonenzymic hydrolysis of cephalosporin C and was not through biochemical reaction (Huber *et al.*, 1968). Recently, one of the cephalosporin C-negative mutants, No. 81, derived from *C. acremonium*, was shown to produce cephalosporin C acetylhydrolase in the late stage of fermentation (Fujisawa *et al.*, 1973, 1975b). Almost all of the cephalosporin C, once formed, was converted to deacetylcephalosporin C by the hydrolase. The enzyme was partially purified from the culture filtrate of the mutant No. 81 and the characteristics of the enzyme were examined. The hydrolase has no activity catalyzing the transformation of deacetylcephalosporin C to cephalosporin C. Therefore, the enzyme is considered to be irrelevant to cephalosporin C biosynthesis.

Cephalosporin C acetylhydrolase activity was detected in the culture broth of *S. clavuligerus* which produced deacetyl-3'-O-carbamoylcephalosporin C (A16886A). Although A16886A was chemically synthesized from deacetylcephalosporin C, this reaction could not be demonstrated with the organism. Therefore, the role of the enzyme in the formation of A16886A was not clear (Brannon *et al.*, 1972).

B. Arylamidase

An arylamidase activity, which inactivated cephalosporin C, was found in spent media of *C. acremonium* (Dennen *et al.*, 1971). The arylamidase existed in three aggregational forms. The exocellular enzyme was purified from the culture filtrate by DEAE-cellulose column chromatography, gel filtration, and gel electrophoresis. The arylamidase catalyzed the hydrolysis of β-lactam bond of many cephalosporins in addition to that of L-leucyl-β-naphthylamide. However, the enzyme failed to hydrolyze the β-lactam bond of penicillins. The molecular weight of the exocellular enzyme is about 60,000, and the Km values for L-leucyl-β-naphthylamide and cephalosporin C are 4.2×10^{-4} M and 9.09×10^{-4} M, respectively. The possible role of the enzyme in the biosynthesis of cephalosporin C remains to be determined.

VII. Conclusions

A variety of cephalosporins and their producers has been found, as described here. But it has not been conclusively determined which are the secondary metabolites in a strict sense of the term. A moderate number of the metabolites listed here are surmised to be intermediates of the biosynthesis and degraded or modified products of secondary metabolites. Nevertheless, the discoveries of 7-methoxycephem derivatives and cephalosporins with the glutaryl group instead of D-α-aminoadipyl residue, which has been recognized to be essential to hydrophilic β-lactam antibiotics, and the discovery of the single β-lactam antibiotic (Aoki et al., 1975) seem of great value. This work must trigger further detection of unpredictable novel families of β-lactam antibiotics.

α-Aminoadipic acid is known to have two phases of metabolism. In fungi, the amino acid is an intermediate of lysine biosynthesis from α-ketoglutarate and acetyl CoA. Apart from this, it is also a catabolite product of lysine in prokaryotes such as streptomycetes. In spite of the differences of metabolism, why do the β-lactam antibiotics require α-aminoadipic acid as a common side chain in fungi and streptomycetes? In addition, the diversity of β-lactam compounds and their producers again arouse our interests in the role of β-lactam antibiotics in the microorganisms.

The features of the biosynthetic pathway of the β-lactam antibiotics, and especially of cephalosporins, have been disclosed through extensive work by several groups. The hypothesis that the tripeptide is a direct precursor of the β-lactam antibiotics was supported by the experiments of Fawcett et al. (1974).

The formation of tripeptide from δ-(L-α-aminoadipyl)-L-cysteine was demonstrated to be catalyzed by 20,000 g particulate fraction. However, the biosynthesis of cephalosporin C after ring closure proceeds through reaction steps catalyzed by deacetoxycephalosporin C hydroxylase and deacetylcephalosporin C acetyltransferase, which are located in the 20,200 g supernatant fraction (Fig. 11). Biochemical studies on the biosynthesis of the β-lactam antibiotics up to this point are not consistent with respect to the site of antibiotic synthesis.

The most important problem about β-lactam antibiotic synthesis now involves the mechanism of the ring closure, i.e., the ability of the oxidation mechanism of the valine moiety to form the five-membered or the six-membered ring and of the cysteine moiety to form the β-lactam ring. The elucidation of this mechanism would simultaneously demonstrate whether the cephem and the penam nucleus are synthesized independently from the tripeptide or whether cephem antibiotics are formed via a penam compound. The employment of skillful approaches such as the use of mutants relating to antibiotic synthesis and isotopes such as ^{13}C, in addition to the exploitation of prepared stable enzyme systems could give rise to fruitful results in this field.

REFERENCES

Abraham, E. P. (1974). In "Biosynthesis and Enzymic Hydrolysis of Penicillins and Cephalosporins" (E. P. Abraham, ed.), pp. 8–11. Univ. of Tokyo Press, Tokyo.

Abraham, E. P., and Newton, G. G. F. (1961). Biochem. J. **79**, 377.

Abraham, E. P., and Newton, G. G. F. (1965). Adv. Chemother. **2**, 23.

Abraham, E. P., and Newton, G. G. F. (1967). In "Antibiotics" (D. Gottlieb and P. D. Shaw, eds.), Vol. 2, pp. 1–16. Springer-Verlag, Berlin and New York.

Abraham, E. P., Newton, G. G. F., and Hale, C. W. (1954). Biochem. J. **58**, 94.

Abraham, E. P., Newton, G. G. F., and Warren, S. C. (1964). Proc. Int. Assoc. Microbiol. Symp. Appl. Microbiol. 1964, Vol. 6, p. 79.

Abraham, E. P., Newton, G. G. F., and Warren, S. C. (1965). In "Biogenesis of Antibiotic Substances" (Z. Vanek and Z. Hostálek, eds.), pp. 169–194. Academic Press, New York.

Aoki, H., Kohsaka, M., Hosoda, J., Komori, T., and Imanaka, H. (1975). Abstr. Intersci. Conf. Antimicrob. Agents Chemother., 15th, 1975, No. 97.

Arnstein, H. R. V. (1958). Annu. Rep. Progr. Chem. **54**, 339.

Arnstein, H. R. V., and Morris, D. (1960). Biochem. J. **76**, 357.

Auden, J. A., Gruner, J., Liersch, M., and Nüesch, J. (1969). Pathol. Microbiol. **34**, 240.

Benz, F., Liersch, M., Nüesch, J., and Treichler, H. J. (1971). Eur. J. Biochem. **20**, 81.

Brannon, D. R., Fukuda, D. S., Mabe, J. A., Huber, F. M., and Whitney, J. G. (1972). Antimicrob. Agents & Chemother. **1**, 237.

Brotzu, G. (1948). Lav. Ist. Ig. Cagliari.

Burton, H. S., and Abraham, E. P. (1951). Biochem. J. **50**, 168.

Caltrider, P. G., and Niss, H. F. (1966). Appl. Microbiol. **14**, 746.

Caltrider, P. G., Huber, F. M., and Day, L. E. (1968). Appl. Microbiol. **16**, 1913.

Day, L. E., and Ellis, L. F. (1971). Appl. Microbiol. **22**, 919.

Demain, A. L. (1963a). Trans. N. Y. Acad. Sci. [2] **25**, 731.

Demain, A. L. (1963b). Clin. Med. **70**, 2045.

Demain, A. L. (1963c). Biochem. Biophys. Res. Commun. **10**, 45.

Demain, A. L. (1966). Biosynth Antibiot. **1**, 29–94.

Demain, A. L., and Newkirk, J. F. (1962). Appl. Microbiol. **10**, 321.

Demain, A. L., Newkirk, J. F., and Hendlin, D. (1963a). J. Bacteriol. **85**, 339.

Demain, A. L., Newkirk, J. F., Davis, G. E., and Harman, R. E. (1963b). Appl. Microbiol. **11**, 58.

Dennen, D. W., and Carver, D. D. (1969). Can. J. Microbiol. **15**, 175.

Dennen, D. W., Allen, C. C., and Carver, D. D. (1971). Appl. Microbiol. **21**, 907.

Drew, S. W., and Demain, A. L. (1973). Biotechnol. Bioeng. **15**, 743.

Drew, S. W., and Demain, A. L. (1974). Abstr. Intersect. Congr. Int. Assoc. Microbiol. Soc., 1st. 1974 p. 141.

Drew, S. W., and Demain, A. L. (1975a). Proc. Int. Congr. Int. Assoc. Microbiol. Soc., 1st, 1975, p. 539.

Drew, S. W., and Demain, A. L. (1975b). Eur. J. Appl. Microbiol. **1**, 121.

Drew, S. W., and Demain, A. L. (1975c). Antimicrob. Agents & Chemother. **8**, 5.

Drew, S. W., and Demain, A. L. (1975d). J. Antibiot. **28**, 889.

Dulaney, E. L. (1947). Mycologia **39**, 570.

Elander, R. P., Stauffer, J. F., and Backus, M. P. (1961). Antimicrob. Agents Annu. pp. 91–102.

Elander, R. P., Corum, C. J., DeValeria, H., and Wilgus, R. M. (1974). *Abstr. Int. Symp. Genet. Ind. Microorg.*, *2nd, 1974*, p. 19.

Fantini, A. A. (1962). *Genetics* **47**, 161.

Fasani, M., Marini, F., and Teatini, L. (1974). *Abstr. Int. Symp. Genet. Ind. Microorg.*, *2nd, 1974*, p. 31.

Fawcett, P., Loder, P. B., Duncan. M. J., Beesley, T. J., and Abraham, E. P. (1973). *J. Gen. Microbiol.* **79**, 293.

Fawcett, P., Usher, J. J., and Abraham, E. P. (1974). *Abstr. Int. Symp. Genet. Ind. Microorg.*, *2nd, 1974*, p. 11.

Feren, C. J., and Squires, R. W. (1969). *Biotechnol. Bioeng.* **11**, 583.

Fleischman, A. I., and Pisano, M. A. (1961). *Antimicrob. Agents Annu.* pp. 48–53.

Fleming, A. (1929). *Br. J. Exp. Pathol.* **10**, 226.

Fujisawa, Y., and Kanzaki, T. (1975a). *J. Antibiot.* **28**, 372.

Fujisawa, Y., and Kanzaki, T. (1975b). *Agric. Biol. Chem.* **39**, 2043.

Fujisawa, Y., Shirafuji, H., Kida, M., Nara, K., Yoneda, M., and Kanzaki, T. (1973). *Nature (London) New Biol.* **246**, 154.

Fujisawa, Y., Shirafuji, H., Kida, M., Nara, K., Yoneda, M., and Kanzaki, T. (1975a). *Agric. Biol. Chem.* **39**, 1295.

Fujisawa, Y., Shirafuji, H., and Kanzaki, T. (1975b). *Agric. Biol. Chem.* **39**, 1303.

Fujisawa, Y., Kitano, K., and Kanzaki, T. (1975c). *Agric. Biol. Chem.* **39**, 2049.

Fujisawa, Y., Kikuchi, M., and Kanzaki, T. (1976). *J. Antibiot.* (in preparation).

Fukase, H., Hasegawa, T., Hatano, K., Iwasaki, H., and Yoneda, M. (1976). *J. Antibiot.* **29**, 113.

Godfrey, O. W. (1973). *Antimicrob. Agents & Chemother.* **4**, 73.

Godfrey, O. W. (1974). *Can. J. Microbiol.* **20**, 1479.

Gorman, M., Hoehn, M. M., Nagarajan, R., Boeck, L. D., Presti, E. A., Whitney, J. G., and Hamill, R. L. (1971). *Abstr. Intersci. Conf. Antimicrob. Agents Chemother., 11th, 1974*, p. 7.

Grosklags, J. H., and Swift, M. E. (1957). *Mycologia* **49**, 305.

Hamill, R. L., Carrell, C. B., Hoehn, M. M., Stark, W. M., Boeck, L. D., and Gorman, M. (1971). *Abstr. Intersci. Conf. Antimicrob. Agents Chemother., 11th, 1971*, p. 7.

Hasegawa, T., Fukase, H., Hatano, K., Iwasaki, H., and Yoneda, M. (1975). *Abstr. Annu. Meet. Agric. Chem. Soc. Jpn.* p. 80.

Higgens, C. E., and Kastner, R. E. (1971). *Int. J. Syst. Bacteriol.* **21**, 326.

Higgens, C. E., Hamill, R. L., Sands, T. H., Hoehn, M. M., Davis, N. E., Nagarajan, R., and Boeck, L. D. (1974). *J. Antibiot.* **27**, 298.

Huang, F., Chan, J. A., Sih, C. J., Fawcett, P., and Abraham, E. P. (1975). *J. Am. Chem. Soc.* **97**, 3858.

Huber, F. M., and Redstone, M. O. (1967). *Can. J. Microbiol.* **13**, 332.

Huber, F. M., Baltz, R. H., and Caltrider, P. G. (1968). *Appl. Microbiol.* **16**, 1011.

Hussey, C., Orsi, B. A., Scott, J., and Spencer, B. (1965). *Nature (London)* **207**, 632.

Inamine, E., and Birnbaum, J. (1972). *Abstr. 72nd Annu. Meet., Am. Soc. Microbiol.* p. 12.

Kanzaki, T., Shirafuji, H., Fujisawa, Y., Fukita, T., Nara, K., Kitano, K., and Kida, M. (1974a). *Abstr. Int. Symp. Genet. Ind. Microorg.*, *2nd, 1974*, p. 32.

Kanzaki, T., Fukita, T., Shirafuji, H., Fujisawa, Y., and Kitano, K. (1974b). *J. Antibiot.* **27**, 361.

Kavanagh, F., Tunin, D., and Wild, G. (1958). *Arch. Biochem. Biophys.* **77**, 268.

Kikuchi, T., and Uyeo, S. (1967). *Chem. Pharm. Bull.* **15**, 549.

200   TOSHIHIKO KANZAKI AND YUKIO FUJISAWA

Kirkpatrick, J. R., Doolin, L. E., and Godfrey, O. W. (1973). *Antimicrob. Agents & Chemother.* 4, 542.

Kitano, K., Kintaka, K., Suzuki, S., Katamoto, K., Nara, K., and Nakao, Y. (1974a). *Agric. Biol. Chem.* 38, 1761.

Kitano, K., Kintaka, K., Suzuki, S., Katamoto, K., Nara, K., and Nakao, Y. (1974b). *J. Ferment. Technol.* 52, 785.

Kitano, K., Kintaka, K., Suzuki, S., Katamoto, K., Nara, K., and Nakao, Y. (1975a). *J. Ferment. Technol.* 53, 327.

Kitano, K., Kintaka, K., Suzuki, S., Katamoto, K., Nara, K., and Nakao, Y. (1975b). *Abstr. Annu. Meet., Soc. Ferment. Technol. Jpn.* p. 244.

Kluender, H., Bradley, C. H., Sih, C. J., Fawcett, P., and Abraham, E. P. (1973). *J. Am. Chem. Soc.* 95, 6149.

Kluender, H., Huang, F., Fritzberg, A., Schnoes, H., Sih, C. J., Fawcett, P., and Abraham, E. P. (1974). *J. Am. Chem. Soc.* 96, 4054.

Komatsu, K., Mizuno, M., and Kodaira, R. (1975). *J. Antibiot.* 28, 881.

Lemke, P. A., and Brannon, D. R. (1972). *In* "Cephalosporins and Penicillins: Chemistry and Biology" (E. H. Flynn, ed.), pp. 370–437. Academic Press, New York.

Lemke, P. A., and Nash, C. H. (1972). *Can. J. Microbiol.* 18, 255.

Liersch, M., Nüesch, J., and Treichler, H. J. (1973). *Pathol. Microbiol.* 39, 39.

Liersch, M., Nüesch, J., and Treichler, H. J. (1974). *Abstr. Int. Symp. Genet. Ind. Microorg., 2nd, 1974,* p. 48.

Loder, P. B., and Abraham, E. P. (1971a). *Biochem. J.* 123, 471.

Loder, P. B., and Abraham, E. P. (1971b). *Biochem. J.* 123, 477.

Loder, P. B., Abraham, E. P., and Newton, G. G. F. (1969). *Biochem. J.* 112, 389.

Metzenberg, R. L., and Parson, J. W. (1966). *Proc. Natl. Acad. Sci. U.S.A.* 55, 629.

Miller, A. K., Celozzi, E., Pelak, B. A., Stapley, E. O., and Hendlin, D. (1972). *Antimicrob. Agents & Chemother.* 2, 281.

Miller, G. A., Kelly, B. K., and Newton, G. G. F. (1956). British Patent 759,624.

Miller, I. M., Stapley, E. O., and Chaiet, L. (1962). *Bacteriol. Proc.* p. 32.

Miller, T. W., Goegelman, R. T., Weston, R. G., Putter, I., and Wolf, F. J. (1972). *Antimicrob. Agents & Chemother.* 2, 132.

Morin, R. B., Jackson, B. G., Flynn, E. H., and Roeske, R. W. (1962). *J. Am. Chem. Soc.* 84, 3400.

Morin, R. B., Jackson, B. G., Mueller, R. A., Lavagnino, E. R., Scanlon, W. B., and Andrews, S. L. (1963). *J. Am. Chem. Soc.* 85, 1896.

Nagarajan, R., Boeck, L. D., Gorman, M., Hamill, R. L., Higgens, C. E., Hoehn, M. M., Stark, W. M., and Whitney, J. G. (1971). *J. Am. Chem. Soc.* 93, 2308.

Nash, C. H., and Huber, F. M. (1971). *Appl. Microbiol.* 22, 6.

Nash, C. H., Higuera, N., Neuss, N., and Lemke, P. A. (1974). *Dev. Ind. Microbiol.* 15, 114.

Neuss, N., Nash, C. H., Lemke, P. A., and Grutzner, J. B. (1971a). *J. Am. Chem. Soc.* 93, 2337.

Neuss, N., Nash, C. H., Lemke, P. A., and Grutzner, J. B. (1971b). *Proc. R. Soc. London, Ser. B* 179, 335.

Neuss, N., Nash, C. H., Baldwin, J. E., Lemke, P. A., and Grutzner, J. B. (1973). *J. Am. Chem. Soc.* 95, 3797.

Newton, G. G. F., and Abraham, E. P. (1954). *Biochem. J.* 58, 103.

Newton, G. G. F., and Abraham, E. P. (1955). *Nature (London)* 175, 548.

Niss, H. F., and Nash, C. H. (1973). *Antimicrob. Agents & Chemother.* 4, 474.

Nüesch, J., Treichler, H. J., and Liersch, M. (1970). *Abstr. Int. Symp. Genet. Ind. Microorg., 1st, 1970,* p. 160.

Nüesch, J., Liersch, M., and Treichler, H. J. (1972). *Abstr. Int. Ferment. Symp.*, *4th, 1972*, p. 228.

Ott, J. L., Godzeski, C. W., Pavey, D., Farran, J. D., and Horton, D. R. (1962). *Appl. Microbiol.* **10**, 515.

Pathak, S. G., and Elander, R. P. (1971). *Appl. Microbiol.* **22**, 366.

Pisano, M. A., and Vellozzi, E. M. (1974). *Antimicrob. Agents & Chemother.* **6**, 447.

Queener, S. W., and Capone, J. J. (1974). *Abstr. Int. Symp. Genet. Ind. Microorg.*, *2nd, 1974*, p. 33.

Queener, S. W., Capone, J. J., Radue, A. B., and Nagarajan, R. (1974). *Antimicrob. Agents & Chemother.* **6**, 334.

Raper, K. B., Alexander, D. F., and Coghill, R. D. (1944). *J. Bacteriol.* **48**, 639.

Roberts, J. M. (1952). *Mycologia* **44**, 292.

Rode, L. J., Foster, J. W., and Schuhardt, V. T. (1947). *J. Bacteriol.* **53**, 565.

Sanders, A. G. (1949). *In* "Antibiotics" (H. W. Florey *et al.*, eds.), Vol. 2, pp. 672–685. Oxford Univ. Press, London and New York.

Segel, I. H., and Johnson, M. J. (1963). *Arch. Biochem. Biophys.* **103**, 216.

Sheehan, J. C., and Henery-Logan, K. R. (1959). *J. Am. Chem. Soc.* **81**, 5838.

Shirafuji, H., Fujisawa, Y., Kida, M., Kanzaki, T., and Yoneda, M. (1975). *Agric. Biol. Chem.* (in preparation).

Smith, B., Warren, S. C., Newton, G. G. F., and Abraham, E. P. (1967). *Biochem. J.* **103**, 877.

Stapley, E. O., Hendlin, D., Hernandez, S., Jackson, M., Mata, J. M., Miller, A. K., Woodruff, H. B., Miller, T. W., Goegelman, R. T., Weston, R. G., Putter, I., Wolf, F. J., Albers-Schönberg, G., Arison, B. H., and Smith, J. L. (1971). *Abstr. Intersci. Conf. Antimicrob. Agents Chemother., 11th, 1971*, p. 8.

Stapley, E. O., Jackson, M., Hernández, S., Zimmerman, S. B., Currie, S. A., Mochales, S., Mata, J. M., Woodruff, H. B., and Hendlin, D. (1972). *Antimicrob. Agents & Chemother.* **2**, 122.

Stevens, C. M., Vohra, P., Inamine, E., and Roholt, O. A., Jr. (1953). *J. Biol. Chem.* **205**, 1001.

Stewart, W. W. (1971). *Nature (London)* **229**, 174.

Takita, T., Muraoka, Y., Yoshioka, T., Fujii, A., Maeda, K., and Umezawa, H. (1972). *J. Antibiot.* **25**, 755.

Traxler, P., Treichler, H. J., and Nüesch, J. (1975). *J. Antibiot.* **28**, 605.

Trown, P. W., Abraham, E. P., Newton, G. G. F., Hale, C. W., and Miller, G. A. (1962). *Biochem. J.* **84**, 157.

Trown, P. W., Sharp, M., and Abraham, E. P. (1963a). *Biochem. J.* **86**, 280.

Trown, P. W., Smith, B., and Abraham, E. P. (1963b). *Biochem. J.* **86**, 284.

Tubaki, K. (1973). *Mycologia* **65**, 938.

Uri, J., Valu, G., and Békési, I. (1963). *Nature (London)* **200**, 896.

Warren, S. C., Newton, G. G. F., and Abraham, E. P. (1967a). *Biochem. J.* **103**, 891.

Warren, S. C., Newton, G. G. F., and Abaham, E. P. (1967b). *Biochem. J.* **103**, 902.

Whitney, J. G., Brannon, D. R., Mabe, J. A., and Wicker, K. J. (1972). *Antimicrob. Agents & Chemother.* **1**, 247.

Wixom, R. L., and Howell, G. B. (1965). *Proc. Soc. Exp. Biol. Med.* **118**, 1145.

Wolfe, S., Godfrey, J. C., Holdrege, C. T., and Perron, Y. G. (1963). *J. Am. Chem. Soc.* **85**, 643.

Woodward, R. B. (1966). *Science* **153**, 487.

Woodward, R. B., Heusler, K., Gosteli, J., Naegeli, P., Oppolzer, W., Ramage, R., Ranganathan, S., and Vorbrüggen, H. (1966). *J. Am. Chem. Soc.* **88**, 852.

Preparation of Pharmaceutical Compounds by Immobilized Enzymes and Cells

BERNARD J. ABBOTT

The Lilly Research Laboratories,
Eli Lilly and Company,
Indianapolis, Indiana

I. Introduction 203
II. Penicillins 206
   A. Introduction 206
   B. Penicillin Deacylation 207
   C. Penicillin Acylation 213
III. Cephalosporins 217
   A. Introduction 217
   B. Cephalosporin Deacylation 218
   C. Cephalosporin Acylation 219
   D. Cephalosporin Deacetylation 225
IV. Amino Acids 225
   A. Introduction 225
   B. Resolving Racemic Mixtures 226
   C. Synthesis 231
V. Other Products 240
   A. Steroids 240
   B. N-Oxidation 242
   C. Poly I:C 242
   D. Protein Hydrolysis 243
   E. Urocanic Acid 243
   F. Sorbosone 246
   G. Malic Acid 247
   H. Coenzyme A 248
VI. Concluding Remarks 249
   References 252

I. Introduction

Industrial exploitation of enzymes and microorganisms traditionally has been accomplished by using intact microorganisms or soluble cell-free enzyme preparations. These processes are not very efficient because the catalysts (enzymes or microorganisms) are used for just one batch reaction or fermentation. Additional uses are not feasible because (a) enzymes and cells are relatively unstable and may lose activity during a fermentation or reaction and (b) conventional recovery methods are either expensive or cause denaturation and loss of catalytic activity. If enzymes and microorganisms are to be reused effectively, their stability must be improved, and inexpensive nondestructive recovery methods must be developed. Immobilization offers a means of achieving both objectives. Immobilization, e.g., attaching an enzyme to an insoluble matrix, facilitates reuse because the large enzyme-matrix complex can be recovered easily and nondestructively by coarse filtration or low-speed centrifugation. The stability of bound enzymes to denaturation is frequently enhanced due to factors such as decreased rotational freedom of the bound enzyme molecules. Other advantages conferred by immobilization are that bound enzymes can be used continuously in a stirred tank or column reactor and that product purification is simplified because of the absence of the soluble extraneous organic material. Among the disadvantages are the additional steps required for immobilization, the cost of the supporting matrix, and the loss of activity during binding. These problems, which are characteristic of many immobilization methods, do not always preclude the economical use of the technology (see Section VI).

Immobilization methods (Fig. 1) can be broadly classified into two types (161): chemical methods and physical methods. Chemical methods involve covalent bond formation as exemplified by the attachment of an enzyme to Sepharose activated with cyanogen bromide (8). Sepharose activated with cyanogen bromide can react in several ways with the free amino groups of enzymes to form a covalent bond (161). Another chemical immobilization method is intermolecular cross-linking of enzyme molecules with a bifunctional reagent such as glutaraldehyde (109). The cross-linked molecules form a large insoluble macromolecular complex that retains catalytic activity.

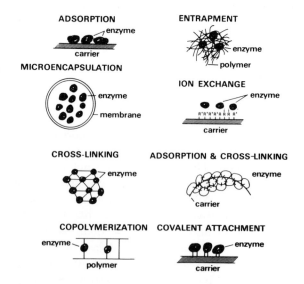

FIG. 1. Methods of enzyme immobilization. From Weetall (155).

Physical immobilization methods do not involve covalent bonding. An example is the adsorption of an enzyme to water-insoluble matrices such as DEAE-cellulose (9) or bentonite (2). Adsorption is due to ionic bonds and/or electrostatic interactions between the enzyme and adsorbent. Other physical methods include the entrapment of an enzyme in a gel such as polyacrylamide (95) or containment of a soluble enzyme by a semipermeable membrane via hollow fibers (52), an ultrafiltration device (22), or microcapsules (26).

Various combinations of physical and chemical methods also can be employed. For example, an enzyme can be adsorbed to a water-insoluble support and then cross-linked with a bifunctional reagent (10). The weak forces responsible for adsorption are not always sufficient to retain the enzyme permanently. The additional step of cross-linking can significantly improve the stability of the complex. Many other chemical and physical immobilization methods have been described. A comprehensive listing of these methods and details of their implementation can be found in various review papers (21,76,127,156,161).

Most immobilization studies involve cell-free enzymes, but in recent years increasing attention has been directed toward the use of immobilized whole cells. These systems obviate the need for a cell separation step prior to immobilization. Immobilized cells may also be more amenable to catalyzing a series of sequential reactions and may provide a means of regenerating in situ the necessary cofactors. The disadvantages are that the integrity of cells is difficult to maintain, and the cell walls and membranes offer permeability and diffusion barriers. Also, the presence of numerous catalytically active enzymes in the cells may lead to unwanted side reactions.

By restricting this review to preparative uses of enzymes and cells,

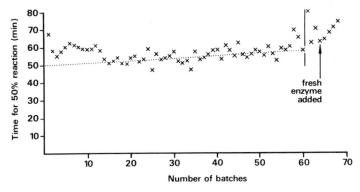

Fig. 2. Hydrolysis of benzylpenicillin by an *E. coli* penicillin acylase immobilized on cyanogen bromide-activated Sephadex G-200. The enzyme complex was recovered by filtration and used in sequential batch reactions. From Ekström *et al.* (48,49).

other pharmaceutical and medical applications of immobilization technology are omitted. Among these are the use of immobilized enzymes to treat metabolic disorders such as lactose intolerance (*102,120*), isomaltose intolerance (*111*), and acatalasia (*27,110*), a rare disease caused by a deficiency of catalase. Other nonsynthetic uses include incorporating immobilized urease into an artificial kidney (*24,133*) and treating acute lymphocytic leukemia with immobilized asparaginase (*6,25,64*). Immobilized enzymes, as components of enzyme electrodes, have clinical use as analytical tools (*16,61,62*). These electrodes couple an enzyme with a pH, dissolved oxygen, or ion-specific electrode and detect changes in reactant concentration due to catalysis. Immobilization technology also has applications in immunology (*16,18,106*) and a wide range of uses as strictly research tools (*17,20,51,60*). The many uses relating to the preparation of pharmaceutical compounds are the subjects of the remainder of this review.

II. Penicillins

A. Introduction

The activity and toxicity of the penicillin molecule are influenced by the side-chain substituent at the C_6–NH_2 position. Substitutions can be made at this position by reacting acyl chloride or anhydride derivatives of the side chain with the penicillin nucleus, 6-aminopenicillanic acid (6-APA) (*74*).

Originally, 6-APA was produced directly by fermentation (*11,12,77*). However, the yields were low and some penicillin biosynthesis invariably occurred which complicated 6-APA purification (*23*). Presently, 6-APA is made commercially from penicillin G or penicillin V since both are readily obtained in high yield by fermentation. Cleavage of penicillin G or penicillin V to 6-APA can be accomplished chemically or enzymically (*67*). Although both methods have been commercially employed, chemical cleavage is now preferred.

The enzyme that hydrolyzes penicillins to 6-APA has been called penicillin amidase, penamidase, etc. The generally accepted designation of penicillin acylase (penicillin amidohydrolase E.C. 3.5.1.11) will be used for this discussion. Penicillin acylases can be classified on the basis of the organisms that produce them and the substrates that they hydrolyze (*63,146,147*). Type I acylases are produced by bacteria and preferentially hydrolyze penicillin G. Type II acylases, which are usually found in actinomycetes and molds, hydrolyze penicillin V. Recently, a third type of acylase was described. This enzyme, which is produced by *Pseudomonas melanogenum* and *Pseudomonas ovalis,* preferentially hydrolyzes ampicillin (97,104,105). Most, but not all, acylases (see 38,98,99,112,149,150 for exceptions) can be placed into one of these categories.

Many of the microorganisms that catalyze the hydrolysis of penicillins to 6-APA also catalyze the reverse reaction, i.e., the synthesis of a penicillin from a side chain and 6-APA (40,41,78,105). The hydrolytic reaction is generally favored at alkaline pH values; whereas, the synthesis (reverse) reaction occurs at acidic pH values (63). Studies on substrate specificities and other studies with a purified *Escherichia coli* acylase (19) suggest that the same enzyme catalyzes both reactions. However, data obtained with an *Erwinia aroideae* acylase (147–149) indicate that in some microorganisms separate enzymes may be involved. The synthetic reaction is greatly enhanced if "energy-rich" derivatives of the chain (e.g., amide or ester derivatives) are employed (40,41,78). Little is known about the mechanism of action of these enzymes but the influence of the derivatives and the reversibility of the reaction suggest that the enzyme catalyzes acyl transfers. In these reactions, the side-chain derivatives serve as acyl donors and 6-APA as an acyl acceptor. For the hydrolysis of penicillins, the penicillin functions as an acyl donor and water serves as the acyl acceptor.

The commercial importance of 6-APA provided the impetus for developing many processes involving penicillin acylase. These processes have utilized fungal spores, mycelia, bacterial cells, crude enzyme preparations, and purified enzymes. More recently, with the development of immobilization technology and recognition of its potential advantages, increasing effort has been devoted to applying the technology to penicillin acylase. The resultant processes (Table I), which are described in the following section, have led to the successful commercial use of immobilized acylases.

B. Penicillin Deacylation

Marconi and co-workers at the Italian company SNAM Progetti immobilized a penicillin acylase obtained from an *E. coli* culture (45,47, 85,128). The enzyme was partially purified by ammonium sulfate fractionation and was immobilized by entrapment in cellulose triacetate fibers. The fibers were formed by adding an enzyme solution dropwise to a solution of cellulose triacetate in methylene chloride to form an emulsion (85). The emulsion was then extruded through a spinneret into a coagulating bath containing toluene. The organic solvents did not denature the acylase (85), but the acylase fibers exhibited less activity than the soluble enzyme. The reduction in activity was attributed to hindered substrate diffusion caused by the fiber matrix. The fibers were packed in a column reactor through which a 12% solution of penicillin G

TABLE I

DEACYLATION OF PENICILLINS BY IMMOBILIZED PENICILLIN ACYLASES

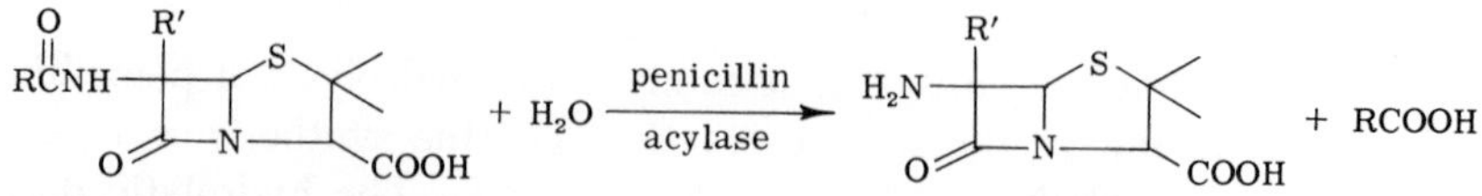

Enzyme source	Immobilization method	Remarks	Company or institution	References
Escherichia coli	Entrapment in cellulose triacetate fibers	79 % of initial activity remained after 30 uses	SNAM Progetti	*45, 48, 85, 128*
E. coli	Bound to bromoacetylcellulose, acidocarboxymethyl cellulose, CM-Sephadex, cation-exchange resins	Enzyme-bromoacetylcellulose complex was re-used 4 times	Beecham	*115*
E. coli	Coupled to a poly-(meth) acrylate resin (XAD-7) with glutaraldehyde and hexamethylenediamine	Enzyme complex was reused 15 times	Beecham	*114*
E. coli	Coupled to sucrose/epichlorohydrin copolymer (M.W. 400,000) and containment in an ultrafiltration (UF) device	Enzyme was bound to water-soluble polymer to increase its molecular weight and minimize losses in UF device	Beecham	*15*
Bacillus megaterium	Adsorption onto bentonite and diatomaceous earth, and containment in an U.F. device	Process employed for commercial production of 6-APA	Squibb	*65, 113*
E. coli	Coupled to a chloro-S-triazine derivative of DEAE-cellulose or cellulose filter paper	No enzyme activity was lost during 11 weeks of use	University College, London	*118*
E. coli	Coupled to cyanogen bromide-activated Sepharose	Process employed for commercial production of 6-APA	Astra Lakemedel	*44, 48, 49*

(Continued)

TABLE I (*Continued*)

Enzyme source	Immobilization method	Remarks	Company or institution	References
Bacillus circulans	Enzyme acylated with succinic anhydride and coupled to DEAE-Sephadex	A column reactor was used and the yield of 6-APA was 93 %	Otsuka Pharmaceutical	*108*
E. coli	Coupled to a co-polymer of acrylamide, *N,N'*-methylenebis-(acrylamide) and maleic acid anhydride	—	Bayer	*68, 69*
—	Coupled to cyanogen bromide-activated dextran (M.W. 500,000)	—	Bayer	*70*
Fusarium moniliforme	Spores of the microorganism	Spores were reused 3 times to hydrolyze phenoxymethyl penicillin	Ayerst Laboratories	*122*
E. coli	Cells entrapped in polyacrylamide gel	A column reactor hydrolyzed 85 % of a 1 % penicillin G solution	Tanabe Seiyaku	*31*
—	Entrapment in polyacrylamide gel	—	Tallin Polytech. Inst., USSR	*81*

was recycled. By recycling and using high flow rates to minimized diffusion limitations, 85% hydrolysis was obtained in 190 minutes. The acylase fibers were quite stable, although some activity losses occurred during initial uses due to leakage of loosely entrapped enzyme. In one series of experiments, the fibers were used 30 times during a 4-month period and only 20% of the initial activity was lost (*45*). In an extension of this study, Dinelli and Morisi found that other polymers such as poly-γ-methylglutamate and ethyl cellulose could be used for immobilization (*128*). The most notable feature of these entrapment systems is the high stability they afford the enzyme. In addition to improving stability to thermal inactivation, the fibers provide protection against bacterial contamination and proteolytic hydrolysis (*45*).

Many immobilization methods involving covalent bond formation have been used with penicillin acylase. Ekström *et al.* (*49*) evaluated several coupling methods involving various polysaccharide or polyacrylamide supports. High specific activity and good coupling yields were obtained when the enzyme was immobilized on cyanogen bromide-activated Sephadex G-200. Cyanogen bromide also was used to bind the acylase to other polysaccharides (Table II). Some of these activated polymers bound more acylase activity than Sephadex G-200, but the latter possessed better filtration properties, lower cost, and greater stability (48). The Sephadex-acylase complex was used by the Swedish company Astra Lakemedel for large-scale production of 6-APA (48). The reactions were catalyzed in stirred tank batch reactors, and filtration was used for enzyme recovery. Use of the complex for at least 30 reactions was necessary to offset immobilization costs. In large-scale reactors, 60 reactions were catalyzed before significant losses of activity occurred (Fig. 2). The rapid loss of activity after 60 uses was attributed to microbial contamination and proteolytic enzymes. A comparison of the antigenicity of the 6-APA produced by the immobilized acylase and that produced by a whole-cell suspension revealed that the 6-APA from the bound enzyme reactor was significantly less allergenic (43,44,48). The allergenicity of penicillins is due in part to the formation of penicilloylated proteins which can cause anaphylactic shock in man (13). Formation of these allergens in the bound acylase reactor was prevented because the enzyme complex was stable and did not release protein into solution.

Savidge et al. (114,117) described an immobilization method in which an acylase was coupled to an XAD resin with glutaraldehyde. This complex produced 16 μmoles of 6-APA/minute/gm complex and was used in 15 successive reactions to hydrolyze 6% penicillin G solutions. High

TABLE II

IMMOBILIZATION OF *E. coli* PENICILLIN ACYLASE TO CYANOGEN
BROMIDE DERIVATIVES OF POLYSACCHARIDES[a]

Support	Specific activity of product (U/gm)	Activity yield (%)
Amylose	1120	1.5
α-Glucan	36,800	47.8
Dextran	17,840	23.0
Carboxymethyl cellulose	859	11.1
Sephadex G 200	23,400	30.3
Carboxymethyl Sephadex	5130	6.6
Sulfoethyl Sephadex	12,440	16.1
DEAE-Sephadex	0	0
Sepharose 4B	34,700	45.1

[a] From Ekström *et al.* (48,49).

specific activities were also reported by Self *et al.* (*118*) using an acylase coupled to triazine cellulose derivatives. The latter complexes were somewhat unusual in that they exhibited apparent Michaelis constant (K_m) values that were less than those measured with the soluble enzyme. An enzyme attached to an insoluble matrix usually exhibits a higher apparent K_m value because of the presence of an unstirred layer (Nernst layer) that surrounds the particles and acts as a diffusion barrier (*76, 100*). The lower K_m with the cellulose complexes may be due to adsorption of the substrate by cellulose. In practical terms, this adsorption is beneficial in two ways: (a). rapid stirring speeds or high-column flow rates are not needed to enhance diffusion, and (b) a lower apparent K_m value results in a less rapid decline in reaction rate as the substrate is depleted.

Huper (*68,69*) described methods for coupling an *E. coli* acylase to copolymers containing maleic anhydride. In initial studies by Savidge *et al.* (*115,116*) a heterogeneous linear ethylenemaleic anhydride copolymer was reacted with an enzyme. Hydrolysis of carboxylic anhydride groups during the reaction increased the water solubility of the polymer. As a result, some of the low molecular weight polymer enzyme complexes were water soluble and so could not be recovered. The insolubility of the complexes was increased by coupling in the presence of higher protein concentrations. Although this procedure reduced the losses due to water solubility, the complexes were so strongly cross-linked by protein that substrate diffusion was hindered. In a modification of this approach (*60a,72a*), the enzyme trypsin was coupled to a copolymer of acrylamide, maleic anhydride, and N,N'-methylenebis(acrylamide). The enzyme was bound covalently through the reaction of its amino groups with the carboxylic anhydride, and the slightly cross-linked polymers minimized losses to water solubility. However, hydrolysis of some of the anhydride functional groups produced carboxylic acids that bound enzyme molecules ionically. These noncovalently attached molecules, representing up to one-half of the total bound enzyme, dissociated from the polymer during catalysis and were lost. With further modifications, Huper (*68,69*) established conditions that resulted in covalent binding of 98% of the penicillin acylase added to the polymer. Maximizing covalent binding prevented activity losses due to dissociation of loosely (ionically) bound enzyme.

A process involving a combination of physical entrapment and covalent immobilization methods was described in a patent assigned to Beecham Group Ltd. (*15*). In this process, an *E. coli* acylase was coupled to a water-soluble sucrose/epichlorohydrin copolymer of molecular weight 400,000. Attachment to the polymer increased the effective molecular weight of the enzyme and thus facilitated its recovery via ultrafiltration. The enzyme-polymer conjugate was recovered on an ultrafiltration membrane with a nominal molecular weight cutoff of 50,000. In a series of batch reactions, 6-APA yields were in excess of 90%. A two-stage continuous process also was devised. In this system unreacted substrate in the ultrafiltrate of the first reactor was transferred to a second stage where hydrolysis was completed.

Adsorption onto an insoluble matrix is one of the simplest ways to immobilize an enzyme. A disadvantage of the method, as illustrated earlier with the maleic anhydride copolymers, is that the enzyme may not be tightly bound. One approach used to improve adsorption to a basic adsorbent was to increase the acidity of the enzyme. This was accomplished with a *Bacillus circulans* penicillin acylase by reacting it with succinic anhydride (*107,108*). The succinoylated enzyme was then adsorbed to DEAE-Sephadex. It was reported that succinoylation did not lower enzyme activity and that the bound acylase retained 90% of the activity of the soluble enzyme (*108*).

A penicillin acylase adsorbed onto bentonite has been used commercially by Squibb to manufacture 6-APA (*123*). In this process, enzyme recovery was simplified by utilizing a species of *Bacillus megaterium* that produced an extracellular acylase (*65,113*). After 48–72-hours incubation the cells were removed by centrifugation and a mixture of bentonite ($AlSiO_4$) and Hyflo-Supercel (a diatomaceous earth filter aid) was added to the supernatant. The suspension was stirred for 1 hour and then the bentonite-acylase-Hyflo complex was recovered by filtration. After washing, the complex was used in batch or continuous reactors to hydrolyze penicillins. The stirred tank enzyme reactors were combined with an ultrafiltration (UF) system to separate enzyme and product (*113*). It is not clear why ultrafiltration was required since the bentonite-enzyme complex can be easily recovered by coarse filtration or low-speed centrifugation. One possibility is that the acylase slowly dissociated from the bentonite and was lost if not contained by the UF membrane. The membrane would prevent continuous dissociation or loss by maintaining some protein in solution in equilibrium with the bulk of the protein on the bentonite.

The incorporation of a UF system in the process should eliminate the need for attaching the enzyme to bentonite. The need for the bentonite-acylase complex is probably attributable to several factors: (a) the UF system operates more effectively because there is very little soluble protein to plug the filter and reduce flow rates, (b) the adsorbed enzyme may exhibit increased stability, (c) the enzyme must be adsorbed onto bentonite to recover it from the fermentation broth (to use the acylase in a soluble form, the additional step of releasing the acylase from bentonite is necessary), (d) kinetic analyses demonstrated that 6-APA productivities may be higher with the adsorbed enzyme than with its soluble counterpart. The differences in kinetic behavior were revealed in computer simulations of the productivities of continuous reactors containing either soluble or immobilized acylase. The simulations were based on rate equations that incorporated experimentally derived kinetic constants (113). Phenylacetic acid and 6-APA, the products of penicillin G hydrolysis, are competitive and noncompetitive inhibitors of penicillin acylase, respectively. The soluble and immobilized acylases exhibited similar K_m values and similar K_i values were measured with phenylacetic acid. In contrast, the K_i value for 6-APA and the bentonite-acylase complex was about 10-fold greater than the K_i measured with the soluble enzyme. This increased K_i indicated that 6-APA inhibited the bound enzyme much less than the soluble enzyme. Predictions of higher productivities with the bound enzyme, based on the lower K_i value for 6-APA, were verified in continuous reactors (113).

Immobilized whole cells also have been used to hydrolyze penicillins. In one study, 12 gm of E. coli cells were entrapped in a gel formed from 9.0 gm acrylamide, 480 mg N,N'-methylenebis(acrylamide), 300 mg di-methylaminopropionitrile, and 150 mg potassium persulfate (31). After polymerization, the gel containing the microorganism was passed through a sieve to obtain particles 3 mm in diameter. These particles (120 ml) were packed into a column (2.1 × 34.8 cm) and 500 ml of a 1% solution of potassium penicillin G were passed through at a rate of 16ml/hr. From the column eluate 2.23 gm ($\approx$85% molar yield) of crystalline 6-APA were recovered.

Escherichia coli cells have also been immobilized by entrapment in cellulose triacetate fibers. The fibers, containing 15 mg wet cells/gm polymer, exhibited 80% of the penicillin acylase activity of a similar amount of cells in free suspension (45).

C. PENICILLIN ACYLATION

The immobilization techniques used for enzymically acylating 6-APA are essentially the same as those employed for producing 6-APA (Table III). There are relatively few reports on the enzymic acylation of 6-APA because the reaction has less commercial utility. Commercial exploitation is hampered by low product yields and the availability of simple chemical methods of acylation. Low yields are attributable to significant product or substrate inhibition and the reversibility of the reaction (63).

The succinoylated penicillin acylase, described earlier for 6-APA production, also catalyzed the synthesis of ampicillin (75). The enzyme, adsorbed on DEAE-Sephadex, was used in a stirred batch reactor and recovered by filtration. In a solution containing 19.7 mg/ml immobilized

TABLE III

ACYLATION OF 6-AMINOPENICILLANIC ACID BY IMMOBILIZED PENICILLIN ACYLASES

Reaction catalyzed	Enzyme source	Immobilization method	Remarks	Company or institution	References
D- penicillin acylase reaction scheme	*Bacillus megaterium* or *Achromobacter* sp.	Whole cells absorbed on DEAE-cellulose	A column reactor was used	Toyo Jozo	*56*
	Bacillus circulans	Succinoylated enzyme was bound to DEAE-Sephadex	Enzyme complex was used for 10 batch reactions	Otsuka Pharmaceutical	*75*
	E. coli	Entrapment in cellulose triacetate	In one experiment 44% of the 6-APA was acylated	SNAM Progetti	*82, 128*
penicillin acylase reaction scheme	*E. coli*	Entrapment in cellulose triacetate	—	SNAM Progetti	*82*

acylase, 34.1 mg/ml ethyl D-phenylglycinate, and 16.0 mg/ml 6-APA, 18.0 mg/ml of ampicillin was produced in 2 hours (75). These yields correspond to about 57% molar incorporation of the side chain into the product. Similar yields were obtained when the acylase complex was packed into a column and methyl D-phenylglycinate was supplied as the side chain. After 10 uses the column reactor lost about 15% of the initial activity (75).

The fiber-entrapped *E. coli* acylase used by Marconi and his colleagues to hydrolyze penicillins also catalyzed the synthesis of penicillins (46, 82). The hydrolysis reaction was conducted at pH 8.0 and 37°C, but penicillin synthesis was favored at pH 7.0 and 25°C. Esters of phenylglycine or *p*-hydroxyphenylglycine were reacted with 6-APA to produce ampicillin or amoxycillin, respectively. The acylase exhibited a preference for the D-configuration of the acyl donor (side chain). The reaction rate with the D-isomer was about 10 times faster than with the L-isomer (82). If the specificity for the D-configuration were absolute, a significant reduction in process cost would accrue since it would not be necessary to resolve the racemic side-chain mixture prior to the acylation step. A time course of amoxycillin synthesis by fiber-entrapped acylase is shown in Fig. 3. The maximum amount of amoxycillin synthesized corresponded to 26.3% molar yield from 6-APA and 58% molar yield from the side chain. In another reaction, containing 92 μmoles/ml 6-APA and 300.0 μmoles/ml phenylglycine methyl ester, 44.5% of the 6-APA and 13.6% of the side chain were converted to ampicillin (82).

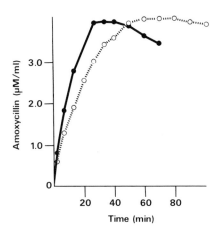

FIG. 3. Effect of the amount of cellulose triacetate-entrapped penicillin acylase on the synthesis of amoxycillin. Initial concentrations of hydroxyphenylglycine ethyl ester and 6-APA were 68.0 and 15.2 μmoles, respectively. ○– – –○, 2.0 gm enzyme fibers; ●——●, 4.0 gm enzyme fibers. From Marconi *et al.* (82).

By exploiting the reversibility of the acylase reaction and by selecting the proper enzymes, a two-step process can be devised to synthesize ampicillin from penicillin G (Fig. 4). In the first step, an *E. coli* acylase hydrolyzes penicillin G to 6-APA and phenylacetic acid. Prior to the acylation reaction, the free side chain normally must be separated from 6-APA to prevent reformation of penicillin G. This separation step is not necessary if the type III acylase (produced by *Pseudomonas melanogenum*) is used (*104,105*). The type III acylase is specific for phenylglycine acyl donors. Thus, this acylase and a phenylglycine acyl donor can be added to the 6-APA/phenylacetic acid mixture and only ampicillin will be produced. Although there are no reports yet on immobilizing the type III acylase, the approach seems worthwhile and undoubtedly has been considered and evaluated.

One problem associated with *P. melanogenum* acylase was that the microorganism produced a penicillinase (*104,105*). Losses to β-lactam hydrolysis were not extensive because the penicillinase had a pH optimum at 7.5 and the acylase pH optimum was 5.5 to 6.0. In addition, it was possible to inhibit penicillinase with 1 mM *o*-phenanthroline or 100 mM thiourea, without inhibiting acylase activity (*104*). Removal of penicillinase also was accomplished by purifying the acylase (*105*) and by isolating penicillinase-deficient mutants (*104*).

In another process, immobilized whole cells were used to synthesize ampicillin from 6-APA and D-phenylglycine methyl ester. Cells of an *Achromobacter* sp. or *Bacillus megaterium* were immobilized by adsorption onto DEAE-cellulose (*56*). The cell complex was used not only to synthesize penicillins but also to acylate the cephalosporin nucleus (*145*).

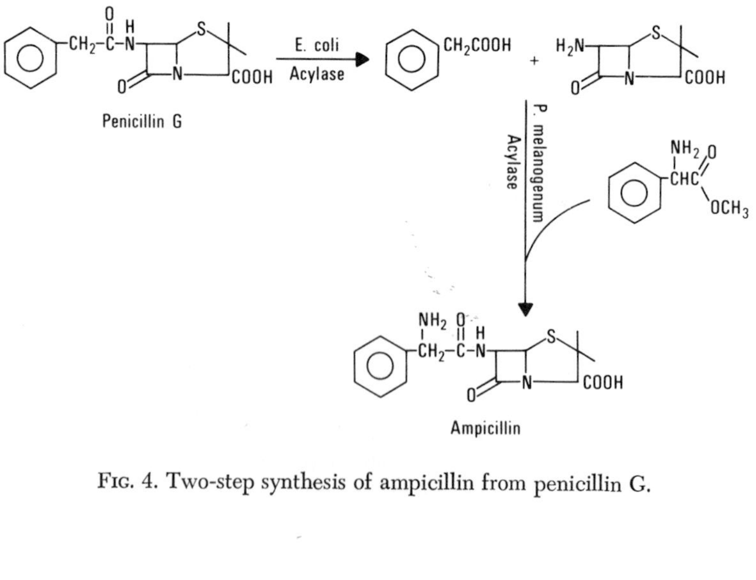

Penicillin G

Ampicillin

FIG. 4. Two-step synthesis of ampicillin from penicillin G.

This dual capability was also exhibited by the fiber-entrapped *E. coli* acylase described by Marconi *et al.* (*82*).

III.   Cephalosporins

A.   INTRODUCTION

Cephalosporin antibiotics can be synthesized from penicillins or made by direct fermentation with *Cephalosporium acremonium*. The major cephalosporin fermentation product is cephalosporin C. This antibiotic contains the nucleus 7-aminocephalosporanic acid (7-ACA) and the side chain α-aminoadipic acid.

$$\underset{\text{HOOCCH(CH}_2)_3}{\overset{\text{NH}_2}{|}}\overset{\text{O}}{\overset{||}{\text{C}}}-\text{N}$$

Cephalosporin C

Cephalosporins containing the 7-ACA nucleus and side chains other than α-aminoadipic acid cannot be made by fermentation. *Cephalosporium acremonium*, unlike *Penicillium chrysogenum*, does not incorporate side-chain precursors from the fermentation medium into the antibiotic. Thus, to make semisynthetic cephalosporins containing the 7-ACA nucleus, the α-aminoadipic acid side chain must be removed. Removal can be accomplished with reagents such as PCl_5 or nitrosyl chloride (*67*), but enzymic hydrolysis has not been unequivocally demonstrated. Walton reported that species of *Achromobacter*, *Brevibacterium*, and *Flavobacterium* removed the α-aminoadipic acid side chain (*152,153*); however, studies by other investigators have not confirmed these results (*67*). After chemical cleavage of α-aminoadipic acid, enzymic or chemical acylation can be used to add other side chains to the nucleus. Acylation of 7-ACA with a 2-thiophene–acetic acid derivative produces the commercially important antibiotic cephalothin.

Cephalosporins can be produced from penicillins by a series of chemical reactions that expand the 5-membered thiazolidine ring of the penicillin to the 6-membered dihydrothiazine ring of a cephalosporin (Fig. 5) (*42*). The resultant cephalosporin contains the penicillin side chain which can be removed enzymically or chemically to obtain the nucleus 7-aminodesacetoxycephalosporanic acid (7-ADCA). Acylation (enzymic or chemical) of 7-ADCA with D-phenylglycine produces the clinically useful antibiotic cephalexin.

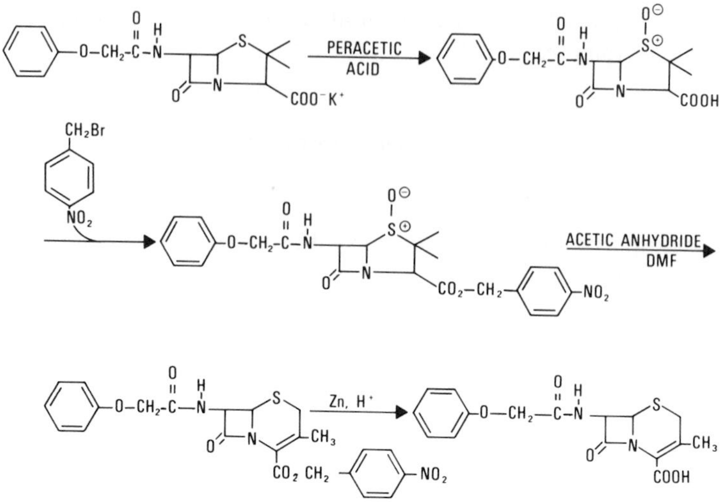

Fig. 5. Chemical synthesis of a cephalosporin from a penicillin (42).

An advantage of enzymic acylation over chemical acylation is that blocking groups are not needed for the reactants (131). Preparation of cephalexin by chemical acylation is accomplished with blocked derivatives such as the trichloroethyl ester of 7-ADCA and the N-tert-butoxycarbonyl derivative of D-phenylglycine. After the acylation reaction, two deblocking steps are required to obtain the final product. Efficient chemical deacylation also requires suitably blocked derivatives of the antibiotic.

B. CEPHALOSPORIN DEACYLATION

Most studies on enzymic deacylation of cephalosporins are directed toward producing the 7-ADCA nucleus. Substrates for the reaction are usually either 7-phenylacetamidodesacetoxycephalosporanic acid or 7-phenoxyacetamidodesacetoxycephalosporanic acid (Table IV). These substrates are used because they are readily obtained by the ring expansion reaction from penicillin G and penicillin V, respectively. The Japanese company of Toyo Jozo has patented several immobilized enzyme processes for conducting the reaction (141,143). In one process, an acylase precipitated from the culture supernatant of B. megaterium was adsorbed onto either celite ("Hyflo"), activated carbon, carboxymethyl cellulose, or an Amberlite ion exchange resin (143). Each carrier bound similar amounts of enzyme except for the Amberlite resin which adsorbed about 20% less. The celite-adsorbed acylase was evaluated on a pilot plant scale. Acylase from 350 liters of culture supernatant was adsorbed onto 3.5 kg of celite. After washing, the enzyme-support complex was packed into a column. A 5 mg/ml solution of 7-phenylacetamidodesacetoxycephalosporanic acid (total amount 1.8 kg/360 liters) was passed through at a flow rate of 5 liters/hour. From the eluate 1.13 kg of 7-ADCA (90% purity) was recovered, corresponding to hydrolysis of 89.2% of the substrate (143). The activity of the immobilized acylase gradually decreased during repeated uses. The decreases were attributed in part to bacterial contamination. With the addition of toluene to minimize contamination, the immobilized acylase had a useful life of more than 10 days.

An extracellular acylase produced by *Proteus rettgeri* was used in a process almost identical to that described above. This enzyme produced 7-ADCA in an overall molar yield of 90% (141).

Glaxo Laboratories immobilized a partially purified acylase from *Erwinia aroideae* by entrapment in cellulose triacetate fibers (53). The fibers were packed into a column through which a substrate solution containing 16.7 mg/ml 7-phenoxyacetamidodesacetoxycephalosporanic acid was circulated. After 3 hours at 37°C, 58% of the substrate was hydrolyzed to 7-ADCA.

C. CEPHALOSPORIN ACYLATION

Marconi *et al.* (82) used an *E. coli* acylase entrapped in cellulose triacetate fibers to synthesize cephalexin (Table V). This immobilized enzyme system was the same one that was used to acylate the penicillin nucleus (see Section II,C). A time course of cephalexin synthesis by the fibers in a stirred batch reactor is shown in Fig. 6. The decline in cephalexin concentration after 100 minutes of reaction time was probably due to the acylase catalyzing the reverse reaction (i.e., cephalexin deacylation). The reaction mixture initially contained 46 μmoles of 7-ADCA and 182 μmoles of D-phenylglycine methyl ester. The maximum amount of cephalexin synthesized corresponded to acylation of 74.5% of the 7-ADCA.

The Japanese company Toyo Jozo developed several immobilized whole cell systems for synthesizing cephalexin (5,145). Acetone-dried *Achromobacter* cells (160 mg) were adsorbed to 10 gm of carboxymethyl cellulose, DEAE-cellulose, or TEAE-cellulose (145). The cellulose-cell complexes were packed into columns and 100 ml solutions containing 0.1% 7-ADCA and 0.5% D-phenylglycine methyl ester were passed through. After 5 hours from 80 to 85% of the 7-ADCA was acylated by the DEAE- and TEAE-cellulose complexes and 40% yields were obtained with the carboxymethyl cellulose complex. When DEAE-cellulose com-

Reaction catalyzed

IV
IMMOBILIZED ACYLASES AND ESTERASES

Enzyme source	Immobilization method	Remarks	Company or institution	References
Bacillus megaterium	Adsorption on diatomaceous earth	A column reactor was operated for 9 days	Toyo Jozo	143
Arthrobacter viscosus	Adsorption on Ca$_3$(PO$_4$)$_2$ and "Dicalite"	—	Banyu Pharmaceutical	132
Bacillus megaterium	—	—	Toyo Jozo	140a
Erwinia aroideae	Cells were entrapped in cellulose triacetate fibers	—	Glaxo Laboratories	53
Bacillus subtilis	Adsorption onto bentonite or containment by ultrafiltration (UF)	UF-contained enzyme was reused 20 times	Eli Lilly and Company	1–4
—	Coupled to methacrylate maleic acid copolymer	—	Bayer	14

TABLE V

SYNTHESIS OF CEPHALOSPORINS BY IMMOBILIZED CEPHALOSPORIN ACYLASES

Reaction catalyzed	Enzyme source	Immobilization method	Remarks	Company or institution	References
D- [cephalosporin acylase-catalyzed reaction of D-phenylglycine methyl ester ($CHC=OCH_3$, NH_2) plus 7-ADCA (H_2N—, CH_3, COOH) → acylated cephalosporin]	Achromobacter sp.	Ionic binding of whole cells to DEAE cellulose	CM-cellulose and TEAE-cellulose were also used as supports	Toyo Jozo	145
	Achromobacter sp.	Adsorption to hydroxyapatite		Toyo Jozo	145
	Bacillus megaterium	Adsorption to diatomaceous earth		Toyo Jozo	145
	E. coli	Entrapment in cellulose triacetate fibers	74.5% of the 7-ADCA in the reaction mixture was acylated	SNAM Progetti	82, 129
[cephalosporin acylase-catalyzed reaction of 7-ACA (H_2N—, CH_2OCCH_3, COOH) plus 2-thienylacetic acid methyl ester ($CH_2C=OCH_3$) → cephalothin-type product (CH_2CNH—, CH_2OCCH_3, COOH)]	Bacillus megaterium	Adsorption onto alumina, diatomaceous earth, or hydroxyapatite	Enzyme was adsorbed directly from the culture supernatant	Toyo Jozo	142

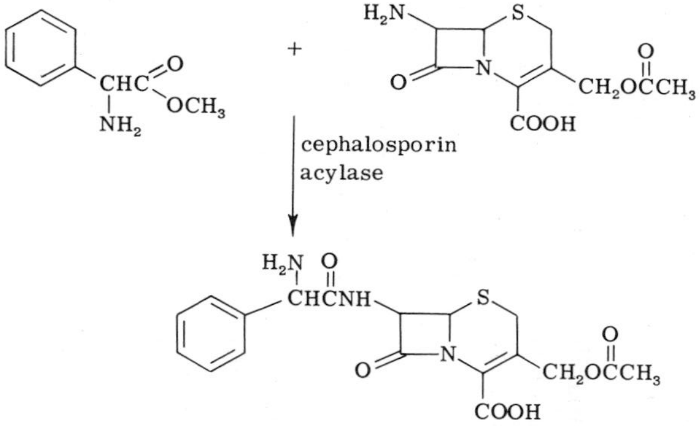

cephalosporin acylase

Bacillus megaterium

Adsorption on diatomaceous earth

—

Toyo Jozo *144*

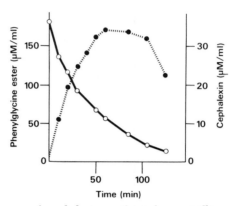

FIG. 6. Time course of cephalexin synthesis by penicillin acylase entrapped in cellulose triacetate fibers. Initial concentration of 7-ADCA was 46.0 μmoles. ○——○, Disappearance of phenylglycine methyl ester; ●– – –●, synthesis of cephalexin. From Marconi *et al.* (*82*).

plexes of four microorganisms were compared, highest cephalexin yields were obtained with cells of *Benecka hiperoptica* and an *Achromobacter* sp. Yields were significantly lower with *Alcaligenes faecalis* and *Flavobacterium aquatile* (*145*). In related experiments, an extracellular acylase from *B. megaterium* (*5*) was adsorbed onto celite and a cell-free *Achromobacter* acylase was adsorbed onto hydroxyapatite (*145*). These immobilized enzymes also produced high yields (80%–85%) of cephalexin. Unfortunately, data were not presented in these studies to illustrate the stability of the complexes to inactivation or dissociation during multiple reuses.

Previous examples illustrated the capability of a *B. megaterium* acylase(s) to catalyze the deacylation of penicillins and cephalosporins and the acylation of 6-APA and 7-ADCA. Immobilized forms of this acylase also have been used to acylate the cephalosporin C nucleus, 7-ACA (*142,144*). Cephaloglycin (Table V) was produced by reacting a 2 mg/ml solution of 7-ACA with D-phenylglycine methyl ester in the presence of 20 mg/ml celite-acylase complex (*144*). Product yields were quite low; only 40 μg/ml of cephaloglycin were produced. Much higher yields were obtained when the celite-enzyme complex was used to synthesize cephalothin by acylating 7-ACA with 2-thiophene acetic acid methyl ester. In a large-scale process, the acylase from 580 liters of *B. megaterium* culture supernatant was adsorbed onto 6 kg of celite and packed into a column (*142*). A buffered solution containing the side chain and 1.0 kg of 7-ACA (10 mg/ml) was passed through. About 85% of the 7-ACA was acylated and 1.34 kg of crystalline cephalothin was recovered.

D. Cephalosporin Deacetylation

The biological activity of cephalosporins is dependent not only on the side chain on the C-7 position, but also on the substituent at the C-3 position. Many substituents can be added to the C-3 position of 7-ACA by nucleophilic displacement of the acetate group (96). Other modifications can be made only after acetate is hydrolyzed from the molecule. Acetate can be removed by chemical hydrolysis but the yields are low because of side reactions such as migration of the Δ^3 double bond to the Δ^2 position (91) or lactone formation between the hydroxymethyl and the adjacent carboxyl group (80). Enzymic hydrolysis is, at present, the only method of preparing deacetyl-7-ACA in high yield.

Several processes that use cell suspensions or soluble enzyme preparations to deacetylate cephalosporins have been described (1,73,103,157). Recently, Abbott and Fukuda (3,4) developed an immobilized enzyme system for catalyzing this reaction. An extracellular esterase produced by B. subtilis was adsorbed onto bentonite and used for multiple batch reaction to deacetylate 5 mg/ml 7-ACA solutions (3). The immobilized esterase lost very little activity when stored in solution for 1 month at 25°C, but the complex was unstable while catalyzing the reaction because the esterase dissociated from the bentonite particles. The complex was partially stabilized by adding divalent cations, aluminum hydroxide gel, or cross-linking with bifunctional reagents. The best stabilization, which was achieved with aluminum hydroxide, enabled the complex to be used nine times before one-half the initial activity was lost.

The B. subtilis esterase has a molecular weight of about 190,000 (1), and it is very resistant to thermal inactivation. These characteristics make the enzyme amenable to immobilization by containment in an ultrafiltration device. With this immobilization method, the enzyme was reused 20 times over an 11-day span (Fig. 7). Solutions containing 4–24 mg/ml of 7-ACA were deacetylated and yields >94% were obtained from each reaction. Some activity losses occurred during these experiments, but they were attributed to loss of protein during handling and filtration and not to inactivation of the enzyme (4).

IV. Amino Acids

A. Introduction

Amino acids are used as food supplements, cosmetic additives, medicinal agents, and as raw materials for the synthesis of other compounds. For food supplementation and for most medicinal applications, only the L-isomers of the amino acid are biologically active. There are efficient

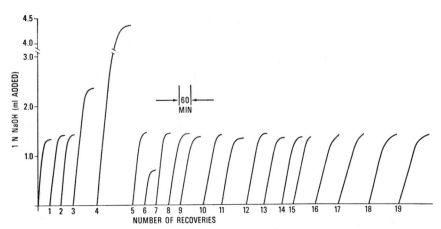

FIG. 7. Multiple reuse of a cephalosporin acetylesterase. Curves represent continuous pH stat titration of the acetic acid liberated by the reaction. Esterase was recovered after each reaction by ultrafiltration. The initial 7-ACA concentration was 4–24 mg/ml. From Abbott *et al.* (*4a*).

chemical syntheses for amino acids, but the products of the reactions are racemic D,L-mixtures. Although these mixtures can be resolved chemically or enzymically, the resolution steps are expensive. Biosynthesis of amino acids by microorganisms or isolated enzyme systems leads exclusively to the L-isomers. For this reason biological synthesis usually is the preferred production method. As illustrated in the following section, immobilization technology has been applied both to the biosynthesis of amino acids and to the resolution of racemic mixtures (Table VI).

B. RESOLVING RACEMIC MIXTURES

The first commercial process that utilized an immobilized enzyme was for the resolution of racemic amino acid mixtures (*32*). The Tanabe Seiyaku Company in Japan developed the process after an extensive investigation of many immobilization methods (*32,134–138*). The enzyme employed was an aminoacylase produced by *Aspergillus oryzae*. This enzyme catalyzes the hydrolysis of N-acetyl-L-amino acids. The N-acetyl derivatives of the racemic acid mixtures are prepared chemically. After preferential hydrolysis of the acetyl group from the L-isomer, the free L-amino acid and unreacted N-acetyl D-amino acid are easily separated by crystallization. The unreacted D-isomer can then be racemized to a D,L-mixture and recycled back to the enzyme reactor.

The immobilization methods that were tested with the aminoacylase were (*32,135*): (a) physical adsorption on silica gel, alumina, and charcoal; (b) ionic binding to cellulose derivatives and ionic exchange resins;

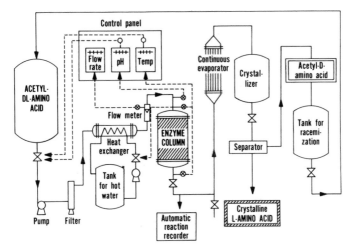

FIG. 8. Flow diagram for the continuous production of L-amino acids by immobilized aminoacylase. From Chibata *et al.* (32).

(c) covalent binding to glass beads and halogenated cellulose derivatives; (d) cross-linking of an enzyme that was ionically bound to cellulose derivatives; and (e) entrapment in gels and encapsulation by microcapsules. These methods were compared on the basis of the amount of activity they bound per unit weight of support. From these comparisons, ionic binding to DEAE-Sephadex, covalent binding to iodoacetylcellulose, and entrapment by polyacrylamide were selected for further evaluation.

A comparison of some physicochemical properties of these immobilized acylases is shown in Table VII. Selection of a support for industrial use was based on the following criteria: (a) the activity and stability of the immobilized acylase, (b) the ease (and cost) of immobilization, and (c) the ability of the support to be regenerated when enzyme is lost or inactivated. From these considerations, ionic binding to DEAE-Sephadex was selected as the most effective immobilization method (32).

The Sephadex-acylase complex was used in a column reactor to hydrolyze a 0.2 M (36.2 mg/ml) solution of acetyl-DL-methionine. Higher substrate concentrations (i.e., higher ionic strengths) were not used because they caused a significant increase in leakage of the acylase from the column. A flow diagram for the process is shown in Fig. 8. The column was operated at 50°C and maintained 60% of its initial activity after 32 days of operation. Full activity was restored simply by adding more aminoacylase to the column. By periodic regeneration, the Sephadex was used over 2 years without loss of binding capacity or physical deterioration (32). An economic analysis of the process revealed that

Reaction catalyzed

$$\text{DL-R}-\underset{\underset{\text{NHCOR}'}{|}}{\text{CHCOOH}} + H_2O \xrightarrow[\text{acylase}]{\text{amino}} \text{L-R}-\underset{\underset{\text{NH}_2}{|}}{\text{CHCOOH}} + \text{D-R}-\underset{\underset{\text{NHCOR}'}{|}}{\text{CHCOOH}}$$

$$\text{HOOCCH}=\text{CHCOOH} + NH_4 \xrightarrow{\text{aspartase}} \text{L-HOOCCH}_2\underset{\underset{\text{NH}_2}{|}}{\text{CHCOOH}}$$

$$\underset{\underset{\text{COOH}}{|}}{\underset{\underset{\text{CHNH}_2}{|}}{(CH_2)_3\text{NHC}-NH_2}} \begin{matrix} | \\ NH_2 \end{matrix} + H_3O^+ \xrightarrow[\text{desiminiase}]{\text{arginine}} \underset{\underset{\text{COOH}}{|}}{\underset{\underset{\text{CHNH}_2}{|}}{\text{L-}(CH_2)_3-\text{NHC}-NH_2}} \begin{matrix} \| \\ O \end{matrix} + NH_4$$

VI
AMINO ACIDS BY IMMOBILIZED ENZYMES

Enzyme source	Immobilization method	Remarks	Company or institution	References
Aspergillus oryzae	Ionic binding to DEAE-Sephadex	A column reactor, with periodic regeneration, was used over 2 years	Tanabe Seiyaku	*32, 134–138*
E. coli	Entrapment in polyacrylamide gel	Whole cells and isolated enzyme were immobilized	Tanabe Seiyaku	*34, 139, 140*
—	Entrapment in hollow fibers	—	Kanebo Ltd.	*88*
Pseudomonas putida	Entrapment of cells in polyacrylamide gel	A column reactor lost no activity during 1 month of operation	Tanabe Seiyaku	*35, 159*
E. coli	Entrapment in cellulose triacetate fibers	Fibers were reused 50 times with little activity loss	SNAM Progetti	*45, 83, 162*
E. coli	Attachment of apotryptophanase to pyridoxal phosphate derivatives of Sepharose	Complex was used for 5 days with no loss of activity	Kyoto University	*58, 59, 71, 72*
E. coli	Cells entrapped in polyacrylamide gel	—	Tanabe Seiyaku	*29*
E. coli	Cells entrapped in polyacrylamide gel	—	Tanabe Seiyaku	*30*
Escherichia intermedia	Attachment to pyridoxal phosphate derivatives or cyanogen bromide derivatives of Sepharose	CNBr complex lost 30 % activity during 7 days use	Kyoto University	*8, 58, 59*

(Continued)

TABLE VI

Reaction catalyzed

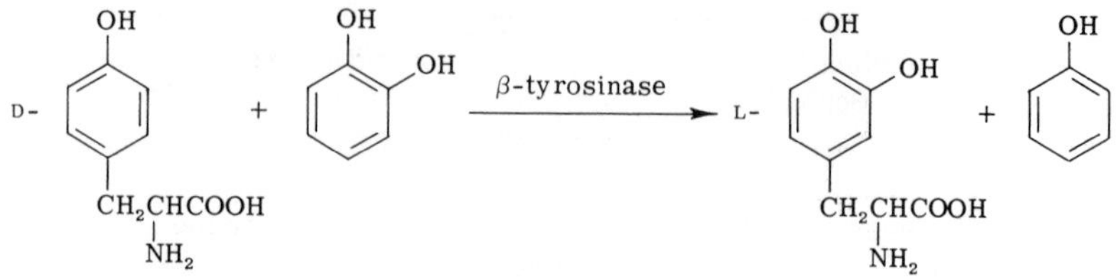

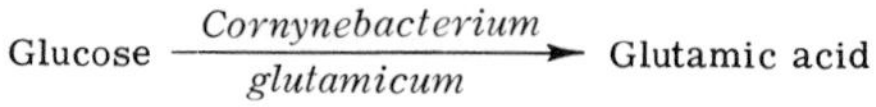

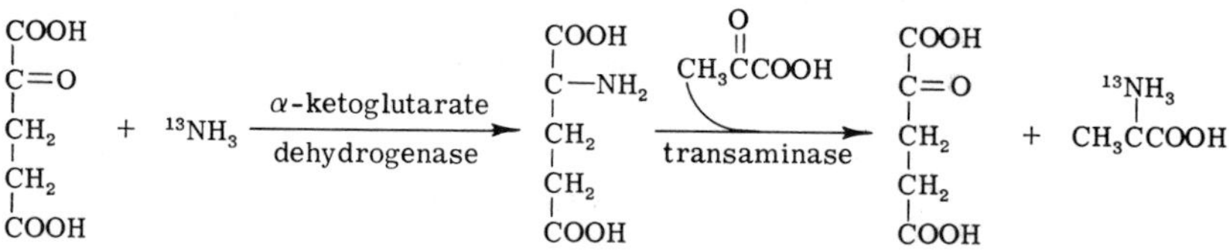

$$\text{Glucose} \xrightarrow[\textit{glutamicum}]{\textit{Cornynebacterium}} \text{Glutamic acid}$$

production costs were 40% less than for a comparable process employing a soluble acylase. The Tanabe Seiyaku Company has the capacity to produce over 20 tons/month of L-amino acids with the Sephadex-acylase process.

TABLE VII

COMPARISON OF ENZYMIC PROPERTIES OF VARIOUS IMMOBILIZED AMINOACYLASES[a,b]

		Immobilized aminoacylases		
Properties	Native aminoacylase	Ionic binding (DEAE-Sephadex)	Covalent binding (iodoacetyl-cellulose)	Entrapped by poly-acrylamide
Optimum pH	7.5–8.0	7.0	7.5–8.5	7.0
Optimum temperature (°C)	60.0	72.0	55.0	65.0
Activation energy (kcal/mole)	6.9	7.0	3.9	5.3
Optimum Co^{2+} (mM)	0.5	0.5	0.5	0.5
K_m (mM)	5.7	8.7	6.7	5.0
V_{max} (μmole/hour)	1.52	3.33	4.65	2.33

[a] Data are from the results with acetyl-DL-methionine as a substrate.
[b] From Chibata et al. (32).

(Continued)

Enzyme source	Immobilization method	Remarks	Company or institution	References
—	Entrapment in cellulose triacetate fibers	Little activity was lost during 15 days of use	SNAM Progetti	*45*
Corynebacterium glutamicum	Cells entrapped in polyacrylamide gel	Cells accumulated 14 mg/ml glutamic acid in 144 hours	New England Enzyme and Tufts University	*126*
—	Enzymes bound to derivatized porous glass beads	Process used for rapid synthesis of ¹³N-alanine	Veterans Administration Hospital	*38*

C. Synthesis

1. Aspartic Acid

The enzyme aspartase, which is produced by many microorganisms, catalyzes the interconversion of L-aspartate and fumarate. Partially purified aspartases usually contain the enzyme fumarase. In processes for converting fumarate to L-aspartate, fumarase catalyzes an unwanted side reaction:

$$\text{fumarate} \xrightarrow[\text{H}_2\text{O}]{\text{fumarase}} \text{malate}$$

The fumarase activity in an aspartase preparation from *E. coli* was selectively destroyed by heating for 4 hours at 37°C (*140*). Tosa *et al.* (*140*) immobilized the furmarase-free aspartase by ionic binding, covalent binding, physical adsorption, and entrapment. A comparative study of these immobilization methods revealed that entrapment in a polyacrylamide gel afforded the highest aspartase activity. Aspartase was inactivated by the entrapment procedure if fumarate or aspartate was not present. These compounds bind to the active site of the enzyme and protect it from inactivation during immobilization.

The gel-entrapped and soluble aspartase exhibited similar pH optima, temperature optima, heat stability, and kinetic behavior. The enzyme was capable of converting fumarate to aspartate in high yield. In one experiment, a 1 M solution of ammonium fumarate (pH 8.5) was passed through a column (5 × 100 cm) containing gel-entrapped aspartase at a flow rate of 160 ml/hour (140). From 1 liter of column effluent, 127 gm of L-aspartic acid was isolated. This yield corresponded to the conversion of about 95% of the fumarate to L-aspartate. The gel columns also were relatively stable and retained about one-half the initial aspartase activity after 15–20 days of operation.

Another approach to L-aspartate production involved the reuse of whole cells from a fermentation for isoleucine production (89). A total of 766 gm of "waste" *Brevibacterium roseum* cells were recovered from a fermentation which produced 16.3 mg/ml of isoleucine. These cells, which contain aspartase, converted fumarate in 83% molar yield to L-asparate.

More extensive use of aspartase-containing cells was made by immobilizing *E. coli* (33,34). Three methods were employed (34): (a) entrapment in polyacrylamide gel, (b) cross-linking with bifunctional reagents, and (c) encapsulation in polyurea microcapsules. The encapsulated cells retained no aspartase activity, but cells entrapped in polyacrylamide gel or cross-linked with *N,N'*-diallyltartardiamide exhibited about 72% of the activity of cells in free suspension. The activity of the gel-entrapped cells increased during incubation in solutions of slightly alkaline pH and high ionic strength. Electron microscopic observation suggested that the "activation" (i.e., activity increase) was due to lysis of the cells. The aspartase activity of cell lysate was much higher than the activity of aspartase enzyme preparations obtained from sonicated cells. A comparison of "activated" immobilized cells and an immobilized aspartase obtained from sonicated cells (Fig. 9) demonstrated that considerably higher activities were obtained with the comparatively simple immobilized whole cell system. In subsequent studies, gel-entrapped *E. coli* cells that were packed into a column reactor formed L-aspartic acid in 95% molar yield from a 1 M fumarate solution (139). At a flow rate of 0.5 ml/ml bed volume/hour, no aspartase activity was lost during 40 days of operation.

2. *Tryptophan*

The enzyme tryptophanase is a multifunctional enzyme that requires the cofactor pyridoxal phosphate. *Escherichia coli* tryptophanase catalyzes α,β-elimination and β-replacement reactions with tryptophan, serine, cysteine, S-methylcysteine, O-methylserine, and other compounds

(*92,101*). The reversible α,β-elimination reaction can be used to synthesize tryptophan from indole, pyruvate, and ammonia (*154*).

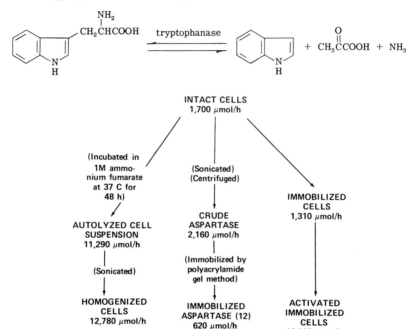

FIG. 9. Comparison of aspartase activities of various enzyme preparations per unit of intact cells. One gram of packed cells corresponds to 0.2 gm dry weight. Numerical values in parentheses are aspartase activities from 1 gm packed cells. From Chibata *et al.* (*34*).

When 5-hydroxyindole is added in place of indole, 5-hydroxytryptophan is formed. Decarboxylation of the latter compound produces serotonin (5-hydroxytryptamine).

Fukui *et al.* (*57–59,71,72*) immobilized tryptophanase by covalent binding to cyanogen bromide or diazotized derivatives of Sepharose, and by attachment of the apoenzyme to pyridoxal phosphate (PLP) derivatives of Sepharose. Three methods were used to bind pyridoxal phosphate to Sepharose (Figs. 10 and 11). One derivative, designated SP-A, was prepared by reacting PLP with diazotized *p*-aminobenzamidohexyl-Sepharose. The other two derivatives (SP-B and SP-C) were formed by reacting PLP with bromoacetamidohexyl-Sepharose. The SP-B derivative contained PLP attached via alkylation of the pyridine nitrogen. The third derivative (SP-C) was coupled at the 3-hydroxy group by alkylation at pH 6.0 where the pyridine nitrogen is predominantly protonated and the 3-hydroxy group is ionized. Apo-tryptophanase was attached to the

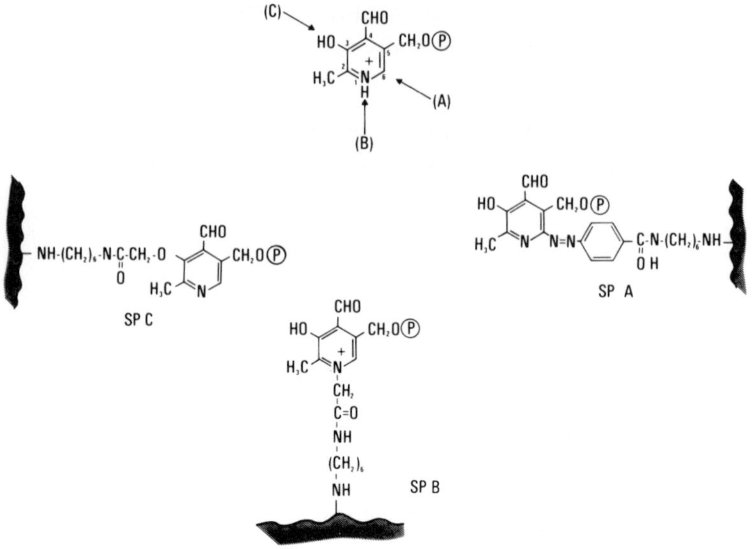

FIG. 10. Three types of Sepharose derivatives containing pyridoxal 5′-phosphate. From Fukui and Ikeda (58).

PLP-Sepharose derivatives by formation of a Schiff base linkage between the 4-formyl group of the bound PLP and the ε-amino group of a lysine residue at an active site of the enzyme. The Schiff base was then stabilized by reduction with $NaBH_4$.

A comparative study was made of tryptophanase immobilized by the three PLP-Sepharose derivatives and by cyanogen bromide and diazotized derivatives of Sepharose (58). The data revealed that the SP-A derivative bound more enzyme than the other Sepharose derivatives. Tryptophanase bound to SP-A also exhibited the highest activity.

Some properties of the SP-A tryptophanase and soluble tryptophanase were compared. Immobilization did not alter the K_m value for tryptophan (0.33 mM), but the pH optimum was shifted 0.5–1.0 pH units to the alkaline side (58). The immobilized tryptophanase exhibited higher thermal stability and both soluble and immobilized enzymes were substantially activated by 5% v/v ethanol (58,59). In a series of batch reactions, it was found that about one-half of the activity of the SP-A tryptophanase was lost after four reactions. The activity losses were attributed to dissociation of PLP from the tryptophanase complex. Full activity was restored by the addition of more PLP. In the presence of adequate PLP, SP-A tryptophanase synthesized tryptophan from indole, pyruvate, and ammonia in sequential batch reactions (at 30–37°C) over a 5-day period with no appreciable loss of activity (59).

In another series of studies, a partially purified tryptophanase from

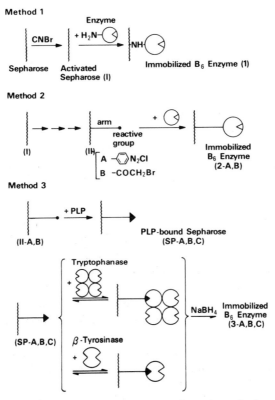

FIG. 11. Methods of immobilization of tryptophanase to Sepharose derivatives. From Fukui and Ikeda (58).

E. coli was immobilized by entrapment in cellulose triacetate fibers (83,162). The fibers were used to synthesize L-tryptophan from indole and D,L-serine. Initial experiments revealed that the rate of indole disappearance was greater than the rate of tryptophan formation. The rapid loss of indole was due to adsorption of indole by the fibers (162). About one-half the indole in a 390 mg/liter indole solution was adsorbed by 1 gm/liter of fibers. Indole adsorption during tryptophan synthesis was accompanied by a progressive decrease in reaction rate, apparently due to a decrease in fiber permeability rather than to enzyme inhibition by locally high indole concentrations. Similar decreases were observed with fiber-entrapped invertase even though soluble invertase was not inhibited by indole. The detrimental effects of indole were minimized by periodically washing indole from the fibers and by stepwise additions of indole to the reaction mixture to maintain a low indole concentration. With these procedures it was possible to use the fibers for more than 50 batch reac-

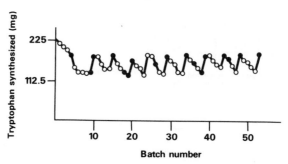

FIG. 12. Sequential batch reactions for the synthesis of L-tryptophan from indole and serine by tryptophanase entrapped in cellulose triacetate fibers. From Zaffaroni *et al.* (*162*).

tions with little loss of activity (Fig. 12). The reaction mixtures in these experiments consisted of 500 mg fibers (75 mg protein) in 50 ml phosphate buffer, pH 7.8 containing 2 μg/ml PLP, 40 μg/ml glutathione, 8 mg/ml D,L-serine, and 0.7 mg/ml indole. Periodic additions of indole were made and after 9 hours of incubation at 25°C, between 120 and 190 mg of tryptophan were synthesized. The fibers were also used in a continuous feed recycle reactor, but the productivity was only one-eighth that of the batch reaction system (*83*).

Tryptophan also has been synthesized by *E. coli* cells entrapped in a polyacrylamide gel. In one reaction 30 gm of immobilized cells were suspended in 100 ml of buffer containing 1 gm indole and 2 gm serine. After 24 hours of incubation at 30°C, 1.5 gm L-tryptophan were isolated (*29*). From a similar reaction mixture containing 1 gm 5-hydroxyindole in place of indole, 1.1 gm of crystalline 5-hydroxytryptophan were obtained (*30*).

3. *Tyrosine*

The enzyme β-tyrosinase catalyzes a series of α,β-elimination and β-replacement reactions analogous to those catalyzed by tryptophanase (*50,158*). The reversible α,β-elimination reaction can be used to synthesize L-tyrosine or L-dopa.

$$\text{HO} \overset{}{\underset{(\text{HO})}{\bigcirc}} - CH_2CHCOOH \underset{}{\overset{\beta\text{-tyrosinase}}{\rightleftarrows}} \text{HO} \overset{}{\underset{(\text{HO})}{\bigcirc}} + CH_3\overset{O}{\overset{\|}{C}}COOH + NH_3$$

L-Tyrosine   Phenol
(L-Dopa)   (pyrocatechol)

β-Tyrosinase requires the cofactor PLP, and Fukui *et al.* immobilized this enzyme by the same methods they employed with tryptophanase

(58,59). In contrast to tryptophanase, β-tyrosinase immobilized on pyridoxal derivatives of Sepharose was much less active than when it was immobilized on cyanogen bromide or diazotized derivatives of Sepharose (58). The differences were attributed to the macromolecular composition of the enzymes. Tryptophanase is a tetramer (containing four subunits) and each subunit contains an active site. Immobilization of the molecule by PLP-Sepharose inactivates one active site, but the other three retain activity. β-Tyrosinase is a smaller molecular and binds only 2 moles of PLP. Immobilization of this enzyme on PLP-Sepharose inactivates one of the PLP binding sites and may, in addition, induce a conformational change that alters the activity of the second site (59).

When apo-β-tryosinase (5.0 mg) was reacted with 1.36 gm CNBr-Sepharose, 98% of the enzyme was immobilized and 40% of the initial enzyme activity was retained (58). Immobilization of β-tyrosinase by this method increased the thermal stability of the enzyme, did not alter the K_m value for tyrosine, and shifted the pH optimum 0.5–1.0 units to the alkaline side. In sequential batch reactions the activity of the immobilized β-tyrosinase gradually declined, but full activity was restored by adding PLP. A column containing the immobilized β-tyrosinase was operated continuously to produce L-tyrosine from phenol, pyruvate, and ammonia. At a flow rate of 0.18 ml/ml bed volume per hour, about 60% of a 13.5 mM phenol solution was converted to L-tyrosine. During 7 days of continuous operation at 25°C and pH 8.0, the column lost only 30% of its initial activity (59).

4. Citrulline and Ornithine

Franks (54,55) studied arginine catabolism by cells of Streptococcus faecalis entrapped in polyacrylamide gel. The study was initiated to determine if a sequential multienzyme system (and associated cofactors) could be effectively immobilized without isolating and immobilizing the indivdual components of the system.

A cell suspenson, derived from a frozen cell paste of S. faecalis cells, produced citrulline (>50%), ornithine, and a small amount of putrescine from L-arginine. The entrapped cell paste catabolized arginine more slowly and produced only citrulline, indicating that ornithine transcarbamylase was lost or inactivated during cell immobilization. Freshly prepared cells (i.e., not previously frozen) entrapped in polyacrylamide gel accumulated ornithine but not citrulline. The interpretation was that entrapment did not destroy the ornithine transcarbamylase activity of fresh cells and that the enzyme rapidly converted citrulline to ornithine, preventing accumulation of the former. The entrapped cells were packed into a column and subjected to alternate periods of use and storage to assess their stability. Over an 11-day period there was little change in rate of arginine catabolism.

Yamamato *et al.* (*35,159*) made a thorough study of arginine catabolism by entrapped cells with the goal of developing an efficient commercial process for L-citrulline production. A strain of *Pseudomonas putida* was employed because this microorganism lacks ornithine transcarbamylase and has been successfully used (in free suspension) for commercial citrulline production. The cells, after entrapment in polyacrylamide gel, exhibited 56% of the arginine desimidase activity of a comparable amount of freely suspended cells. A comparative study revealed that citrulline production by free and entrapped cells had similar pH optima, but the entrapped cells exhibited a higher temperature optimum and a greater thermal stability. L-Citrulline was produced by the entrapped cells in a continuously operated column. At 37°C and a flow rate of 0.19 ml/ml bed volume/hour, the column converted 100% of a 0.5 M L-arginine solution to L-citrulline and lost no activity during 30 days of operation.

The surface-active agent cetyltrimethylammonium bromide (CTAB) was essential for a high rate of citrulline production be free cells (*159*). The surfactant had no effect on citrulline production by entrapped cells. Enhanced production in the presence of CTAB has been attributed to an increase in the permeability of the cells to substrate and/or product. The lack of an effect of CTAB on entrapped cells suggested that the permeability barrier was destroyed by the entrapment process.

5. Glutamic Acid

Glutamic acid can be obtained by extraction of vegetable protein, chemical synthesis, or fermentation. Most of the total world production, estimated at 400 million pounds in 1972, is made by fermentation. The microorganism commonly employed is a biotin-requiring strain of *Corynebacterium glutamicum* that accumulates as much as 75 mg/ml of glutamic acid in the culture medium.

Slowinski and Charm (*126*) immobilized *C. glutamicum* cells by entrapment in polyacrylamide gels. The entrapped cells were tested for their ability to synthesize glutamic acid from glucose. Two types of gel were prepared. One contained more N,N-methylenebis(acrylamide) than the other and, thus, was more highly cross-linked. The gels (0.20–0.22 gm dry weight), containing *C. glutamicum* (74 mg dry weight), were added to 40 ml of fermentation media. After 144 hours they produced 13–14 mg/ml glutamic acid. The gels then were recovered, washed, and resuspended in fresh fermentation medium. During this second incubation, the cells entrapped in the more highly cross-linked gel produced about the same amount of glutamic acid they produced previously. Cells entrapped in the loosely cross-linked gel accumulated about 3 mg/ml less.

Freely suspended cells also produced less glutamic acid when they were reused. Unfortunately, more extensive evaluations of the stability and productivities of the immobilized cells were not made. However, the results that were obtained are noteworthy because they demonstrate the feasibility of immobilizing a microorganism and maintaining intact a complex multienzyme pathway that produces relatively large amounts of fermentation product.

Venkatasubramanian *et al.* (*151*) made an economic analysis of glutamic acid by a hypothetical immobilized cell system. The hypothetical system consisted of *C. glutamicum* entrapped in collagen membranes and used in continuously operated reactors. The membranes were assumed to contain 30% (w/w) cells and to produce glutamic acid at a rate of 0.05 g/g cell/hour. This rate is comparable to production rates in the conventional glutamic acid fermentation. An additional assumption, which seems overly optimistic, was that the cell-collagen complex would have an operating life of 3 months.

The economic analyses indicated that production costs by the immobilized cells would be about 20% higher than in the conventional fermentation. Capital investment, however, would be about 40% less, due mainly to the elimination of the costly fermentation equipment. Because of the lower capital costs, it was estimated that the immobilized cell process would provide a 50% return on investment as opposed to 36% from the conventional process (*151*).

6. ^{13}N-Alanine

The development of the biomedium cyclotron has led to a resurgence of interest in the short-lived nuclides ^{11}C (half-life = 20 minutes) and ^{13}N (half-life = 10 minutes). These isotopes are useful because they can be detected by external (i.e., noninvasive) methods. Short-lived labeled compounds are usually prepared by chemical synthesis, but these methods are relatively slow and yield racemic mixtures. Enzymic synthesis is usually more rapid and specific, but the end products may contain antigenic and pyrogenic substances. Cohen *et al.* (*38*) utilized immobilized enzymes to minimize product contamination and shorten reaction times. ^{13}N-L-Glutamic acid was synthesized from ^{13}N-NH$_3$ and α-ketoglutarate using glutamate dehydrogenase immobilized on an N-hydroxysuccinimide derivative of silica beads. The glutamate was formed by rapidly passing the substrate solution through a column reactor. Pyruvate was added to the eluate which was then rapidly passed through a second column containing an immobilized transaminase. The eluate from the second column contained ^{13}N-L-glutamate and ^{13}N-L-alanine which were separated by subsequent ion-exchange chromatography. The entire two-step alanine synthesis was performed in about 4 minutes as opposed to 10 minutes

240   BERNARD J. ABBOTT

when soluble enzymes were used. Product yields from the column reactor also were higher. Higher yields and shorter reactions were undoubtedly due to the relatively large amounts of immobilized enzymes that were used. The advantage conferred by immobilization was that these large amounts could be used without increasing the pyrogenicity and antigenicity of the product.

V.   Other Products

A.   STEROIDS

Microorganisms catalyze numerous biotransformation reactions with steriod substrates (28). Many of the reaction products are very difficult to obtain through chemical modifications and can only be made in high yield by fermentation with the appropriate microorganism. Most of the products formed in these fermentations arise by a one-step enzymic modification. Despite the involvement of only one enzyme, large-scale steroid transformations do not employ cell-free enzyme preparations. One reason is that many steroid-modifying enzymes are oxidoreductases that require a cofactor which must be continuously regenerated to sustain the reaction. Thus, successful implementation of an immobilized enzyme for steroid transformation must include a mechanism for cofactor regeneration. One mechanism was examined in a steroid reduction reaction catalyzed by a 20-β-steroid dehydrogenase. This enzyme utilizes NADH to reduce a C-20 keto group of a steroid to a C-20β-alcohol. Kaplan *et al.* (7) proposed regenerating NADH by coupling steroid reduction with the enzymic oxidation of ethanol to acetate (Fig. 13). The three

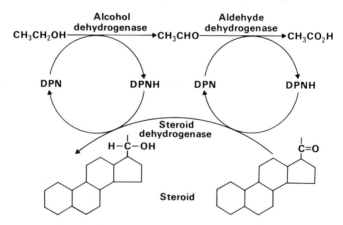

FIG. 13. Coupled immobilized enzyme reactions for steroid reductions. From (7). Reprinted with permission of the copyright owner, The American Chemical Society, from the February 25, 1974 issue of *Chemical and Engineering News*.

enzymes, 20-β-dehydrogenase, ethanol dehydrogenase, and acetaldehyde dehydrogenase, were coupled to glass beads with glutaraldehyde. A preliminary report (7) indicated that these enzymes, in a small batch reactor containing NAD and ethanol, catalyzed the reduction of prednisolone.

Another approach to the cofactor problem is to immobilize the intact cell of the microorganisms and utilize intracellular enzyme systems for cofactor regeneration. Soviet investigators reported that they immobilized cells of *Mycobacterium globiforme* and *Tieghemella orchedis* in polyacrylamide gel (124,125). The entrapped microorganisms catalyzed $\Delta^{1,2}$-dehydrogenations, C-20 ketone reductions, deacetylations, and 11α- or 11β-hydroxylations.

Mosbach and Larsson (94,95) used cells of *Curvularia lunata* entrapped in polyacrylamide gel to produce hydrocortisone by 11β-hydroxylation of Reichstein's Compound S (11-deoxy-17-hydrocorticosterone). The entrapped cells lost part of their hydroxylase activity during storage. The lost activity was recovered by reactivating (incubating) the entrapped cells for 16 hours in a nutrient solution containing Compound S. The restoration of activity was probably due to growth of the microorganisms in the gel and/or the induction by Compound S of higher levels of 11β-hydroxylase.

The rate of hydrocortisone production by the entrapped cells was slow. In a 20 ml solution containing 5 gm wet gel granules, 2.0–2.9 μmoles of cortisone was formed per hour. Another limitation of the system was that concentrations of substrate (Reichstein's Compound S) above 0.21 mg/ml lowered the reaction rate.

In the same report (95) Mosbach and Larsson described the entrapment of a Δ^1 steroid dehydrogenase in polyacrylamide gel. The enzyme was obtained from cells of *Corynebacterium simplex* and partially purified by ammonium sulfate fractionation. The entrapped enzyme catalyzed the conversion of hydrocortisone to prednisolone. Phenazine methosulfate was added to the reaction mixture as an electron acceptor. The gel-entrapped enzyme, which exhibited only 7% of the activity of the free enzyme, produced prednisolone at a rate of 9 μmoles/hour/gm dry polymer.

Vieth and co-workers studied prednisolone production by cells of *C. simplex* entrapped in a collagen matrix (151). The matrix, which contained 52% by weight cells, was cut into small chips and packed into a column. Operating conditions were varied to maximize prednisolone production. The presence of 15% ethanol in the substrate solution greatly increased product yields; the increase was probably due to higher solubility of the substrate (hydrocortisone) in the ethanol solution. Optimum residence time in the column was reported to be 0.5 hour, and maximum conversion to prednisolone was obtained with 0.7 mg/ml hydrocortisone. The column could be operated continuously for 5.5 days before losing one-half of the initial activity.

B. N-OXIDATION

Many clinically useful drugs that contain amine- or hydrazine-functional groups are enzymically oxidized by a flavoprotein from hepatic microsomes. The enzyme is a mixed function oxidase that requires NADPH and oxygen. It exhibits broad specificity and catalyzes the N-oxidation of lipid-soluble primary and secondary amines that do not contain a polar functional group near the nitrogen. Sofer et al. (130) immobilized a flavoprotein oxidase in order to produce N-oxidized drug metabolites. These metabolites are difficult to obtain by chemical syntheses and only small amounts can be isolated after in vivo administration of a drug.

The enzyme was immobilized by binding to nylon tubing, Sepharose, or glass beads (130). In addition to the substrate, the reaction mixtures contained NADP, isocitrate, and isocitrate dehydrogenase. The latter enzyme regenerated the NADPH consumed by the flavoprotein oxidase. The rate of ethylmorphine oxidation by the nylon-immobilized enzyme increased at high flow rates, suggesting the reaction was limited by mass transfer. The highest reaction rate was 120 μmoles/minute/meter of tubing. The Sepharose-immobilized enzyme and the enzyme immobilized on glass beads catalyzed ethylmorphine oxidation at rates of 30 μmoles/minute/ml gel and 60 μmoles/minute/gm dry beads, respectively. Among the other substrates oxidized were chlorpromazine, prochlorperzine, bromphenylamine, and N,N-dimethylaniline. Almost complete conversion of the substrates was obtained by making multiple passes through the reactors. In one experiment 45 μmoles ($\approx$16 mg) of crystalline ethylmorphine-N-oxide was obtained from 20 ml of reaction solution.

C. POLY I:C

Poly I:C is a synthetic ribonucleic acid that is a potent inducer of interferon. The polymer consists of paired homopolymer strands of polyriboinosinic acid and polyribocytidylic acid. Hoffman et al. (66) attempted to produce poly I:C in relatively large quantities for pharmacological and clinical testing. The homopolymer strands were synthesized from nucleotide-5'-diphosphates with a polynucleotide phosphorylase. However, difficulties were encountered in separating the polynucleotide from the enzyme. In addition, the enzyme, which is expensive, was destroyed by the isolation process. To circumvent these problems the enzyme was bound to cyanogen bromide-activated Sepharose. The immobilized phosphorylase retained 26% of the activity of the free enzyme and lost no activity over a period of 6 months of storage and intermittent use. The intermittent uses totaled more than 40 consecutive polymerization cycles. Since the enzyme was easily recovered after each use by centrifugation, it did not interfere with the isolation of the product.

D. Protein Hydrolysis

The presence of protein contaminants in antibiotic preparations may cause allergenic reactions. Protein can be removed from antibiotic solutions by treatment with proteases. The proteases themselves are antigenic, but only small amounts are needed and the enzymes destroy themselves by their own proteolytic activity. A more effective way of removing unwanted protein is by treatment with immobilized proteases. Immobilized proteases offer several advantages: (a) large amounts of protease can be used to achieve short reaction times without concern for residual enzyme remaining in solution after hydrolysis; (b) in fixed bed reactors, the protease molecules do not hydrolyze one another; and (c) the immobilized proteases may be reused many times.

Shaltiel and Sela (119) utilized immobilized pronase to reduce the antigenicity of penicillin solutions. Pronase, a protease mixture produced by *Streptomyces griseus*, was selected because of its broad specificity. The enzymes were coupled to bromoacetylcellulose and used to hydrolyze protein in commercial preparations of benzylpenicillin, ampicillin, and 6-APA. *In vivo* testing in rabbits revealed that treatment with immobilized pronase greatly reduced the immunogenicity of the antibiotic.

E. Urocanic Acid

Urocanic acid is useful as a pharmaceutical and cosmetic agent because of its sun-screening properties. The compound can be produced from L-histidine by the enzyme L-histidine ammonia lyase (Table VIII). Many bacteria produce this enzyme, but *Achromobacter liquidum* is the most suitable for commercial production (160). Yamamoto et al. (36,160) attempted to improve the process for urocanic acid production by utilizing immobilized cells. Seven microorganisms that produce histidine ammonia lyase were entrapped in polyacrylamide gels (160). A comparison of urocanic acid production by the entrapped cells revealed that the highest lyase activity was obtained with *A. liquidum*. The entrapped cells displayed 64% of the lyase activity of freely suspended microorganisms. *Achromobacter liquidum* also produced the enzyme urocanase, but

TABLE
SYNTHESIS OF VARIOUS PHARMACEUTICAL

Reaction catalyzed
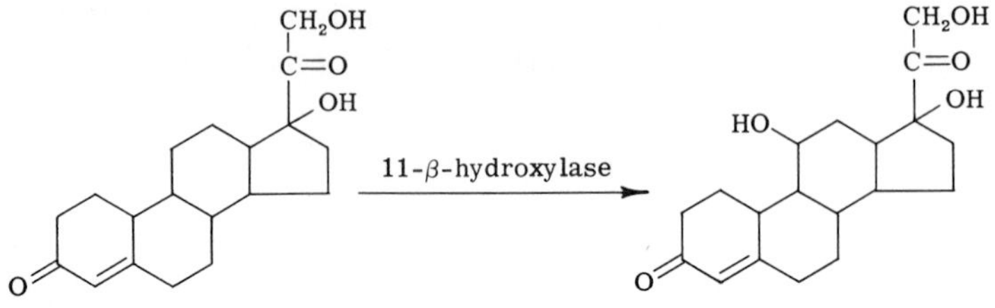 CH₂OH C=O OH 11-β-hydroxylase HO O
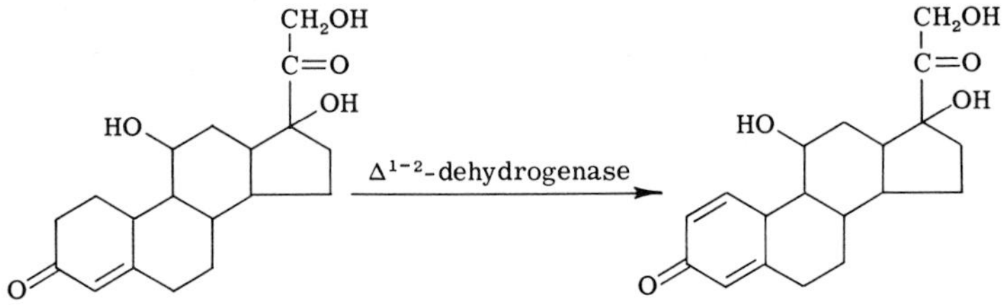 CH₂OH C=O OH HO Δ1-2-dehydrogenase O
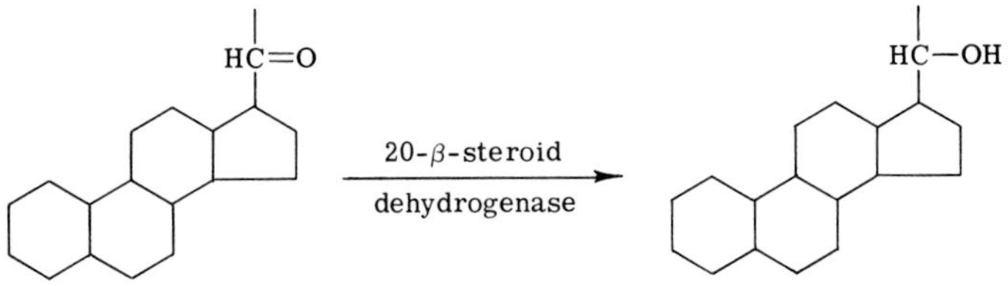 HC=O 20-β-steroid dehydrogenase HC—OH
Inosine diphosphate, Cytosine diphosphate —polynucleotide phosphorylase→ Polyriboinosinic acid, Polyribocytidylic acid
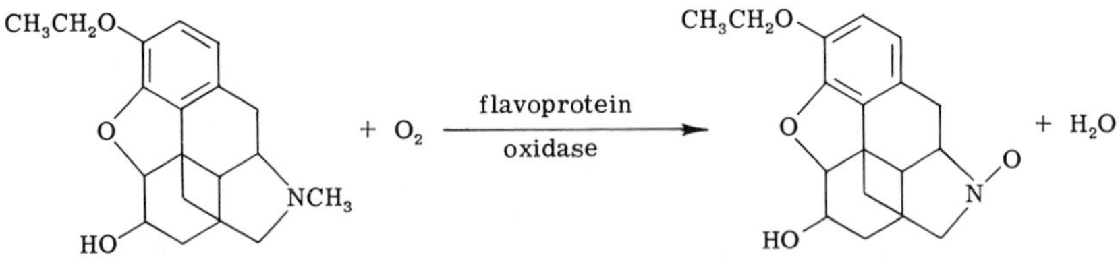 CH₃CH₂O O HO NCH₃ + O₂ flavoprotein oxidase N O + H₂O
Protein —pronase→ Amino acids

VIII
COMPOUNDS BY IMMOBILIZED ENZYMES

Enzyme source	Immobilization method	Remarks	Company or institution	References
Curvularia lunata	Entrapment of whole cells in polyacrylamide gel	Activity was 6 μmoles hydroxylated/hour/gm polymer	University of Lund, Sweden	*94, 95*
Corynebacterium simplex	Entrapment of enzyme or whole cells in polyacryl amide gel	Activity was 9 μmoles dehydrogenated/hour/gm polymer	University of Lund, Sweden	*94, 95*
Corynebacterium simplex	Entrapment of cells in a collagen matrix	A column reactor lost $\frac{1}{2}$ its activity during 5 days of operation	Rutgers University	*151*
—	Coupled to glass beads with glutaraldehyde	—	University of California	*7*
Micrococcus luteus	Coupling to cyanogen bromide activated cellulose	Complex was used for 40 polymerization cycles	Merck Sharp & Dohme	*66*
Pig liver microsomes	Binding to nylon, sepharose, or glass	The enzyme also catalyzed the N-oxidation of other compounds	University of Texas	*130*
Streptomyces griseus	Coupling to bromoacetyl-cellulose	Enzyme was used to lower the antigenicity of penicillins	Beecham	*119*

(Continued)

TABLE VIII

Reaction catalyzed

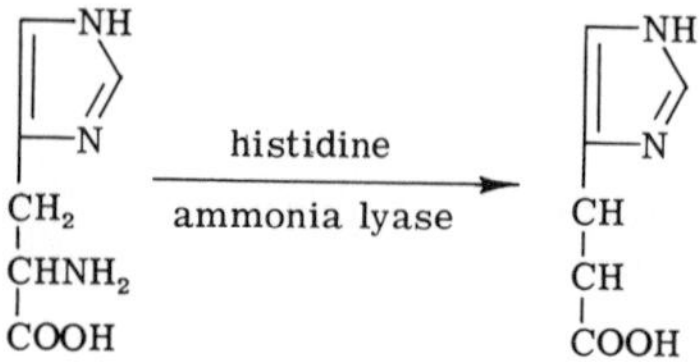

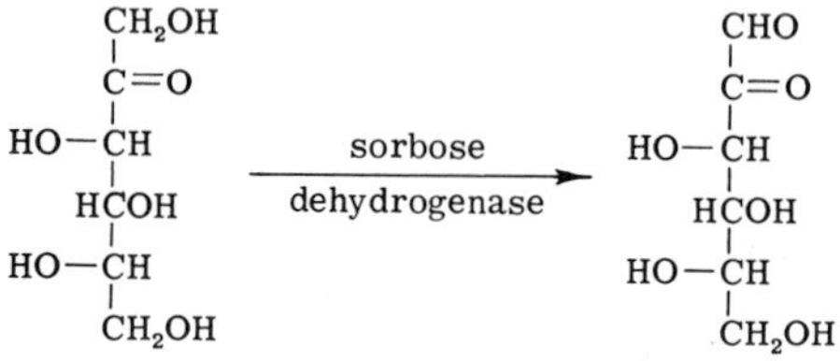

$$\text{HOOCCH}=\text{CHCOOH} \ + \ H_2O \xrightarrow{\text{fumerase}} \text{L-HOOCCH}_2\overset{\overset{\displaystyle OH}{|}}{\text{CHCOOH}}$$

it was possible to selectively inactivate this enzyme by heating the cells for 30 minutes at 70°C. Cells lacking urocanase activity produced urocanic acid in 100% molar yield.

The gel-entrapped microorganisms were packed into a column and a 0.25 M L-histidine solution was passed through at a flow rate of 0.06 ml/ml bed volume/hour (160). The column lost no lyase activity during 40 days of continuous operation. About 91% of the L-histidine passed through was isolated from the effluent as crystalline urocanic acid.

F. Sorbosone

Sorbosone a potential intermediate for the synthesis of vitamin C, can be made from sorbose by *Gluconobacter melanogenus* (79). Martin and Perlman studied the immobilization of *G. melanogenus* cells as a method to achieve continuous production of L-sorbosone (86,87). The cells were entrapped in polyacrylamide gels of different composition and tested for their ability to produce sorbosone (86). Entrapment resulted in a loss of at least 50% of the sorbose dehydrogenase activity. Immobilization did not alter the pH or temperature optima, but the entrapped cells dis-

(*Continued*)

Enzyme source	Immobilization method	Remarks	Company or institution	Refer-ences
Achromobacter liquidum	Cell entrapped in polyacrylamide gel	A column was used for 40 days without loss of activity	Tanabe Seiyaku	*36, 160*
Gluconobacter melanogenus	Cells entrapped in polyacrylamide gel	–	University of Wisconsin	*86, 87*
Brevibacterium ammonia-genes	Cells entrapped in polyacrylamide gel	Column half-life was 2 months	Tanabe Seiyaku	*37, 90*
Pseudomonas sp.	Enzyme entrappd in polyacrylamide gel	Column lost no activity for 60 days	SNAM Progetti	*84,93*

played improved thermal stability when heated in the presence of sorbose. Sorbosone production by the immobilized cells increased during the first 40 hours of incubation and then rapidly decreased so that by 100 hours the sorbosone production rate was quite low. In column reactors, the initial rate of sorbosone production increased when the flow rate was increased. The higher production rates were attributed to an increased availability of dissolved oxygen (*87*). Higher flow rates also caused a more rapid loss of enzyme activity. Sorbose dehydrogenase was inactivated by low concentrations (0.2%) of sorbosone, but inactivation was prevented by high concentrations (20%) of sorbose. In the presence of 20% sorbose, 310 mg of entrapped cells accumulated >450 mg of sorbosone during 360 hours of incubation (*87*).

G. Malic Acid

Malic acid is used for intravenous infusion mixtures and for the treatment of hepatic malfunctions. The Tanabe Seiyaku Company has developed a commercially operating immobilized cell process for making

L-malic acid from fumaric acid (*90,123*). The reaction is catalyzed by the enzyme fumarase which is an intracellular enzyme found in many microorganisms. The Tanabe Seiyaku process (Table VIII) employs a strain of *Brevibacterium ammoniagenes* that is entrapped in poly-acrylamide gel (*37,90*). The microorganism also contains the enzyme succinic dehydrogenase, which catalyzes an unwanted side reaction: the reduction of fumarate to succinate. The latter reaction could be inhibited by pretreating the cells with bile extract or deoxycholic acid, and the treatment also stimulated the rate of L-malic acid production. A large increase in production rate ($\sim$10-fold) was achieved by adding surfactants to the reaction mixture. This increase was probably due to increased cell permeability caused by disruption of the cell membrane. By treatment with bile extract and a surfactant, production rates as high as 7.7 mmoles/hour/gm cells were obtained (*90*). In continuously operated column reactors the entrapped cells had a half-life of about 2 months. The columns produced L-malic acid in about 70% yield from 1 M sodium fumarate solutions (*90*).

Morisi *et al.* (*84,93*) developed several immobilized enzyme processes for producing L-malic acid. These processes utilized enzyme preparations from swine tissue, a *Pseudomonas* sp., or from *Paracolibacterium aerogenides* (*93*). In one example, a partially purified fumarase preparation from a pseudomonad was entrapped in cellulose triacetate fibers. The fibers (70 gm) were packed into a 3 $\times$ 60-cm column through which a 0.5 M ammonium fumarate solution was passed at 50 ml/hour. The column converted about 60% of the fumarate to L-malic acid and operated continuously for 60 days with no noticeable loss of activity (*93*).

H. Coenzyme A

Shimizu *et al.* (*121*) studied the production of Coenzyme A (CoA) from pantothenic acid by immobilized cells of *Brevibacterium ammoniagenes*. The study was undertaken to evaluate the utility of immobilized cells for catalyzing sequential multienzyme reactions. The reaction sequence consisted of five steps: pantothenic acid → phosphopantothenic acid → phosphopantothenoylcysteine → phosphopantetheine → dephospho-CoA → CoA. Air-dried cells were immobilized by (a) entrapping in polyacrylamide gel, (b) containment within cellophane tubing, (c) entrapping in a silastic resin, (d) covalent attachment to a copolymer of ethylene-maleic anhydride, and (e) cross-linking with glutaraldehyde. Only the cells in cellophane tubing or entrapped in polyacrylamide gel exhibited activity after immobilization (*121*). The cells in the tubing were used for 7 sequential 10-hour batch reactions. The amount of CoA synthesized increased after the first reaction and then steadily declined during subsequent uses. The initial increase was probably due to lysis of the microorganism.

The gel-entrapped cells (7.1 gm dried cells) were incubated in a 100 ml solution containing pantothenate, cysteine, ATP, MgSO₄, and phosphate buffer. After 8 hours the gel was transferred to a fresh reaction mixture and incubated an additional 8 hours. The process was repeated until the gel had been used for four batch reactions. A total of 213 mg of CoA was produced in these experiments. When the gel was packed into a column, the rate of CoA production declined about 50% after 5 days of continuous operation (121).

VI. Concluding Remarks

Immobilized enzymes and cells can be used to prepare many pharmaceutical compounds. Of the examples cited, at least six have been conducted on a commercial scale. These six are the bentonite-immobilized penicillin acylase process developed by Squibb, the Sepharose-immobilized penicillin acylase used by Astra Lakemedel to produce 6-APA, the Sepharose-immobilized amino acid acylase used by the Tanabe Seiyaku Company to resolve racemic amino acid mixtures, the Tanabe Seiyaku Company's immobilized cell processes for synthesizing L-aspartic and L-malic acid, and a Kyowa Hakko Kogyo process for L-aspartic acid. There is also speculation that the Toyo Jozo Company operates an immobilized cell process for producing cephalexin by acylation of 7-ADCA. Three of these processes use immobilized cell systems. The vast potential of immobilized cells is just being fully recognized. This potential is illustrated by the production of glutamic acid from glucose by *C. glutamicum* entrapped in polyacrylamide gel and by the production of CoA by gel-entrapped *B. ammoniagenes* cells. Results with these systems demonstrate that a complex metabolic pathway can be maintained intact and that the cells can continuously regenerate the cofactors needed to sustain the operation of the pathways. If immobilized cells can synthesize glutamic acid and CoA, perhaps other immobilized microorganisms can make other amino acids, nucleotides, or anaerobic products, e.g., ethanol, or even secondary metabolites such as antibiotics. More extensive optimization studies are needed and more effective immobilization methods must be developed to fully evaluate those possibilities.

A compilation of immobilized cell systems (Table IX) reveals that very few immobilization methods were used relative to the large number that have been employed with cell-free enzymes. Notably absent are methods for the attachment of cells to insoluble supports by covalent bonds. Cells covalently immobilized should be more stable to dissociation than ionically bound cells and less restricted by substrate and product diffusion

TABLE IX

IMMOBILIZED CELL SYSTEMS FOR SYNTHESIZING PHARMACEUTICAL COMPOUNDS

Substrate	Product	Microorganism	Immobilization method	Company or institution	References
Penicillin G	6-APA	E. coli	Entrapment in PA[a] gel	Tanabe Seiyaku	31
6-APA	Ampicillin	B. megaterium	Adsorption on DEAE-cellulose	Toyo Jozo	56
7-ADCA	Cephalexin	Achromobacter sp.	Adsorption to DEAE-cellulose	Toyo Jozo	145
Fumarate	Aspartate	E. coli	Entrapment in PA gel	Tanabe Seiyaku	34, 139, 140
Arginine	Citrulline	P. putida	Entrapment in PA gel	Tanabe Seiyaku	35, 159
Indole	Tryptophan	E. coli	Entrapment in PA gel	Tanabe Seiyaku	29
Compound S	Hydrocortisone	C. lunata	Entrapment in PA gel	University of Lund	94, 95
Hydrocortisone	Prednisolone	C. simplex	Entrapment in collagen	Rutgers University	151
Glucose	Glutamic acid	C. glutamicum	Entrapment in PA gel	New England Enzyme Institute and Tufts University	126
Histidine	Urocanic acid	A. liquidum	Entrapment in PA gel	Tanabe Seiyaku	36, 160
Sorbose	Sorbosone	G. melanogenus	Entrapment in PA gel	University of Wisconsin	86, 87
Fumarate	Malate	B. ammoniagenes	Entrapment in PA gel	Tanabe Seiyaku	37, 90

[a] PA = polyacrylamide.

than cells entrapped in a polymer. Development of good methods of covalent cell immobilization may be lagging because, ideally, these methods should attach cells by their walls without interaction with other cell constituents. To accomplish this objective, methods for chemical or enzymic attachment must be developed that depend on a specific interaction with a unique wall component or functional group.

Some future directions of immobilization technology can be seen from research projects currently under way at the University of Pennsylvania and at MIT. Both projects involve immobilization of enzyme systems that require cofactors. At the University of Pennsylvania, a process is being developed for the synthesis of chenodesoxycholic acid from cholic acid (E. K. Pye, personal communication). Chenodesoxycholic acid, which may be used for the dissolution of gallstones, can be made from cholic acid by a complex 8-step chemical synthesis. The synthesis can be simplified by a coupling of two chemical reactions with two enzymic reductions. The proposed sequence starts with a chromic acid oxidation of cholic acid to 3,7,12-dehydrocholic acid. The latter compound is reduced with immobilized 3-keto and 7-ketoreductases. The resultant 12-dehydrocholic acid is then subjected to a Wolff–Kishner reaction to produce chenodesoxycholic acid. Microorganisms that produce the reductases have been isolated, and methods for immobilizing the enzymes involved are being developed. Since both enzymes require reduced pyridine nucleotides, methods for regenerating the cofactor are also being investigated.

A group at MIT is developing an immobilized enzyme system for the synthesis of the cyclic decapeptide antibiotic gramicidin S. This antibiotic is synthesized by two enzyme fractions from *Bacillus brevis*. The synthesis of 1 mole of gramicidin S requires 10 moles of ATP. Thus, a major part of the research program involves the regeneration of ATP consumed in the reaction. One approach being studied is to enzymically react AMP with ATP to produce ADP. The ADP can then be converted to ATP with acetyl kinase using chemically prepared acetyl phosphate as the phosphate donor. The successful development of this technology may provide an important impetus for the commercial exploitation of other ATP-requiring enzymes.

Of the many enzymes that have been immobilized, there are only two examples of the immobilization of mixed-function oxygenases. One of these is the flavoprotein oxygenase used by Sofer *et al.* (*130*) for the N-oxidation of drugs, and the other is the 11β-hydroxylase immobilized by gel entrapment of *C. lunata cells*. Mixed-function oxygenases may have wide utility in synthetic bioorganic chemistry because they catalyze a variety of oxidative transformations. These oxidations include aliphatic and aromatic hydroxylations, N- and O-dealkylations, sulfoxidations, and epoxidations. Exploitation of these enzymes via immobilization is hampered by the complexity of the reaction mechanism, the instability of the enzymes, and cofactor requirements. A complete system consists of the oxygenase, one or two electron transport proteins, a reduced cofactor, substrate, and molecular oxygen. As an additional complexity, immobilization of this system must be done in such a way as to permit the interaction of proper juxtaposition of the three protein constituents for electron transfer to occur. If the enzymes were sufficiently stable, they could be immobilized by containment in semipermeable membranes. More likely, commercial exploitation of mixed-function oxygenases will only be achieved by immobilization of the entire cell or cell constituents (e.g., membrane fragments or microsomes) that contain the enzyme.

ACKNOWLEDGMENTS

The author gratefully acknowledges the helpful comments of Dr. R. Fuller, Dr. O. Zaborsky, and Dr. S. May, and the excellent secretarial assistance of Ms. P. Hager.

REFERENCES

1. Abbott, B. J., and Fukuda, D. (1975). *Appl. Microbiol.* **28**, 58–63.
2. Abbott, B. J., and Fukuda, D. (1975). *Abstr., 75th Annu. Meet., Am. Soc. Microbiol.* p. 195.
3. Abbott, B. J., and Fukuda, D. (1975). *Antimicrob. Agents & Chemother.* **8**, 282–288.
4. Abbott, B. J., Fukuda, D., and Cerimeli, B. (1974). *Acad. Pharm. Sci., 17th Natl. Meet., 1974* p. 88.
4a. Abbott, B. J., Cerimele, B., and Fukuda, D. (1976). *Biotechnol. Bioeng.* **18**, 1033–1042.
5. Abe, J., Watanabe, T., Yamaguchi, T., and Matsumoto, K. (1973). U.S. Patent 3,761,354.
6. Allison, J. P., Davidson, L., Gutierrez-Hartman, A., and Kitto, G. B. (1972). *Biochem. Biophys. Res. Commun.* **47**, 66–71.
7. Anonymous. (1974). *Chem. & Eng. News* **52**, 19–20.
8. Axén, R., and Ernback, S. (1971). *Eur. J. Biochem.* **18**, 351–360.
9. Bachler, M. J., Stranberg, G. W., and Smiley, K. L. (1970). *Biotechnol. Bioeng.* **12**, 85–93.
10. Balcom, J., Foulkes, P., Olson, N. F., and Richardson, T. (1971). *Process Biochem.* **6**, 42–44.
11. Ballio, A., Chain, E. B., Dentice di Accadia, F., Mauri, M., Rauer, K., Schlesinger, K. J., and Schlesinger, S. (1961). *Nature (London)* **191**, 909–910.
12. Batchelor, F. R., Doyle, F. P., Nayler, J. H. C., and Rolinson, G. N. (1959). *Nature (London)* **183**, 257–258.
13. Batchelor, F. R., Dewedney, J. M., Feinberg, J. G., and Weston, R. D. (1967). *Lancet* **1**, 1175–1177.
14. Bayer AG. (1975). German Patent 2,355,079.
15. Beecham Group Ltd. (1974). German Patent 2,356,630.
16. Bergmeyer, H. U., and Michal, G. (1974). *In* "Industrial Aspects of Biochemistry" (B. Spencer, ed.), Vol. 30, pp. 187–211. Elsevier, New York.
17. Blatt, L. M., and Kim, K. H. (1971). *J. Biol. Chem.* **246**, 4895–4898.

18. Boguslaski, R. C., Smith, R. S., and Mhatre, N. S. (1972). *Curr. Top. Microbiol. Immunol.* **58**, 1–68.
19. Bondareua, N. S., Levitov, M. M., and Rabinovich, M. S. (1969). *Biokhimiya* **34**, 478–482.
20. Brown, H. D., Chattopadhyay, S. K., and Patel, A. (1966). *Biochem. Biophys. Res. Commun.* **25**, 304–308.
21. Brown, H. D., and Hasselberger, F. X. (1971). *In* "Chemistry of the Cell Interface" (H. D. Brown, ed.), Part B, pp. 85–258. Academic Press, New York.
22. Butterworth, T. A., Wang, O. I. C., and Sinskey, A. J. (1970). *Biotechnol. Bioeng.* **12**, 615–631.
23. Carrington, T. R. (1971). *Proc. R. Soc. London, Ser. B* **179**, 321–333.
24. Chang, T. M. S. (1966). *Trans. Am. Soc. Artif. Intern. Organs* **12**, 13–19.
25. Chang, T. M. S. (1971). *Nature (London)* **229**, 117–118.
26. Chang, T. M. S. (1972). *In* "Enzyme Engineering" (L. B. Wingard, ed.), pp. 395–401. Wiley (Interscience), New York.
27. Chang, T. M. S., and Pozansky, M. J. (1968). *Nature (London)* **218**, 243–245.
28. Charney, W., and Herzog, H. L. (1967). "Microbial Transformations of Steroids." Academic Press, New York.
29. Chibata, I., Kakimoto, T., and Nabe, K. (1974). Japanese Patent 7481591.
30. Chibata, I., Kakimoto, T., and Nabe, K. (1974). Japanese Patent 7481590.
31. Chibata, I., Osaka, S., Tosa, T., Sato, T., and Tadashi, T. (1974). German Patent 2,414,128.
32. Chibata, I., Tosa, T., Sato, T., Mori, T., and Matsuo, Y. (1972). *In* "Fermentation Technology Today" (G. Terui, ed.), pp. 383–389. Soc. Ferment. Technol. Osaka, Japan.
33. Chibata, I., Tosa, T., and Sato, T. (1974). Japanese Patent 7425189.
34. Chibata, I., Tosa, T., and Sato, T. (1974). *Appl. Microbiol.* **27**, 878–885.
35. Chibata, I., Tosa, T., Sato, T., and Yamamoto, K. (1974). Japanese Patent 7426486.
36. Chibata, I., Tosa, T., Sato, T., and Yamamoto, K. (1974). Japanese Patent 7461390.
37. Chibata, I., Tosa, T., Sato, T., and Yamamoto, K. (1975). German Patent 2,450,137.
38. Cohen, M. B., Spolter, L., Chang, C. C., MacDonald, N. S., Takahashi, J., and Bobinet, D. D. (1974). *J. Nucl. Med.* **15**, 1192–1195.
39. Cole, M. (1966). *Appl. Microbiol.* **14**, 98–104.
40. Cole, M. (1969). *Biochem. J.* **115**, 747–756.
41. Cole, M. (1969). *Biochem. J.* **115**, 757–764.
42. Cooper, R. D. G., and Spry, D. O. (1973). *In* "Cephalosporin and Penicillin Compounds: Their Chemistry and Biology" (E. H. Flynn, ed.), pp. 183–254. Academic Press, New York.
43. Delin, P. S., Ekström, B., Sjöberg, B., Thelin, K. H., and Nathorst-Westfelt, L. S. (1972). German Patent 1,966,428.
44. Delin, P. S., Ekström, B., Sjöberg, B., Thelin, K. H., and Nathorst-Westfelt, L. S. (1973). U.S. Patent 3,736,230.
45. Dinelli, D. (1972). *Process Biochem.* **7**, 9–12.
46. Dinelli, D., and Morisi, F. (1971). Belgian Patent 782,646.
47. Dinelli, D., and Morisi, F. (1973). French Patent 2,134,530.
48. Ekström, B., Lagerlof, E., Nathorst-Westfelt, L., and Sjöberg, B. (1973). *Int. Congr. Pharm. Sci., 1973.*

49. Ekström, B., Lagerlof, E., Nathorst-Westfelt, L., and Sjöberg, B. (1974). *Sven. Farm. Tidskr.* **78**, 531–535.

50. Enei, H., Nakazawa, H., Matsui, H., Okumura, S., and Yamada, H. (1972). *FEBS Lett.* **21**, 39–41.

51. Falb, R. A., and Grode, G. A. (1971). *Fed. Proc., Fed. Am. Soc. Exp. Biol.* **30**, 1688.

52. Fink, D. J., and Rodwell, V. W. (1975). *Biotechnol. Bioeng.* **17**, 1029–1050.

53. Fleming, I. D., Turner, M., and Napier, E. J. (1974). German Patent 2,422,374.

54. Franks, N. T. (1971). *Biochim. Biophys. Acta* **252**, 246–254.

55. Franks, N. E. (1972). *Biotechnol. Bioeng. Symp.* **3**, 327–339.

56. Fujii, T., Hanamitsu, K., Izumi, R., Yamaguchi, T., and Watanabe, T. (1973). Japanese Patent 7399393.

57. Fukui, S., and Ikeda, S. (1975). *Process Biochem.* June, 3–8.

58. Fukui, S., Ikeda. S., Fujimura, M., Yamada, H., and Kumagai, H. (1975). *Eur. J. Biochem.* **51**, 155–164.

59. Fukui, S., Ikeda, S., Fujimura, M., Yamada, H., and Kumagai, H. (1975). *Eur. J. Appl. Microbiol.* **1**, 25–39.

60. Gassen, H. G., and Nolte, R. (1971). *Biochem. Biophys. Res. Commun.* **44**, 1410–1415.

60a. Gruber, W., and Bergmeyer, H. U. (1974). German Patent 1,935,711.

61. Guilbault, G. C.(1971). *Pure Appl. Chem.* **25**, 727–740.

62. Guilbault, G. G. (1974). *In* "Enzyme Engineering" (E. K. Pye and L. B. Wingard, eds.), Vol. II, pp. 377–383. Plenum, New York.

63. Hamilton-Miller, J. M. T. (1966). *Bacteriol. Rev.* **30**, 761–771.

64. Hasselberger, F. X., Brown, H. D., Chaitopadhyay, S. K. Mather, A. N. Stasiw, R. O., Patel, A. B., and Pennington, S. N. (1970). *Cancer Res.* **30**, 2736–2738.

65. Heuser, L. J., Chiang, C., and Anderson, C. F. (1969). U.S. Patent 3,446,705.

66. Hoffman, C. H., Harris, E., Chodroff, S., Micholson, S., Rothrock, J. W., Peterson, E., and Reuter, W. (1970). *Biochem. Biophys. Res. Commun.* **41**, 710–714.

67. Huber, F. M., Chauvette, R. R., and Jackson, B. G. (1973). *In* "Cephalosporin and Penicillin Compounds: Their Chemistry and Biology" (E. H. Flynn, ed.), pp. 27–73. Academic Press, New York.

68. Huper, F. (1973). German Patent 2,157,970.

69. Huper, F. (1973). German Patent 2,157,972.

70. Huper, F. (1974). German Patent 2,312,824.

71. Ikeda, S., and Fukui, S. (1973). *Biochem. Biophys. Res. Commun.* **52**, 482–488.

72. Ikeda, S., Hara, H., and Fukui, S. (1974). *Biochim. Biophys. Acta* **372**, 400–406.

72a. Jaworek, D., Gruber, W., and Bergmeyer, H. U. (1970). German Patent 1,908,290.

73. Jeffery, J. D'A., Abraham, E. P., and Newton, G. G. F. (1961). *Biochem. J.* **81**, 591–596.

74. Kaiser, G. V., and Kukolja, S. (1973). *In* "Cephalosporin and Penicillin Compounds: Their Chemistry and Biology" (E. H. Flynn, ed.), pp. 74–134. Academic Press, New York.

75. Kamogashira, T., Kawaguchi, T., Mivazaki, W., and Doi, T. (1972). Japanese Patent 7228190.

76. Katchalski, E., Silman, I., and Goldman, R. (1971). *Adv. Enzymol.* **34**, 445–536.

77. Kato, K. (1953). *Kagaku (Tokyo)* **23**, 217–218.

78. Kaufman, W., Bauer, K., and Offe, H. A. (1961). *Antimicrob. Agents Annu.* pp. 1–5.
79. Kitamura, I., and Perlman, D. (1975). *Biotechnol. Bioeng.* 17, 349–359.
80. Kukolja, S. (1968). *J. Med. Chem.* 11, 1067–1069.
81. Mandel, M., Koestner, A., Silmer, E., Kleiner, G. I., Elizarovskaya, L. M., and Stamer, V. (1975). *Prikl. Biokhim. Mikrobiol.* 11, 219–225.
82. Marconi, W., Bartoli, F., Cecere, F., Galli, G., and Morisi, F. (1975). *Agric. Biol. Chem.* 39, 277–279.
83. Marconi, W., Bartoli, F., Cecere, F., and Morisi, F. (1974). *Agric. Biol. Chem.* 38, 1343–1349.
84. Marconi, W., Morisi, F., and Mosti, R. (1975). *Agric. Biol. Chem.* 39, 1323–1324.
85. Marconi, W., Cecere, F., Morisi, F., Della Penna, G., and Rappuoli, B. (1973). *J. Antibiot.* 26, 228–232.
86. Martin, C. K. A., and Perlman, D. (1975). *Biotechnol. Bioeng.* 17, 1473–1485.
87. Martin, C. K. A., and Perlman, D. (1975). *Biotechnol. Bioeng.* (in press).
88. Matsui, M., Yoneya, T., and Nagura, M. (1974). Japanese Patent 74101614.
89. Matsushima, H., Murata, K., and Mase, Y. (1974). *J. Ferment. Technol.* 52, 431–437.
90. McAbee, M. K. (1975). *Chem. & Eng. News* 53, 32.
91. Morin, R. B., Jackson, B. G., Mueller, R. A., Lavagnino, E. R., Scanlon, W. B., and Andrews, S. L. (1969). *J. Am. Chem. Soc.* 91, 1401–1407.
92. Morino, Y., and Snell, E. E. (1967). *J. Biol. Chem.* 242, 2800–2809.
93. Morisi, F., Mosti, R., and Bettionte, M. (1974). German Patent 2,415,310.
94. Mosbach, K. (1971). *Sci. Am.* 224, 26–33.
95. Mosbach, K., and Larsson, P. (1970). *Biotechnol. Bioeng.* 12, 19–27.
96. Murphy, C. F., and Webber, J. A. (1973). In "Cephalosporin and Penicillin Compounds: Their Chemistry and Biology" (E. H. Flynn, ed.), pp. 134–182. Academic Press, New York.
97. Nara, T., Okachi, R., and Kato, F. (1962). *Proc. Int. Fermentation Symp.,* 4th, 1972 Abstracts, p. 207.
98. Nara, T., Okachi, R., and Misawa, M. (1971). *J. Antibiot.* 24, 321–323.
99. Nara, T., Misawa, M., Okachi, R., and Yamamoto, M. (1971). *Agric. Biol. Chem.* 35, 1676–1682.
100. Nernst, W. Z. (1904). *Phys. Chem.* 47, 52–62.
101. Newton, W. A., Morino, Y., and Snell, E. E. (1965). *J. Biol. Chem.* 240, 1211–1218.
102. Nilsson, H., Mosbach, R., and Mosbach, K. (1972). *Biochim. Biophys. Acta* 268, 253–256.
103. Nuesch, J., Riehen, and Bickel, H. (1967). U.S. Patent 3,304,236.
104. Okachi, R., Kato, F., Miyamura, Y., and Nara, T. (1973). *Agric. Biol. Chem.* 37, 1953–1957.
105. Okachi, R., and Nara, T. (1973). *Agric. Biol. Chem.* 37, 2797–2804.
106. Orth, H. D., and Brummer, W. (1972). *Angew. Chem., Int. Ed. Engl.* 11, 249–260.
107. Otsuka Seiyaku Company. (1972). Japanese Patent 7228187.
108. Otsuka Seiyaku Company. (1972). Japanese Patent 7228183.
109. Payne, J. W. (1973). *Biochem. J.* 135, 867–873.
110. Poznansfy, M. J., and Chang, T. M. S. (1974). *Biochim. Biophys. Acta* 334, 103–115.
111. Prader, A., and Auriccho, S. (1965). *Annu. Rev. Med.* 16, 345–358.

112. Rossi, D., Romeo, A., Lucente, G., and Tinti, O. (1973). *Farmaco, Ed. Sci.* **28**, 262–264.

113. Ryu, D. Y., Bruno, C. F., Lee, B. K., and Venkatasubramanian, K. (1972). *In* "Fermentation Technology Today" (G. Terui, ed.), pp. 307–314. Soc. Ferment. Technol., Osaka, Japan.

114. Savidge, T., Powell, L. W., and Warren, K. B. (1974). German Patent 2,336,829.

115. Savidge, T., Powell, L. W., and Carrington, T. R. (1969). German Patent 1,917,057.

116. Savidge, T., Powell, L. W., and Carrington, T. R. (1972). Canadian Patent 911353.

117. Savidge, T., Powell, L. W., and Warren, K. B. (1975). United States Patent 3883394.

118. Self, D. A., Kay, G., and Lilly, M. D. (1969). *Biotechnol. Bioeng.* **11**, 337–348.

119. Shaltiel, S., and Sela, M. (1973). U.S. Patent 3,770,584.

120. Sharp, A. K., Kay, G., and Lilly, M. D. (1969). *Biotechnol. Bioeng.* **11**, 363–380.

121. Shimizu, S., Morioka, H., Tani, Y., and Ogata, K. (1975). *J. Ferment. Technol.* **53**, 77–83.

122. Singh, K., Sehgal, S. N., and Vezina, C. (1969). *Appl. Microbiol.* **17**, 643–644.

123. Skinner, K. J. (1975). *Chem. & Eng. News* **53**, 22–41.

124. Skryabin, G. K., Koshchenko, K. A., Mogilnitski, G. M., Surovtsev, V. I., Turin, V. S., and Fikhte, B. A. (1974). *Izv. Akad. Nauk SSSR, Ser. Biol.* pp. 857–866.

125. Skryabin, G. K., Koshchenko, K. A., Surovtsev, V. I., Mogilnitski, G. M., Fikhte, B. A., and Tyurin, V. I. (1974). *J. Steroid Biochem.* **5**, 397.

126. Slowinski, W., and Charm, S. E. (1973). *Biotechnol. Bioeng.* **15**, 973–979.

127. Smiley, K. L., and Strandberg, C. W. (1972). *Adv. Appl. Microbiol.* **15**, 13–18.

128. SNAM Progetti. (1972). Belgian Patent 782,646.

129. SNAM Progetti. (1975). Belgian Patent 823,393.

130. Sofer, S. S., Ziegler, D. M., and Popovich, R. P. (1974). *Biochem. Biophys. Res. Commun.* **57**, 183–189.

131. Takahashi, T., Yamazaki, Y., Kato, K., and Isono, M. (1972). *J. Am. Chem. Soc.* **94**, 4035–4037.

132. Takeda, H., Matsumoto, I., and Matsuda, K. (1975). Japanese Patent 7513588.

133. Tanabe Seiyaku Company Limited. (1974). German Patent 2,252,888.

134. Tosa, T., Mori, T., Fuse, N., and Chibata, I. (1966). *Enzymologia* **31**, 214–224.

135. Tosa, T., Mori, T., Fuse, N., and Chibata, I. (1966). *Enzymologia* **31**, 225–238.

136. Tosa, T., Mori, T., Fuse, N., and Chibata, I. (1967). *Biotechnol. Bioeng.* **9**, 603–615.

137. Tosa, T., Mori, T., Fuse, N., and Chibata, I. (1967). *Enzymologia* **32**, 153–168.

138. Tosa, T., Mori, T., Fuse, N., and Chibata, I. (1969). *Agric. Biol. Chem.* **33**, 1047–1052.

139. Tosa, T., Sato, T., Mori, T., and Chibata, I. (1974). *Appl. Microbiol.* **27**, 886–889.

140. Tosa, T., Sato, T., Mori, T., Matuo, Y., and Chibata, I. (1973). *Biotechnol. Bioeng.* **15**, 69–84.

140a. Toyo Jozo Company. (1972). Belgian Patent 780,676.

141. Toyo Jozo Company. (1973). Belgian Patent 801,044.

142. Toyo Jozo Company. (1973). Belgian Patent 803, 832.

143. Toyo Jozo Company. (1973). British Patent 1,324,159.

144. Toyo Jozo Company. (1974). Japanese Patent 7447594.

145. Toyo Jozo Company. (1974). British Patent 1,347,665.
146. Vandamme, E. J., and Voets, J. P. (1974). *Meded. Fac. Landbouwwet., Rijks. Univ. Gent* **39**, 1463–1470.
147. Vandamme, E. J., and Voets, J. P. (1974). *Adv. Appl. Microbiol.* **17**, 311–369.
148. Vandamme, E. J., and Voets, J. P. (1975). *Experientia* **31**, 140–143.
149. Vandamme, E. J., Voets, J. P., and Dhaese, A. (1971). *Ann. Inst. Pasteur, Paris* **121**, 435–446.
150. Vanderhaeghe, H., Claesen, M., Vlietinck, A., and Parmentier, G. (1968). *Appl. Microbiol.* **16**, 1557–1563.
151. Venkatasubramanian, K., Vieth, W. R., and Constantinides, A. (1975). *75th Annu Meet. Am. Soc. for Microbiol.* Session 116.
152. Walton, R. B. (1964). *Dev. Ind. Microbiol.* **5**, 349–353.
153. Walton, R. B. (1966). U.S. Patent 3,239,394.
154. Watanabe, T., and Snell, E. E. (1972). *Proc. Natl. Acad. Sci. U.S.A.* **69**, 1086–1090.
155. Weetall, H. H. (1973). *Food Prod. Dev.* **7**, 40–52.
156. Weetall, H. H., and Messing, R. A. (1972). *In* "The Chemistry of Biosurfaces" (M. L. Hair, ed.), Vol. II, pp. 563–595. Dekker, New York.
157. Willner, D., Rossomano, V. Z., and Sprancmanis, V. (1973). *J. Antibiot.* **26**, 179–180.
158. Yamada, H., Kumagai, H., Kashima, N., Torii, H., Enei, H., and Okumura, S. (1972). *Biochem. Biophys. Res. Commun.* **46**, 370–374.
159. Yamamoto, K., Sato, T., Tosa, T., and Chibata, I. (1974). *Biotechnol. Bioeng.* **16**, 1589–1599.
160. Yamamoto, K., Sato, T., Tosa, T., and Chibata, I. (1974). *Biotechnol. Bioeng.* **16**, 1601–1610.
161. Zaborsky, O. R. (1973). "Immobilized Enzymes." CRC Press, Cleveland, Ohio.
162. Zaffaroni, P., Vitobello, V., Cecere, F., Giacomozzi, E., and Morisi, F. (1974). *Agric. Biol. Chem.* **38**, 1335–1342.

Cytotoxic and Antitumor Antibiotics Produced by Microorganisms

J. Fuska

Department of Microbiology and Biochemistry, Faculty of Chemistry, Slovak Polytechnical University, Bratislava, Czechoslovakia

AND

B. Proksa

Institute of Experimental Pharmacology, Slovak Academy of Sciences, Bratislava, Czechoslovakia

I.	Introduction	259
II.	Antibiotics and Their Synonyms	260
III.	List of Tumors	274
	A. Mouse	274
	B. Rat	275
	C. Chicken	275
	D. Man	275
	References	335

I. Introduction

Since many antibiotics are able to suppress or retard the growth of tumors, an extensive effort has been made in many laboratories to find new antibiotics with antineoplastic properties either by screening new soil isolates or by chemical modification of the existing antibiotics. The study of the mechanisms of action of known cancerostatics with different structures revealed that certain functional groups of molecules interfere with the multiplication of tumor cells. Many of these groups also have immunosuppressive and antiviral properties.

Microorganisms are among the most prolific producers of new biologically active compounds. To date more than 400 cytotoxic and/or cancerostatic compounds produced by microorganisms were isolated.

This review compiles information on 373 antibiotics possessing antineoplastic properties for which reasonably complete data were available. When an antibiotic was produced by different microorganisms reference was made only to the original publication.

The review consists of four sections. In the first section the names (or numbers) and synonyms of 373 antibiotics with reported antineoplastic activities are listed in alphabetical order. The second section is a list of the most common experimental tumors.

The third section consists of Table I, which contains 7 columns briefly giving basic information on each of the antibiotics and a list of pertinent literature (numbers in parentheses refer to the numbered Reference section). In the column "Chemical nature" structures are listed and references to pertinent publications are presented. In the column headed "Biological effects" several important properties of each antibiotic are briefly described: the first part broadly indicates any antibacterial or antiviral activity. The second part refers to *in vivo* activity against commonly used animal tumors and gives references to key publications. The third part of the column briefly indicates the principal biochemical effect of the drug and again quotes the pertinent literature on the subject. The last column gives additional critical information (usually in one word) about compounds and/or the toxicity in mice (LD_{50} in milligrams per kilogram).

The fourth section presents the numbered list of 1183 publications.

It is expected that this review will be useful as a source of basic information on antibiotics with antineoplastic activities and on their principal biological effects.

II.   Antibiotics and Their Synonyms

1. ABURAMYCIN
    4-Acetoxycycloheximide = E-73
    Acetylene dicarboxamide = LENAMYCIN
    Achromycin = PUROMYCIN
    Actidione = CYCLOHEXIMIDE
2. ACTININ = Mycetine
3. ACTINOBOLIN = Actinovorin
    Actinochrysin = ACTINOMYCIN C
    Actinoflavin = ACTINOMYCIN J
4. ACTINOFLOCIN = Kikuchi's 3rd substance
5. ACTINOGAN = C-1291
6. ACTINOLEUKIN = MA-537 A_1
    Actinomycin A_{II} = ACTINOMYCIN II
    Actinomycin A_{III} = ACTINOMYCIN III
    Actinomycin A_{IV} = ACTINOMYCIN D
7. ACTINOMYCIN B = X-45
    Actinomycin B_I = ACTINOMYCIN D
    Actinomycin B_{II} = ACTINOMYCIN II
    Actinomycin B_{III} = ACTINOMYCIN III
    Actinomycin B_{IV} = ACTINOMYCIN D
8. ACTINOMYCIN C = Actinomycin S-67 = Actinochrysin = HBF-386 = Cactinomycin = Sanamycin = Oncostatin C
9. ACTINOMYCIN D = Actinomycin $AA, A_{IV}, B_I, B_{IV}, C_I, D_{IV}, I_I, X_I$ = 1779
10. ACTINOMYCIN E (mixture)
11. ACTINOMYCIN F (mixture)
    Actinomycin F_8 = ACTINOMYCIN II
    Actinomycin F_9 = ACTINOMYCIN III
12. ACTINOMYCIN J = Actinoflavin

13. ACTINOMYCIN K
14. ACTINOMYCIN L
15. ACTINOMYCIN P = PA-126P$_2$
16. ACTINOMYCIN S (mixture)
    Actinomycin S-67 = ACTINOMYCIN C
    Actinomycin X$_0$ = ACTINOMYCIN III
    Actinomycin X$_I$ = ACTINOMYCIN D
17. ACTINOMYCIN II = Actinomycin A$_{II}$ = B$_{II}$ = F$_8$
18. ACTINOMYCIN III = Actinomycin A$_{III}$ = B$_{III}$ = X$_0$ = F$_9$
    Actinomycin IV = ACTINOMYCIN D
    Actinovorin = ACTINOBOLIN
19. ACTINOXANTHINE
20. ADRIAMYCIN = Doxorubicin
21. AFLATOXIN B$_1$
22. ALANOSINE
23. ALAZOPEPTIN
    Albofungin = KANCHANOMYCIN
24. ALBONOURSIN = B-73 = P-42-2
25. ALIOMYCIN
    Allomycin = AMICETIN
26. ALPHASARCIN
    Ambomycin = DUAZOMYCIN C
27. AMICETIN = Allomycin = Sacromycin = D-13
28. 3-AMINO-3'-DEOXYADENOSINE
29. 2'-AMINO-2'-DEOXYPENTOFURANOSYL GUANINE
    (αS,5S)-α-Amino-3-chloro-4,5-dihydro-5-izoxazole acetic acid =
    U-42,126
30. AMPHOMYCIN = Glumamycin = Crystallomycin
    Angstomycin = Angustomycin (mixture of ANGUSTMYCINS)
31. ANGUIDIN = Diacetoxyscirpenol = ANG-66
    Angustmycin A = DECOYININ
    Angustmycin C = PSICOFURANINE
32. ANHYDROTETRACYCLINE
33. ANISOMYCIN = Flagecidin = PA-106
34. ANTHRACIDIN A
35. ANTHRAMYCIN (active principle of Refuine) = Ro 5-9000
    ANTHRAMYCIN — METHYLETHER
    Anthraquinone = BIKAVERIN
    Antimycin A$_3$ = BLASTMYCIN
    Aquamycin = LENAMYCIN
36. AQUAYAMYCIN
    Arabinofuranosyladenine = Spongoadenosine = ARA-A
37. ARANOFLAVIN
    Aristelomycin = ARISTEROMYCIN

38. ARISTEROMYCIN = Aristelomycin
39. ASCOCHLORIN = Ilicicolin D = LL-Z-1272γ
    Ascotoxin = BREFELDIN A
40. L-ASPARAGINASE
41. ASPERLINE = U-13,933
    Asteromycin = GOUGEROTIN
42. AURANTIN (mixture)
    Aureolic acid = MITHRAMYCIN
43. AYAMYCIN A_2 (A,A_1,A_3)
    Ayamycins (mixture of Ayamycin A,A_1,A_2,A_3)
44. 5-AZACYTIDINE = U-18,496 = 5-aza-CR = Ladakamycin
    8-Azaguanine = PATHOCIDIN
45. AZASERINE
    Azotomycin = DUAZOMYCIN B
46. BEROMYCIN
47. BIKAVERIN = Lycopersin = Anthraquinon
48. BLASTICIDIN S
    3,6-bis (5-chloro-2-piperidyl)-2,5-piperazinedione = 593 A
49. BLASTMYCIN = Antimycin A_3
50. BLEOMYCIN
    Bleomycin B_2 = Phleomycin D_2
    Bleomycin B_4 = Phleomycin F
51. BORRELIDIN
52. BOSTRYCIN
53. BOTRYODIPLODIN = 2-Hydroxy-3-methyl-4-acetyltetrahydro-furan
54. BREDININ
55. BREFELDIN A = Cyanein = Decumbin = Ascotoxin
    Bruneomycin = STREPTONIGRIN
    Cactinomycin = ACTINOMYCIN C
56. CALIOMYCIN
57. CALVACIN
58. CANARIUS = U-13,714
    Cancidin = GANCIDIN
59. 3-CARBOXY-2,4-PENTADIENALLACTOL = A-415 Z_3 = PA-147
60. CARCINOMYCIN = Carzinomycin = Gannmycin
61. CARMINOMYCIN
62. CARYOMYCIN
63. CARZINOPHILLIN A
64. CARZINOCIDIN
65. CARZINOSTATIN
    Cellocidin = LENAMYCIN
66. CELLOSTATIN = E-150 = G-253 B
    Cephalothecin = CROTOCIN

Cerubidin = DAUNOMYCIN
67. CERVICARCIN
68. CHAETOCIN
69. CHETOMIN
70. CHLAMYDOCIN
    CHROMOMYCIN A$_2$
71. CHROMOMYCIN A$_3$ = Toyomycin
72. CHROTHIOMYCIN
73. CHRYSOMALLIN = 2703
74. CICLACIDIN
75. CICLAMYCIN
    Cinerubin (mixture) = Tauromycetin
76. CINERUBIN A = Ryemycin B$_2$
77. CINERUBIN B = Ryemycin B$_1$
78. CIROLEROSUS = U-12, 241
    cis-β-Carboxyacrylamidine = U-20,904
    Clavacin = PATULIN
    Claviformin = PATULIN
    Clitocybine = NEBULARINE
79. COCCOMYCIN
    Concocidin = NETROPSIN
80. CORDYCEPIN
81. CORIOLIN
82. CROTOCIN = Cephalothecin
    Crystallomycin = AMPHOMYCIN
    Cyanein = BREFELDIN A
83. CYCLOCHLOROTINE
84. CYCLOHEXIMIDE = Naramycin A = Actidione = R — 42
85. CYLINDROCLADIN
86. CYTOCHALASIN B = Phomin
87. CYTOCHALASIN D = Zygosporin A$_1$ = Zygosporin F
88. CYTOMYCIN
89. CYTOTETRIN (mixture)
    Dactinomycin = ACTINOMYCIN D
    Danubomycin = DAUNOMYCIN
90. DAUNOMYCIN = Cerubidin = Daunorubicin = Rubidomycin =
    Rubomycin C = Danubomycin = RP-13057 = FI-1762 Dauno-
    rubicin = DAUNOMYCIN
91. DECOYININ = Angustmycin A = A-14
    Decumbin = BREFELDIN A
92. 6-DEMETHYL-6-DEOXYTETRACYCLINE
93. DEMETRIC ACID
94. DENAMYCIN
    14-Deoxylagosin = FILIPIN

95. DERMADIN = U-21,963
96. DESERTOMYCIN
      Desideus probably = NONACTIN
97. DEUTOMYCIN
98. DEXTROCHRYSIN
      Diacetoxyscirpenol = ANGUIDIN
      Diazomycin A,B,C, = DUAZOMYCIN A,B,C
      6-Diazo-5-oxo-L-norleucine = DON
99. DIKETOCORIOLIN B
100. DISTAMYCIN A
      Doxorubicin = ADRIAMYCIN
      9-(β-D-Ribofuranosyl)purine = NEBULARINE
101. DUAZOMYCIN = Diazomycin A = N-acetyl — DON
102. DUAZOMYCIN B = Diazomycin B = Azotomycin = BA — 8509 = 1719
103. DUAZOMYCIN C = Ambomycin = Diazomycin C
104. DUBORIMYCIN = RP-20798
105. DUCLAUXIN = PSX$_2$
      Echanomycin = ECHINOMYCIN
106. ECHINOMYCIN = Echanomycin = Levomycin = Quinomycin A = X-53 III = X-948
107. EKATETRON
108. ENCALINE (mixture)
109. ENOMYCIN
110. ERICAMYCIN
111. ERIZOMYCIN
112. ETABETACIN
      Etamycin probably = 6613
      Etruscomycin = LUCENSOMYCIN
113. EUCALINE
      Expansin = PATULIN
114. FATTY ACIDS
115. FERVENULIN = Planomycin
116. FILIPIN = 14-Deoxylagosin
      Flagecidin = ANISOMYCIN
117. FLAMMULIN
118. FLUOPSIN = 4601 = YC-63
      Flavomycelin = LUTEOSKYRIN
119. FORMYCIN
      Foromacidin A,B,C = SPIRAMYCIN I,II,III,
120. FRAGIN
121. FREQUENTIN
      Fugillin = FUMAGILLIN
122. FUMAGILLIN = Fugillin = Fumidil = Phagopedin σ

Fumidil = FUMAGILLIN
123. FUNICULOSIN = Islandicin = Rhodomycelin
124. FUSARENON-X
125. FUSARIOCIN
126. GANCIDIN A,W = Cancidin A,W,
    Gannmycin = CARCINOMYCIN
127. GELDANAMYCIN
128. GLIOTOXIN
129. GLOBIMYCIN
130. GLUCAN
    Glumamycin = AMPHOMYCIN
131. GLUTAMINASE-A
132. GOUGEROTIN = Asteromycin
133. GRAMINOMYCIN
134. GRANATICIN = Litmomycin
135. GRISEOFAGIN
136. GRISEOLUTEIN B
137. GRISORIXIN
138. GUANAMYCIN
139. HADACIDIN
140. HEDAMYCIN
141. HILAMYCIN A,B
    4e-Hydroxycycloheximide = STREPTOVITACIN A
142. HYGROSCOPIN A,B,C
    Ilicicolin D = ASCOCHLORIN
143. ILLUDIN M
144. ILLUDIN S = Lampterol
145. INOMYCIN
146. IODININ = 1,6-Phenazinediol 5,10-dioxide
    Islandicin = FUNICULOSIN
147. IYOMYCIN A,B_1,B_2,B_3,B_4,B_5
148. JAWAHARENE
149. JULIMYCIN B-II
150. KANCHANOMYCIN = Albofungin = BA-180265 A = P-42-1
151. KIDAMYCIN
152. KIKUMYCIN A,B
    Kikuchi's 3rd substance = ACTINOFLOCIN
153. KINAMYCIN
154. KINOMYCIN B
155. KUNDRYMYCIN
156. KUNOMYCIN
157. LABILOMYCIN
    Ladakamycin = 5-AZACYTIDINE
    Lampterol = ILLUDIN S

158. LARGOMYCIN (mixture)
159. LATERIOMYCIN A,B,F
160. LENAMYCIN = Aquamycin = Cellocidin = Renamycin = Acetyl-enedicarboxamide
161. LENTINAN
162. LEUCINE DEHYDROGENASE
163. LEUCINOSTATIN
 Levomycin = ECHINOMYCIN
164. LIENOMYCIN = 2995
165. LIPIDS
 Litmomycin = GRANATICIN
166. LUCENSOMYCIN = Etruscomycin = FI-1163
167. LUTEOMYCIN = H-2053
168. LUTEOSKYRIN = Flavomycelin
169. LUTEOSTATIN
 Lycopersin = BIKAVERIN
170. LYMPHOMYCIN
171. MACROMYCIN
172. MALEIMYCIN
173. MANNAN
174. MARINAMYCIN
175. MARINAMYCIN—like substance
176. MELANOMYCIN
177. MELINACIDIN I,II,III,IV
 Melrosporus = U-22,956
 Methylmitomycin C = PORFIROMYCIN
178. MEZZANOMYCIN
179. MINIATOMYCIN
180. MINIMYCIN
181. MINOMYCIN
 Mithracin = MITHRAMYCIN
182. MITRAMYCIN = Mithracin = Aureolic acid = LA-7017 = PA-144
183. MITIROMYCIN = Mitiromycin A = Mitiromycin A_2
 Mitiromycin B = MITOMYCIN A
 Mitiromycin C = MITOMYCIN B
 Mitiromycin D = PORIFROMYCIN
 Mitiromycin E = MITOMYCIN C
184. MITOCROMIN = B-35251 = NSC-77471
185. MITOMALCIN
186. MITOMYCIN A = Mitiromycin B
 MITOMYCIN B = Mitiromycin C
187. MITOMYCIN C = Mitiromycin E
 MITOMYCIN R,Y

188. MONAZOMYCIN
189. MONENSIN = Monensic acid
190. MONODEACETYLANGUIDIN
    Muconomycin = VERRUCARIN J
191. MUTAMYCIN
192. MUTOMYCIN = 4305
193. MYCETIN D-17
    Mycoin = PATULIN
194. MYCOPHENOLIC ACID
    Mycospodicin = 323/58
195. MYRORIDIN
    N-Acetyl-DON = DUAZOMYCIN A
196. NAEMATOLIN
    Naramycin A = CYCLOHEXIMIDE
197. NARANGOMYCIN = BA-6903
    Naritheracin = Naritherashin = TOYOCAMYCIN
198. NEBULARINE = 9-(β-D-ribofuranosyl)purine = Clitocybine
199. NEOCARZINOSTATIN
200. NEOCIDE
201. NEOPLURAMYCIN
202. NETROPSIN = Concocidin = Sinamomycin = IA-887 = K-117 = T-1384
203. NIVALENOL
204. NOGALAMYCIN
205. NONACTIN = FH-3582 A = N-329 A = probably Desideus
206. OLEFICIN
207. OLIVOMYCIN A,B,C,D = 16749 (mixture)
208. ONCOSTATNIN B (x)
209. ORYZACHLORIN
210. OSSAMYCIN
211. OXAZINOMYCIN
212. 5-OXO-1H-PYRROLO[2,1-C][1,4]BENZODIAZEPIN-2-ACYLA-MIDES
213. PACHYMAN
    Pactacin = PACTAMYCIN
214. PACTAMYCIN = Pactacin = U-15,800 = NSC 52947
215. PANEPOXYDION, PANEPOXYDON
216. PATHOCIDIN = 8-Azaguanine = B-28
217. PATULIN = Clavacin = Claviformin = Expansin = Mycoin C = Penicidin
218. PELIOMYCIN = NSC 76455 D
    Penicidin = PATULIN
219. PENICILLIC ACID

220. PEPTIMYCIN
221. PERLIMYCIN
      Phagopedin σ = FUMAGILLIN
222. PHENOMYCIN
223. PHLEOMYCIN A,B,C,D_1,D_2,E,F,G,H,I,J,K
      Phleomycin D_2 = Bleomycin B_2
      Phleomycin F = Bleomycin B_4
      Phomin = CYTOCHALASIN B
224. PIGMENT *Streptomyces padanus*
225. PILLAROMYCIN A,B,B_4,C
      Planomycin = FERVENULIN
226. PLURALLIN
227. PLURAMYCIN A,B
228. POINE
229. POLYMYCIN
230. POLYSACCHARIDES
231. PORFIROMYCIN = Methylmitomycin C = Mitiromycin D
232. PORICIN
233. PRIMOCARCIN
      Provomycin = SPIRAMYCIN (mixture)
234. PRUNACETIN
235. PSICOFURANINE = Angustmycin C = U-9586
236. PUROMYCIN = Achromycin = Stylomycin
237. PURPUROMYCIN
238. PYRAZOMYCIN = A-23813
239. PYRENOPHORIN
240. 5*H*-PYRROLO [2,1-*c*][1,4] BENZODIAZEPIN-5-ONE
      Quinomycin A = ECHINOMYCIN
241. QUINOMYCIN B
242. QUINOMYCIN C
      Quinoxaline = TRIOSTIN C
243. RACTINOMYCIN A,B
244. RAROMYCIN
245. REQUINOMYCIN
      Renamycin = LENAMYCIN
246. RETAMYCIN
247. REUMYCIN
      Rhodomycelin = FUNICULOSIN
248. RIFAMYCIN SV
249. RORIDIN A
250. ROSEOLIC ACID
251. ROSSIMYCIN
      Rovamycin = SPIRAMYCIN (mixture)

Rubidomycin = DAUNOMYCIN
252. RUBIFLAVIN
253. RUBOMYCIN B
    Rubomycin C = DAUNOMYCIN
    Rubromycin B = TUOROMYCIN
254. RUFOCHROMOMYCIN = Rufocromycin = Rufocromomycin = RP-5278
    Rufocromycin = RUFOCHROMOMYCIN
255. RUGULOSIN
256. RYEMYCIN A
    Ryemycin B_1, B_2 = CINERUBIN B,A
    Sacromycin = AMICETIN
    Sanamycin = ACTINOMYCIN C
257. SANGIVAMYCIN = BA-90912
258. SARKOMYCIN
    Selectomycin = SPIRAMYCIN (mixture)
259. SEPTACIDIN
    Sequamycin = SPIRAMYCIN (mixture)
260. SHOWDOMYCIN = MSD-125A
261. SIBIROMYCIN
    Sinamomycin = NETROPSIN
    Siromycin = TOYOCAMYCIN
262. SOEDOMYCIN = M3
    Sparsogenin = SPARSOMYCIN
263. SPARSOMYCIN = Sparsogenin
    Sparsomycin A = TUBERCIDIN
264. SPINAMYCIN
265. SPIRAMYCIN (mixture) = Foromacidin = Provomycin = RP-5337 = Rovamycin = Selectomycin = Sequamycin
    SPIRAMYCIN I,II,III = Foromacidin A,B,C
266. SPORIDESMIN
    SPORIDESMIN B,C,D,E,F,G,H
267. STATOLON = 1758 = 8450
268. STEFFIMYCIN = U-20,661
269. STEFFIMYCIN B
270. STREPTOGAN
271. STREPTONIGRIN = Bruneomycin = BA-163 = BA-4721
272. STREPTORUBIN A,B
273. STREPTOVARICIN = B-44P
274. STREPTOVITACIN A, C_1, C_2, D, E = Hydroxycycloheximides
275. STREPTOZOTOCIN = NSC 37917
    Stylomycin = PUROMYCIN
276. SULFOCIDIN

Tauromycetin = CINERUBIN (mixture)

277. TENUAZONIC ACID
278. TERREIC ACID
279. TOMAYAMYCIN
280. TOYOCAMYCIN = Unamycin B = Siromycin = Naritheracin = Nariterashin = Vengicide = E-212 = 9-27 = 9-48 = 1037
Toyomycin = CHROMOMYCIN A3
281. TRICHOMYCIN
282. TRICHOTHECIN
283. TRIENINE
284. TRIOSTIN C = Quinoxaline
285. TUBERCIDIN = Sparsomycin A = U-19183
286. TUOROMYCIN = Rubromycin B
Unamycin B = TOYOCAMYCIN
287. USSAMYCIN
288. VARIAMYCIN = N-6604-9A
289. VARIOMYCIN
Vengicide = TOYOCAMYCIN
290. VERMICULIN
291. VERRUCARIN A,B,H,J,K
292. VERTICILLIN A,B
293. VERTIMYCIN
294. VIBRIOCIN
295. VIOLAMYCIN
296. VIROCYTIN = 13
297. VIROLITIN
298. XANTOCILLIN X METHYLETHER
299. ZEARALENONE
300. ZEDALAN = U-15774 = *Trans*-3-(hydroxyimino)acetamido acrylamide
301. ZORBAMYCIN = U-30604E = YA-56X
Zygosporin A = CYTOCHALASIN D
Zygosporin F = CYTOCHALASIN D MONOACETATE
A-14 = DECOYININ
302. A-114
303. A-216
304. A-280
A-415Z_3 = 3-CARBOXY-2,4-PENTADIENALLACTOL
A-2410 = NSC A-649
A-6413 = M-259
A-23813 = PYRAZOMYCIN
ANG-66 = ANGUIDIN
305. ARA-A = Arabinofuranosyladenine = Spongoadenosine

B-28 = PATHOCIDIN
B-44P = STREPTOVARICIN
B-73 = ALBONOURSIN
B-35251 = MITOCROMIN
BA-163 = STREPTONIGRIN
BA-4721 = STREPTONIGRIN
BA-6903 = NARANGOMYCIN
BA-8509 = DUAZOMYCIN B
306. BA-17039A
BA-90912 = SANGIVAMYCIN
BA-180265A = KANCHANOMYCIN
307. BA-181314
308. BU-306
C-1291 = ACTINOGAN
309. C-1292
D-13 = AMICETIN
310. D-180
311. D-284
312. D-440
313. DCS-
314. DON = 6-Diazo-5-oxo-L-norleucine
315. E-73 = 4-Acetoxycycloheximide
E-150 = CELLOSTATIN
E-212 = TOYOCAMYCIN
FH-3582A = NONACTIN
FI-1163 = LUCENSOMYCIN
FI-1762 = DAUNOMYCIN
G-253B = CELLOSTATIN
316. G-253C
H-2053 = LUTEOMYCIN
317. HA-135
HBF-386 = ACTINOMYCIN C
IA-887 = NETROPSIN
K-117 = NETROPSIN
LA-7017 = MITHRAMYCIN
LL-Z-1272γ = ASCOCHLORIN
M 3 = SOEDOMYCIN
318. M-259 = A-6413
319. M-741
320. M-6672
M 5-18903 = NSC A-649
MSD-125A = SHOWDOMYCIN
MA-537A$_1$ = ACTINOLEUKIN

321. MT-10
322. N-294
     N-329A = NONACTIN
     N-6604-9A = VARIAMYCIN
323. NSC-A-649 = M 5-18903 = A-2410
324. O-2
325. O-5
326. OS-3256-B
     P-42-1 = KANCHANOMYCIN
     P-42-2 = ALBONOURSIN
     PA-126-P_2 = ACTINOMYCIN P
     PA-106 = ANISOMYCIN
     PA-144 = MITHRAMYCIN
     PA-147 = 3-CARBOXY-2,4-PENTADIENALLACTOL
327. PBF = 1683
328. PS-X_1
     PS-X_2 = DUCLAUXIN
329. P-42-13
330. R-12
     R-42 = CYCLOHEXIMIDE
     Ro 5-9000 = ANTHRAMYCIN
331. RP-9768
     RP-5278 = RUFOCHROMOMYCIN
     RP-5337 = SPIRAMYCIN
332. RP-9865 mixture of RP-13057, RP-13213, RP-13330
333. RP-9971
     RP-13057 = DAUNOMYCIN
334. RP-13213 fraction of RP-9865
335. RP-13330 fraction of RP-9865
336. RP-16978
337. RP-17967
     RP-20798 = DUBORIMYCIN
338. S-339-3
339. SF-1306
340. SK-229 A,B
341. SL-1846
342. SL-2266
343. SL-3364
344. SL-3440
345. SQ-15859
346. SR-1768 A,B,C,D,E,F,G
347. T 12/3
     T-1384 = NETROPSIN

348. TA-435A
349. TA-2590
350. TS-885
  U-9586 = PSICOFURANINE
  U-12241 = CIROLEROSUS
  U-13714 = CANARIUS
  U-13933 = ASPERLINE
  U-15774 = ZEDALAN
  U-15800 = PACTAMYCIN
  U-18496 = 5-AZACYTIDINE
  U-19183 = TUBERCIDIN
  U-20,661 = STEFFIMYCIN
351. U-20904 = cis-β-Carboxyacrylamidine
352. U-22,956 = Melrosporus
  U-30,604E = ZORBAMYCIN
353. U-42,126 = (αS,5S)-α-Amino-3-chloro-4,5-dihydro-5-isoxazoleacetic acid
354. U-43,795 = Hydroxyderivat U-42,126
355. WR-142
  X-45 = ACTINOMYCIN B
  X-53-III = ECHINOMYCIN
  X-948 = ECHINOMYCIN
356. YA-56 (complex)
  YA-56X = ZORBAMYCIN
  YC-63 = FLUOPSIN
  13 = VIROCYTIN
357. 23-21
358. 173-t
359. 289E
360. 289Fo
361. 323/58 = Mycospodicin
362. 593A = 3,6-bis(5-chloro-2-piperidyl)-2,5-piperazinedione
363. 661
364. 757
  9-27 = TOYOCAMYCIN
  9-48 = TOYOCAMYCIN
  10-37 = TOYOCAMYCIN
365. 1418-A$_1$
  1683 = PBF
  1719 = DUAZOMYCIN B
  1758 = STATOLON
  1779 = ACTINOMYCIN D
  2703 = CHRYSOMALLIN

2995 = LIENOMYCIN
366. 3002
367. 3853
368. 4035
369. 4418
370. 4695
4601 = FLUOPSIN C
371. 5888 I,II,
372. 6270
373. 6613 = probably Etamycin
8450 = STATOLON
16749 = OLIVOMYCIN A,B,C,D

III. List of Tumors

A. Mouse

AKR	Leukemia AKR
B 82	Leukemia B82
BC63	Bashford carcinoma 63
Ca755	Mammary carcinoma 755
Ca1025	Carcinoma 1025
C3HBA	Mammary carcinoma C3HBA
EC	Ehrlich carcinoma
ECA	Ehrlich carcinoma ascitic form
EO 771	Mammary carcinoma EO 771
FVL	Friend leukemia
G 26	Glioma 26
GL	Lymphosarcoma 6C3HED
HPM	Harding–Passey melanoma
K 2	Krebs II carcinoma
K2A	Krebs II carcinoma ascites
L 1210	Leukemia L 1210
L 4946	Leukemia L 4946
L 5178	Leukemia L 5178
LIO 1	Lymphosarcoma LIO 1
MC	Myiom carcinoma
Mecca	Mecca lymphosarcoma
NF	Sarcoma N-F
NK/Ly	Lymphoma NK/Ly
P388	Leukemia P 388
P815	Leukemia P 815
P1534	Leukemia P 1534

RC   Mammary carcinoma RC
ROS   Osteosarcoma Ridgeway
S 37   Sarcoma 37
S 91   Melanoma S 91
S 180   Sarcoma 180
S 180A   Sarcoma 180 ascitic form
SN 36   Leukemia SN 36
T 241   Sarcoma T 241
WOS   Osteosarcoma Wagner

B. RAT

AH 130   Hepatoma AH 130
AH 130A   Hepatoma AH 130 ascitic form
FCJ   Flexner–Jobling carcinoma
GC   Guerin carcinoma
JS   Jensen sarcoma
MH 134   Hepatoma MH 134
MSL   Murphy–Sturm lymphosarcoma
NovH   Novikoff hepatoma
S 45   Sarcoma 45
SSK   Sarcoma SSK
US   Usubuchi sarcoma
W 256   Walker 256 carcinosarcoma
YS   Yoshida sarcoma

C. CHICKEN

RS   Rous sarcoma

D. MAN

H.Ad.1
HEp-1
KB
HS 1, HS 2.

TABLE I

No.	Antibiotic	Produced by[a]	Chemical nature	Micro-organisms	Tumors	Mechanism of action	Footnotes or LD$_{50}$ (mg/kg)
					Biological effects		
				Effect on			
1	ABUROMYCIN	S. aburaviensis	Aureolic acid (1,3,5)	Bacteria	ECA, S-180 (1-3,6)		2.5 sc
2	ACTININ	Streptothrix felis	Peptide (1,3)	Bacteria (1,2)	ECA (1-3,6)		
3	ACTINOBOLIN	S. griseoviridis var. atrofaciens	Lactone (1,2,4,5,27,28)	Bacteria (1,2,26)	HS-1, HEp-3, G-26, Ca-1025, W-256, ECA, S-180 (3,6,29,30)	Protein and DNA synthesis (31)	800 iv
4	ACTINOFLOCIN	Streptomyces sp., E-212, S. albus (1,2)	Actinomycin (3)	Bacteria (1-3,6)	EC, S-180 (32)		3 ip
5	ACTINOGAN	Streptomyces sp., Nocardia humifera	Peptide (1-3,33)		S-180, EG MA-387, Ca-755 (1-3,34)		
6	ACTINOLEUKIN	S. aureus, Streptomyces sp., S. abikoensis	Quinoxaline (1-3,5,35)	Bacteria (1-3)	EC, HeLa, ECA, YS, AH-130 (6,36,37)		1.5 ip
7	ACTINOMYCIN B	S. Antibioticus	Actinomycin (1-3,5,8)	Bacteria	ECA, GL (1-3,25)		

276

No.	Name	Source		Active against	Test systems	Mechanism / Remarks
8	ACTINOMYCIN C	*S. chrysomallus*	Actinomycin (1-3,5,8,41)	Bacteria	EC, S-37, W-256, RC, L-4946, GL, S-180, YS (1-3,6,25)	Transformation in bacteria (38), Mixture surface and structure of the cells (39), DNA (40), Immunosuppressant. See also (9,11-13, 16,22).
9	ACTINOMYCIN D	*S. parvullus*, *S. chrysomallus*, *S. antibioticus*, *A. olivobrunensis*, *A. michiganensis* (42-47)	Actinomycin (1-3,5,6,8,48)	Bacteria, viruses	S-91, C3HBA, S-180, EC, ECA, E0-771, RS, Ca-1025, K2A (6,25,49,50)	Complexes with DNA (53,54,64, 81,93), breakage of DNA (100), inhibition of DNA synthesis (65,78), RNA synthesis (55,56, 65-67,79,86), cellular RNA synthesis (58,68,87), ribosomal RNA synthesis (61,84), nucleolar and cytoplasmic RNA synthesis (60), viruses RNA formation (69), DNA-dependent RNA synthesis (51,52,58,59), DNA-dependent DNA synthesis (58), RNA-dependent DNA synthesis (82), DNA-RNA polymerase (71), DNA-dependent RNA polymerase (57). Affected DNase (85), lipase (73), malate and lactate dehydrogenase (89), respiration and anaerobic glycolysis (62,88-90), amino acids transport (63,83), lipids and phospholipids synthesis (79,83), ribonucleotide synthesis (76), β-carotene

(Continued)

TABLE I (Continued)

No.	Antibiotic	Produced by	Chemical nature	Biological effects			Footnotes or LD$_{50}$ (mg/kg)
				Effect on		Mechanism of action	
				Micro-organisms	Tumors		
						synthesis (101), uptake of thymidine (86), virus transformed cells (96), cell cycle (95), cell adhesion (74), mitosis (94), chromosome aberration (99), transfections (75), transcription (71), decay of pulses-labeled RNA (80), changes in nuclei and nucleoli (72,91,92). Binding to spermatozoa (77,98). Immunosuppressant, teratogen (70), mutagen (97). See also (7,9,11-13,16,17,22,23).	
10	ACTINOMYCIN E	*S. chrysomallus*	Actinomycin (1-3,5)		JS, W-256, L-1210, P-388 (1,3,25)		1 ip
11	ACTINOMYCIN F	*S. chrysomallus*	Actinomycin (1-3,5)		W-256, L-1210, P-1081, B-82 (1,2,5,25)		8 ip
12	ACTINOMYCIN J	*S. flaveolus*	Actinomycin (1-3,5)		ECA, K2, S-180, YS, Ca-1025 (1-3,6,25)		
13	ACTINOMYCIN K	*S. melanochromogenes*	Actinomycin (1-3)	Bacteria	ECA, S-180 (1-3)	RNA synthesis in ECA (25).	

14	ACTINOMYCIN L	*Streptomyces* sp. 2104L	Actinomycin (1,2)	Tumor cells (1,2)		
15	ACTINOMYCIN P	*S. aureofaciens*	Actinomycin (1,2)	Ca-755, HPM (1,2,25)		
16	ACTINOMYCIN S	*Streptomyces* sp.	Actinomycin (1,3)	ECA (1-3,6,25)	Formation of complexes with DNA (102). Inhibition of RSV reverse transcriptase (103).	
17	ACTINOMYCIN II	*S. antibioticus*	Actinomycin (1-3,5)	GL (1-3,25)		6 ip
18	ACTINOMYCIN III	*S. antibioticus*	Actinomycin (1-3,5)	GL (1-3,25)		1.5 ip
19	ACTINOXANTHINE	*A. globisporus* 1131 (1-3)	Peptide (1-3)	Bacteria	Nucleic acids and mitosis in ECA (105).	
20	ADRIAMYCIN	*S. peucetius* var. *caesius* (2,106)	Anthracycline (122,123)	Bacteria, viruses, protozoa (2)	Complexes with DNA (118), breaking of DNA (126). Affects RNA and DNA synthesis (108,109), macromolecular synthesis (111), DNA polymerase (114), (125), RNA-dependent DNA polymerase (110), uridine kinase activity (115), mitochondrial respiration (119), proliferation of tumor cells (112), cell cycle (116), cell mitosis (111), cell kinetics	

(Continued)

279

TABLE I *(Continued)*

No.	Antibiotic	Produced by	Chemical nature	Effect on Micro-organisms	Effect on Tumors	Mechanism of action	Footnotes or LD$_{50}$ (mg/kg)
						(113). Binding in nuclei (117), to antibodies (124). Antitransformation effect (127), immunosuppressant (120), mutagen (121). See also (9,10,19,20).	
21	AFLATOXIN B$_1$	*Aspergillus flavus*, *Aspergillus* sp., *Penicillium* sp., (Biosynthesis) (128,147)	(4,14,15)		YS (130)	Binding to DNA, RNA, proteins (133,141). Affects RNA synthesis (131), protein synthesis (132, 140), RNA polymerase (131,142), lipid synthesis (138), formation of subribosomal particles (136), translocation mechanism (132), breaking of chromosomes in mammalian cells (129), germination and growth of seelings (144), histones (143), membrane (134), binding to microsomes (133,137). Immunosuppressant (135), carcinogen (139), teratogen, mutagen (146), mycotoxin (14). See also (11,14).	
22	ALANOSINE	*S. alanosinicus*	Amino acid (2,24,148)	Viruses, protozoa, fungi (149)	Sarcoma (2)	RNA synthesis (148), purine and thymidine synthesis in ECA and *C. albicans* (150), conversion inosine monophosphate to adenosine monophosphate (151). See also (22).	

No.	Name	Source	Chemical class	Sensitive organisms	Tumor systems		Mechanism / Notes
23	ALAZOPEPTIN	*S. griseoplanus*, *S. candidus* var. *azaticus* (1-3,152)	Azaamino acid (1,2,5)	Bacteria	S-180, EC, Ca-1025, L-1210, WOS (1-3,6,153)		Purine synthesis (154).
24	ALBONOURSIN	*S. tumemacerans*, *A. albus* var. *fungatus*, *S. noursei* (155,156)	Diketopipera-zine (1-4)	Bacteria	EC (1-3)	45 ip	
25	ALIOMYCIN	*S. acidomyceticus* (1,2)	Polyene (1,3,5,24)	Fungi, protozoa	YS (1-3)		
26	ALPHA SARCIN	*Aspergillus giganteus*	Peptide (157)		S-180, Ca-755, G-26, T-241, EO-771 (158)		
27	AMICETIN	*S. vinaceus-drappus*, *S. sarcomyceticus*, *S. fasciculatus*, *S. sindenensis*	4-Aminohexose-pyrimidine nucleoside (1-3,5,160)	Bacteria (159)	KB, L-82 (1,3,6,161)		Protein synthesis (163,164), translocation and peptide formation (162). See also (9,12,16).
28	3'-AMINO-3-DEOXYADENOSINE	*Cordyceps militaris*, *Helmintosporium* sp. (215,166)	3'-Deoxypurine nucleoside (16,165,167)		GL, S-180, S3A (168)		Cytoplasmic RNA (170), RNA and DNA synthesis, incorporation of adenine and uridine into RNA and DNA (169), RNA polymerase (10). See also (10,16).

(Continued)

TABLE I *(Continued)*

No.	Antibiotic	Produced by	Chemical nature	Biological effects			Footnotes or LD_{50} (mg/kg)
				Effect on			
				Micro-organisms	Tumors	Mechanism of action	
29	2'-AMINO-2'-DEOXYPENTO-FURANOSYL GUANINE	*Aerobacter cloacae*			Tumor cells (171)		
30	AMPHOMYCIN	*S. canus*, *S. violaceus*, *S. laverdulae* (175)	Peptide (1-3,5,172,173)	Bacteria, protozoa	Tumor cells (1-3,174)		120 iv
31	ANGUIDIN	*F. equisetti*, *F. scirpi*, *F. sambucinum* (176)	Trichothecene (15,177)	Fungi	ECA, P-815, KB, S-37, W-256	DNA synthesis and mitosis (176,178). Immunosuppressant (176), toxin (15). See also (15)	
32	ANHYDRO-TETRACYCLINE	*S. aureofaciens* (179)	Tetracycline (3,8)		ECA (180)	Incorporation of adenine, L-valine, uridine, thymidine into ECA cells (180). Effective also 12a-deoxydedimethyl-aminotetracycline, 4a,12a-anhydrodedimethylaminotetra-cycline (180).	
33	ANISOMYCIN	*S. griseolus*, *S. roseochromogenes* (1-3)	(1,5)	Bacteria, fungi, protozoa (1-3)	HeLa (1-3)	Binding to ribosomes (183,185), affects protein and DNA synthesis in HeLa and yeast (181,182), aminoacyl-tRNA (184). See also (9,12).	140 iv
34	ANTHRACIDIN A	*S. hygroscopicus*, *Streptomyces* sp. (1,2,24)		Bacteria	HeLa (1,2)		

35	ANTHRAMYCIN	*S. refuineus* var. *thermotolerans* (1,2,24,186)	Pyrrolobenzodi-azepine (1,187)	Bacteria, fungi (1,2)	P-388, L-1210 (188)	RNA synthesis in L-1210 cells (189), reaction with DNA (190-192), segregation of the nucleus (193), uncoupled oxydative phosphorylation (195). Chemosterilant (194).	
	ANTHRAMYCIN METHYLETHER				S-180, W-256, L-1210 (196, 197)	RNA and DNA synthesis (196), affected incorporation of uridine into L-1210 cells (197), erythropoiesis (198). See also (9-11,16,17,22).	
36	AQUAYAMYCIN	*S. misawaensis*	Anthracycline (2,21,24)	Bacteria, fungi (2)	ECA, YS, S-180 (2)	Tyrosine hydrolase and dopamine-β-hydrolase (2,199,200), tryptophan-2,3-dioxy-enase (201). See also (12).	11 iv
37	ARANOFLAVIN	*Arachniotus flaveoluteus*	Macrolide	Bacteria, protozoa	YS, L-1210, S-180 (202)		4.6 ip
38	ARISTEROMYCIN	*S. citricolor* (203)	Adenine nucleoside	Bacteria, fungi (2)	HEp-2 (204)	*De novo* purine synthesis, adenosine kinase, calf intestinal deaminase and nucleotide kinase (204). See also (16).	

(Continued)

TABLE I (Continued)

| No. | Antibiotic | Chemical nature | Produced by | Biological effects | | Mechanism of action | Footnotes or LD$_{50}$ (mg/kg) |
| | | | | Effect on | | | |
				Micro-organisms	Tumors		
39	ASCOCHLORIN	Terpene (4,206,208)	*Ascochyta viciae*, Libert (205); *Cylindrocladium ilicicola*, *Nectaria coccinea* (207)	Bacteria, fungi	ECA, HeLa	Plaque-forming in DNA and RNA viruses (205), chromosome aggregation in HeLa (209).	20 ip
40	L-ASPARAGINASE	Peptide (218,241)	Bacteria, Bacilli, *Candida*, *Erwinia*, *Pseudomonas*, *E. coli*, Yeast, *Fusaria*, *Vibrio succinogenes*, (210–215,233). Guinea pig serum (216) (Biosynthesis) (217,232,234)	Viruses (219)	6C3HED, P-1798, RU/3, leukemia (220–223)	Affects protein synthesis (238), incorporation of uridine into sRNA (225), asparaginyl-tRNA (224) incorporation of thymidine (226,236), asparagine metabolism (230), thymidine and amino acid metabolism (235), β-aspartohydroxamic acid (227), depletion of extracellular L-asparagine (228), cell membrane (229). Binding on substrate (239). Immunosuppressant (231,237). See also (19, 22,240,241).	
41	ASPERLINE	Lactone (242)	*Aspergillus nidulans*		KB	DNA, RNA, and protein synthesis (243).	
42	AURANTINS	Actinomycin (21)	*A. citriofluorescens*, *A. fluorescens*, *Streptomyces* sp. (1–3,244)	Bacteria	S-180, LIO-1, ECA (1–3,245)	RNA synthesis (7,246), binding to DNA (247,248).	Mixture

43	AYAMYCIN A$_2$	S. flaveolus, Streptomyces sp. C-80 (2)	Anthracycline (Indicator) (1-3)	Bacteria	ECA, YS, W-256 (1-3,249,250)	40 ip Active also Ayamycin A, A$_1$,A$_3$.
44	5-AZACYTIDINE	S. ladakanus (2,251)	Azapirimidine nucleoside	Bacteria, viruses (252)	ECA, W-256, AKR, L-1210 (2,251)	Ribosomal RNA (260,268), nucleolar RNA (267). Affects protein synthesis (263), RNA synthesis (254), RNA metabolism (264), macromolecular synthesis (255,266), polyamine synthesis (258), ribosomal precursors (256), polyribosomes and tyrosine transaminase (253), uridine kinase (261,262), mitosis and chromosomal damage of L-1210 cells (257), nuclei of calf thymus cells (263. Mutagen (259). See also (13,16,18,19).
45	AZASERINE	S. fragilis	Azaamino acid (1,2,5)	Bacteria, fungi, protozoa	S-180, W-256, RS, FJC (2,3, 6,269,270)	Affects purine synthesis (272), riboside intermediates (271), inosinic acid (273), L-glutamine amidoligase, incorporation of glycine and formiate into NA (275), phosphoribosyl formylglycine amidine synthetase (276).

(Continued)

TABLE I (Continued)

No.	Antibiotic	Produced by	Chemical nature	Biological effects		Footnotes or LD$_{50}$ (mg/kg)
				Effect on	Mechanism of action	
				Micro-organisms	Tumors	

No.	Antibiotic	Produced by	Chemical nature	Micro-organisms	Tumors	Mechanism of action	Footnotes or LD$_{50}$ (mg/kg)
46	BEROMYCIN	A. griseoruber var. bruneomycini n.sp., (271)	(280)		LIO-1, NK/Ly, HPM, L-5178 (281-283)	NAD, NADH (277). Immunosuppressant (274), mutagen (278). See also (7,9-13,18,22,23).	8 iv
47	BIKAVERIN	Gibberella fujikuroi, F. lycopersici (285,286)	Benzoxanthone (287)	Protozoa (285)	ECA (288)	Replication of influenzareproduction of RNA and DNA containing viruses (284). See also (20).	
48	BLASTICIDIN S	S. globiter, S. griseochromogenes, S. morookaensis (291)	4-Aminohexose pyrimidine nucleoside (1-3,5,292)	Bacteria, fungi	ECA, S-180, W-256, Ca-755 (3,293)	Incorporation of precursors into nucleic acids and proteins (288, 290), energetic metabolism (289). Ribosomal RNA synthesis (298), aminoacyl-tRNase (295), peptidyl transferase (297), ribosomes (294,296). See also (1,7,9,11, 12,16,18).	2.8 iv
49	BLASTMYCIN	S. blastmyceticus, S. kitasavaensis	Nonpolyene macrolide (1-3,5,299)	Fungi	RC (1)	Induction of carotenoid in M. marinum (300).	1.87 ip
50	BLEOMYCIN	S. verticillus (NIHJ) (424), (24, 301) (Biosynthesis) (302-306)	Peptide (1,2,307-309)	Bacteria (301)	ECA, Ca-755, RS, Squamous (310-312)	RNA (337), DNA fragmentation (329), Tm of DNA (303,323), degradation and scission of DNA (313-319). Affects DNA synthesis (320-322,343), polyribonucleotides (303), protein	Mixture

No.	Name	Source	Class	Susceptible organisms	Tumor system	Mode of action	Dose
51	BORRELIDIN	S. rochei, S. griseus, Streptomyces sp., C 2898, (1,24)	Macrolide (1-3,5)	Bacteria, viruses, protozoa	K2A (1,2)	synthesis (320), liberating of DNA from DNA-membrane complex (303,336,341), DNase (315,324, 325), DNA polymerase (324), DNA-dependent DNA polymerase (303,327), DNA-dependent RNA polymerase (328), ligase (324, 326), chromosome aberration (333-335,340,342), transcription (328), cell cycle of tumor cells (330-333,344), SH groups. Immunosuppressant (303,338,339). See also (9,10,16,19,20,22).	36 iv
52	BOSTRYCIN	Bostrychonema alpestre	Anthraquinone (21,351)	Bacteria, fungi	S-180, FVL (350)	Incorporation of threonine into tRNA (345), Threonyl-tRNA synthetase (346-348), amino acid synthesis (349). See also (11,22).	109 ip
53	BOTRYODIPLODIN	Botriodiplodia theobromae (352)	Tetrahydrofuran (353)		Leukemia (353)		
54	BREDININ	Eupenicillium brefeldianum		Viruses	Tumor cells (354,355)		

(Continued)

287

TABLE I *(Continued)*

No.	Antibiotic	Produced by	Chemical nature	Biological effects: Effect on Micro-organisms	Biological effects: Effect on Tumors	Biological effects: Mechanism of action	Footnotes or LD_{50} (mg/kg)
55	BREFELDIN A	*P. cyaneum*, *Pennicillium* sp., *P. brefeldianum*, *P. simplicissimum*, *Ascochyta imperfecta* (3,356-358). (Biosynthesis) (361)	Macrolide (4,14,359)	Fungi, viruses	ECA, YS, HeLa (360)	Nucleic acids and protein synthesis in microbial and animal cells (360), incorporation of ^{32}P into ECA, mitosis (360), disturbance of lipid metabolism (362). Mycotoxin (2). See also (14).	200 ip
56	CALIOMYCIN	*S. filamentosus*			YS (24,363)		
57	CALVACIN	*Calvatia gigantea* (3)	Peptide (3,6,4)		EC, S-180, EO-771, ROS, W-256, FVL (3,365)		
58	CANARIUS	*S. canarius* var. *canarius* (366)	(1,2,367)	Viruses, yeast	Tumor cells (2)		14 ip
59	3-CARBOXY-2,4-PENTADIENAL LACTOL	*Streptomyces* sp., A 415-Z3 (2)	(1,5)	Bacteria	Tumor cells (1,368)		250 iv
60	CARCINOMYCIN	*S. carcinomyceticus*, *S. ganmycicus* (2)	Peptide (1,5,6)	Bacteria, protozoa	Tumor cells (6,369)		500 iv
61	CARMINOMYCIN	*Actinomadura carminata* (370)	Anthracycline (371)		L-1210, S-180, LIO-1 (372)	DNA synthesis, complexes with pyrimidine nucleotide, template activity of DNA (373), reproduction of RNA and DNA containing viruses (375), phagocytic activity (374). See also (10).	

No.	Name	Source	Type (refs)	Susceptible organisms	Tumor cells (6)	Mode of action	Dose
62	CARYOMYCIN	*S. filamentosus*					
63	CARZINOPHILLIN A	*S. sahachiroi* n. sp.	Peptide (1,2,5,376)	Bacteria	ECA, S-180 (1-3,6,377)	DNA synthesis, induction of cross-linking in DNA (378,379). See also (11).	
64	CARZINOCIDIN	*S. carcinocidicus*, *S. kitasawaensis*	Peptide (1-3,5,382)	Yeast	ECA, EC, YS (1-2)		4.7 iv
65	CARZINOSTATIN	*Streptomyces* sp.	(1-3)	Bacteria	ECA, S-180, Sn-36, Ah-130 (1-3,380,381)	DNA synthesis, cytological effect (381), immunological effect (19).	Mixture
66	CELLOSTATIN	*S. cellostaticus*	(1-3)	Bacteria, fungi, protozoa	EC (1-3,6)		15 ip
67	CERVICARCIN	*S. ogaensis*	(1-3)		ECA, S-180, S-37, NF, FVL (1-3,383,384)		60 ip
68	CHAETOCIN	*Chaetomium minutum*	Dithiodioxo-piperazine	Bacteria	P-815 (385,386). See also (15)		1.7 ip
69	CHETOMIN	*Chaetomium cochliodes*, *Chaetomium globosum*	Dithiodioxo-piperazine (3,5,387)	Bacteria	HeLa (388)		

(Continued)

289

TABLE I *(Continued)*

No.	Antibiotic	Produced by	Chemical nature	Biological effects		Mechanism of action	Footnotes or LD$_{50}$ (mg/kg)
				Effect on			
				Micro-organisms	Tumors		
70	CHLAMYDOCIN	*Diheterospora chlamidosporia*	Tetrapeptide (389)		P-815 (390)		
71	CHROMOMYCIN A$_3$ (A$_2$, A$_5$)	*S. griseus, S. olivochromogenes*	Aureolic acid (1-3,5)	Bacteria	EC, S-180, SN-36, AH-130 (1-3,6,391)	DNA interaction (394-396), decay of RNA (397). Affects RNA synthesis (392,393,400), DNA-dependent DNA and RNA synthesis (398,399), incorporation of adenine and uracil into RNA (403), *in vitro* transcription (401,402), RSV-genome (404). See also (7,9-11,17,19, 22).	2.1 ip
72	CHROTHIOMYCIN	*S. pluricolorescens*	(2)		YS	Inhibition thyrosine hydrolase and dopamine-β-hydrolase (2, 405)	25 ip
73	CHRYSOMALLIN	*A. chrysomallus* 2703	Actinomycin (1,2,24)		LIO-1, S-180, W-256, S-37, SSK, GC (1,2,406)		
74	CICLACIDIN	*S. capoamus* (2)	Indicator pH		Tumor cells (2,24)		
75	CICLAMYCIN (A)	*S. capoamus* 4670-1A-37	Anthracycline	Bacteria	S-180, W-256 (1,2,24)		Coproduct Ciclacidin

No.	Organism	Class	Susceptible organisms	Tumor systems	Mechanism / Remarks
76	CINERUBIN A	*S. antibioticus*, *S. galilaceus*, *S. niveoruber*	Anthracycline (1-3,5,407)	Bacteria, fungi, protozoa, viruses (2)	ECA, FJC, S-180, W-256, EO-771 (1-3)	Complexes with DNA (408), RNA synthesis (409), inhibition of RSV RNA-directed RNA (410,411), cell nuclei (408). See also (7,9-11).
77	CINERUBIN B	*S. antibioticus*	Anthracycline (3,5)	Bacteria, viruses	ECA, FJC, S-180, EO-771 (1,2,412)	See also (9,11).
78	CIROLEROSUS	*S. bellus* var. *cirolerosus* NRRL 3107	(1,2)	Bacteria	KB (1,2,413)	Binding to DNA (414).
79	COCOMYCIN	*Streptomyces* sp.		Bacteria	HeLa (1,2)	
80	CORDYCEPIN	*Cordyceps militaris* Link, *Aspergillus nidulans* (415)	3'-Deoxypurine nucleoside (3-5,8,416)	Bacteria, protozoa	ECA, KB (417)	DNA and RNA synthesis (423, 428,429), cytoplasmic RNA and nucleochromosomal RNA (418, 424), purine synthesis (422), incorporation of formate into purines of RNA and DNA (419), glycine and hypoxanthine and adenine into RNA of ECA (420) 32P into RNA and DNA of ECA (421). Affects phosphoribosyl pyrophosphate amidotransferase and ribose pyrophosphate kinase (7,8), thyrosine

(Continued)

TABLE I (Continued)

No.	Antibiotic	Produced by	Chemical nature	Biological effects — Effect on — Micro-organisms	Biological effects — Effect on — Tumors	Mechanism of action	Footnotes or LD$_{50}$ (mg/kg)
						aminotransferase (425), nuclear poly(A) (426), collapse in chromosome (427). See also (7,9-11,13,16,18).	
81	CORIOLIN	*Coriolus consor*	Sesquiterpene (4)	Bacteria, protozoa	L-1210, YS (430)		
82	CROTOCIN	*Cephalosporium crotocinigenum*, *Trichothecium roseum*	Trichothecane (4,15)	Fungi, viruses	YS (431)		
83	CYCLOCHLORO-TINE	*P. islandicum Sopp.*	Peptide		Leukemia, HeLa (432)	Glycogen neogenesis *in vivo* (433), incorporation of glycine into the protein fraction (434).	
84	CYCLOHEXIMIDE	*S. griseus*, *S. noursei*, *S. naraensis*, *S. viridochromogenes*, *S. chrysomallus*	Glutarimide (1-5,8)	Fungi, protozoa, viruses	Ca-1025, EO-771, ROS, Mecca (1,2,6,435)	RNA synthesis (436,438), DNA synthesis (439-431), protein synthesis in yeast, mammalian cells and plant roots (436, 437,449), virus specific mRNA (444), incorporation of leucine into mitochondrial and cytoplasmic proteins (443), hypoxanthine transport (450), uridine kinase activity (448), porphyrin synthesis (445), interaction between respiration	190 iv Mutagen

292

No.	Name	Source	Type (refs)	Cell lines	Effects
85	CYLINDROCLADIN	*Cylindrocladium ilicicola* ATCC 20229		HeLa (454)	and gluconeogenesis (446), ornithine decarboxylase (452), cell adhesion (447), antigen concentration (442), vacuole formation and exocytosis in *T. pyriformis* (453), induction of xenotropic type C virus (451). See also (7,9,11,12,16-18,22,23).
86	CYTOCHALASIN B	*Phoma* sp., *Helminthosporium dematoideus* (455) (456) (Biosynthesis)	Macrolide (4,22)	AH-130, MSL, Hep-1 (457)	DNA synthesis (469,482). Affects cytokinase in NovH (459), lymphocytes lysosomal hydrolases (470,478), sugar transport (463-466,479,480), glycolyse in human leukocytes (460), mitochondrial ATP (468), pigment movement in the chromatophores (484), interaction with lipid molecules (483), liver plasma membranes (472), phagocytosis and changes in cell shape (461,462), cell cycle (467), cellular adhesion (473,481), epidermal cell migration (458), leukocyte locomotion (460),

(Continued)

293

TABLE I (Continued)

No.	Antibiotic	Produced by	Chemical nature	Biological effects			Footnotes or LD$_{50}$ (mg/kg)
				Effect on			
				Micro-organisms	Tumors	Mechanism of action	
						nuclear division (476), morphological changes in cells (474, 475,477), binding to cells (471). See also (22).	
87	CYTOCHALASIN D	*Metarhisium anisopliae*, *Zygosporium masonii*	Macrolide (4,485,486)		AH-130, W-256, MSL (457)	Protuberances at the cell surface (zeiosis) (487). Antiinflammatory effect (2).	
88	CYTOMYCIN	*S. griseochromogenes*	Pyrimidine group (1-3)		ECA, W-256, S-180 (1-3,488)		
89	CYTOTETRIN	*S. griseoflavus*	Anthrocycline	Bacteria	Tumor cells (489)		Mixture
90	DAUNOMYCIN	*S. peucetius*, *S. coeruleorubidus*	Anthracycline (1-4,8,21)	Bacteria, fungi, protozoa, viruses	W-256, HeLa, KB, ECA, S-180, L-1210 (1-3, 492)	DNA and RNA interaction (497-500). Affects DNA and RNA synthesis (503), incorporation of nucleotide precursors (511), DNA polymerase (501,517), RNA polymerase (513), DNA-dependent DNA polymerase (502,514), mitochondrial ATPase activity (518), uridine kinase (504), transcription and replication (505), cell cycle (509,510,516), chromosome aberrations (506,508). Immunosuppressant (512). See also (7,9-11,17,19,20,22).	15-20 iv Mutagen

No.	Name	Source	Class	Spectrum	Tumor systems	Mode of action/Remarks	LD_{50}
91	DECOYININ	*S. hygroscopicus* var. *angustmyceticus*, *S. hygroscopicus* var. *decoicus*	Ketohexose nucleoside (1,2,7,24)	Bacteria, fungi (2)	W-256, Ca-755, S-180, ECA (1,519)	Conversion of xanthosine mono-phosphate to guanosine mono-phosphate (520-522). Immuno-suppressant (16). See also (8,9,16,18).	2.5 po
92	6-DEMETHYL-6-DEOXYTETRA-CYCLINE	*S. rimosus*			Ca-755, S-180, ECA, L-1210 (24,523)		
93	DEMETRIC ACID	*S. umbrosus* var. *suragaoensis*	Polyene (1,2)	Bacteria, fungi, protozoa (2,24)	HeLa, Ca-755, Ca-1498, L-1210 (1,2,524)		255 ip
94	DENAMYCIN	*Streptomyces* sp. 8756-CC$_2$	(2)	Bacteria	ECA (2)	Incorporation of thymidine and uridine into ECA cells, RNA synthesis (525).	
95	DERMADIN	*Trichoderma viridae*			Tumor cells (526)		
96	DESERTOMYCIN	*S. flavofungini*	(5)	Bacteria	ECA, S-180 (1-3,6)	Permeability changes in cells (527).	1.35 iv
97	DEUTOMYCIN	*S. flavochromogenes*, *S. deutoensis*	Peptide (2)		Tumor cells (2,528)		

(Continued)

TABLE I (Continued)

No.	Antibiotic	Produced by	Chemical nature	Effect on Micro-organisms	Effect on Tumors	Mechanism of action	Footnotes or LD$_{50}$ (mg/kg)
					Biological effects		
98	DEXTROCHRYSIN	S. calvus var. dextrochrysus	Pyrrolobenzodiazepine	Bacteria	ECA (529)		0.75
99	DIKETOCORIOLIN B	Coriolus consor	Sesquiterpene	Bacteria, protozoa	ECA, YS, L-1210 (530)	Efflux of amino acids and K$^+$, Na$^+$-K$^+$-ATPase from Yoshida sarcoma cells (530). Membrane enzymes, transport of glycine and mitochondrial ATPase (531). Immunosuppressant (532). See also (19).	40 ip
100	DISTAMYCIN A	S. distallicus NCIB 893	(1-3,5)	Bacteria, fungi, viruses (2,24)	EC, S-180, W-256, GC (1,3)	Complexes with DNA (533,534, 538). Affects DNA synthesis (536,539), protein synthesis (546), initiation of new RNA (540), binding to poly(dA-dT) (535,537), RNA polymerase (543,545), DNA-dependent DNA and RNA polymerase (536, 544), reverse transcriptase (541), galactooxydase (542). Immunosuppressant (539), mutagen. See also (9-11,22).	76 iv Active also Distamycin B,C
101	DUAZOMYCIN A	S. ambofaciens	Azaamino acid (1-3)	Bacteria, fungi (547)	S-180, Ca-750, L-1210 (1,2,548)	Purine synthesis (549,550). See also (8,22).	

296

No.	Name	Source	Type	Spectrum	Test systems	Notes	Dose
102	DUAZOMYCIN B	*S. ambofaciens*	Azaamino acid (1-3)	Bacteria, fungi	S-180, Ca-755, L-1210 (1-3,6, 551)	Purine synthesis (552,555), glutamine antagonist (554). Immunosuppressant (553,554). See also (10,22).	
103	DUAZOMYCIN C	*S. ambofaciens*	Azaamino acid (1-3)	Bacteria	S-180, Ca-755 (1-3,6)	Competitive inhibition of glutamine (22).	
104	DUBORIMYCIN	*S. coeruleorubidus*	Anthracycline		S-180, L-1210 (556)		6.5 ip
105	DUCLAUXIN	*P. duclauxi, P. stipitatum Thom* (4,557)			ECA, HeLa (558)	Incorporation of uridine into ECA cells (558).	
106	ECHINOMYCIN	*S. echinatus, Streptomyces* sp.	Quinoxaline (1-3,5,8)	Bacteria, protozoa, viruses	AH-130, HeLa (1,2)	RNA and DNA synthesis (559,560), 0.4 ip phage T2 production (561,562). See also (9,18).	
107	EKATETRON	*S. aureofaciens* (563)			ECA	L-Valine incorporation into ECA cells (564).	
108	ENCALINE	*S. halstedtii, Streptomyces* sp.	Fatty acids (1,2)		LIO-1, NK/LY (24,565)	Mixture of fatty acids	

(Continued)

297

TABLE I *(Continued)*

No.	Antibiotic	Produced by	Chemical nature	Biological effects			Footnotes or LD_{50} (mg/kg)
				Effect on		Mechanism of action	
				Micro-organisms	Tumors		
109	ENOMYCIN	*S. maueolor*	Peptide (1-3)		ECA, S-180, YS (1-3)	Incorporation of L-lysine and proline into ECA and HeLa cells (566), aminoacyl-sRNA with polyribosomes (567).	
110	ERICAMYCIN	*S. varius* SE-548	Indicator pH (1,2)	Bacteria	Tumor cells (1,2)		0.5 ip
111	ERIZOMYCIN	*S. griseus* var. *erisensis*			KB (2,568)		
112	ETABETACIN	*Streptomyces* sp.EP-7	(2)	Bacteria	ECA, leukemia (2,24,569)		
113	EUCALINE	*Actinomyces* sp.13363			NK/Ly (2)		
114	FATTY ACIDS	*P. crustosum*, *P. tardum*, *P. stipitatum*, *Sepedonium ampulosporum*, *Cephalosporium diospyri* (570)	Fatty acids (5,4)	Bacteria	ECA, L-1210, SN-36 (570,572)	Inhibition of adenine incorporation (571). See also (11,12,14).	59 ip Mixture
115	FERVENULIN	*S. fervens*, *S. rubrreticuli*	Pyrimidotriazine (2,5,6)	Bacteria, fungi, protozoa	Tumor cells (1,2,6)		65 ip

No.	Name	Organism	Type	Source/Class	Active against	Mechanism	Dose
116	FILIPIN	*S. filipinensis*	Polyene (1-5)	Fungi	ECA, NovH (574)	Thymidine incorporation into DNA of ECA cells (574), interaction with sterol (575,576), induced K⁺ leak and 6-phosphate dehydrogenase (577), oxygen consumption in NovH (578,579), function of cytoplasmic membrane (580,581), induction of hypercholesterolemia in insects (582). See also (8,9).	17 ip
117	FLAMMULIN	*Flammulina velutipes*	Polyene		ECA, S-180 (3)	Mitosis and cytopathogenic effect in tumor cells (583).	25 ip
118	FLUOPSIN	*Pseudomonas* sp.		Bacteria	Tumor cells (584)		
119	FORMYCIN	*Nocardia interforma*, *S. lavendulae* (585)	Pyrazolopyrimidine nucleoside (1-3)	Bacteria, viruses (586)	ECA, L-1210 (1-3)	RNA (592,594), complexes with polyribonucleotides (593), purine synthesis *de novo* in tumor cells (587,588), nucleotide synthesis (589), nucleoside transport (599), incorporation into RNA in place of the corresponding normal purine nucleosides (591). Mitosis (595). See also (9-11,13,16,18).	250 iv

(Continued)

TABLE I *(Continued)*

No.	Antibiotic	Produced by	Chemical nature	Micro-organisms	Tumors	Mechanism of action	Footnotes or LD$_{50}$ (mg/kg)
				Effect on ← Biological effects			
120	FRAGIN	*Pseudomonas fragi*		Bacteria, fungi	YS (596)		
121	FREQUENTIN	*P. frequentans Westling, P. palitans, P. brefeldianum*	Carbocyclic compound (3-5)	Bacteria, fungi	ECA, YS (597)	Adenine and L-valine incorporation into ECA cells (597).	
122	FUMAGILLIN	*Aspergillus fumigatus*	Polyene (3,4,7,598)	Bacteria, fungi, protozoa	S-180, RC, C3HBA, Ca-755 (599)	RNA in ECA cells (600), inhibition of RNA synthesis (601). See also (14,22).	800 sc
123	FUNICULOSIN	*P. funiculosum Thom*	Anthraquinone (5)	Bacteria	ECA (602)		4 ip
124	FUSARENON X	*F. nivale*	Trichothecane	Fungi (2,14,15)	ECA	Uptake of leucine and thymidine in ECA, interference with poly(U)-directed synthesis of polyphenylalanine (603).	Toxin
125	FUSARIOCIN	*F. moniliforme*		Fungi	ECA	RNA synthesis in tumor cells (604).	Toxin
126	GANCIDIN A,W	*Streptomyces* sp. AK-82	(1-3,6)	Bacteria	ECA (1-3,6)		80 iv
127	GELDANAMYCIN	*S. hygroscopicus* var. *geldanus*	Ansamycine group (2,10)	Bacteria, fungi, protozoa	KB, L-1210 (2,605)		1 ip

300

128	GLIOTOXIN	*Aspergillus tereus,* *Aspergillus* *chevalieri,* *Trichoderma viridae,* *Gliocladium* *fimbriatum*	Dithiodiketo-piperazine (3-7)	Bacteria, fungi, viruses (606)	RC, S-180 (3,6,607)	Induction of ATP-energized mitochondrial volume changes (608), RNA in viruses (609), viral RNA-dependent RNA polymerase (610,611), inhibition of viral RNA replication (611, 612). See also (11,15).	25 ip Toxin
129	GLOBIMYCIN	*A. globisporus*	Peptide		Tumor cells (613)	Affected template function of DNA transcription process (614).	
130	GLUCAN	*Coriolus versicolor,* *Aspergillus niger,* *P. crustosum*	(617)		Sarcoma cells, S-180 (615-617)		
131	GLUTAMINASE A	*Pseudomonas* *aeruginosa* (618,620)			ECA (619)		
132	GOUGEROTIN	*Streptomyces* sp. S-514, *S. gougerotii*	4-Aminohexose pyrimidine nucleoside (1,2)	Bacteria	HeLa, S-180 (621)	Peptidyl transferase, protein synthesis, inhibition of peptide chain elongation, liver uridine kinase (622-624). See also (8,9,11,12,16).	250 ip
133	GRAMINOMYCIN	*Streptomyces* sp.	(1)	Bacteria	HeLa (1,2)		

(Continued)

TABLE I *(Continued)*

No.	Antibiotic	Produced by	Chemical nature	Biological effects				Footnotes or LD$_{50}$ (mg/kg)
				Effect on				
				Micro-organisms	Tumors	Mechanism of action		
134	GRANATICIN	*S. olivaceus*, *S. litmogenes*	Indicator pH	Bacteria	Tumor cells (625)	Inhibition of RNA and protein synthesis, interaction with DNA template and with RNA polymerase, interference with phosphorylation of RNA precursors (10,11).		
135	GRISEOFAGIN	*S. griseus*			Tumor cells (626)			
136	GRISEOLUTEIN B	*S. griseoluteus*	(1-3,5)	Bacteria, protozoa	ECA, EC (3,627,628)			Antiinflammatory (2)
137	GRISORIXIN	*Streptomyces* sp., *S. griseus*	(2,629)	Bacteria, fungi	HeLa (630)			
138	GUANAMYCIN	*Streptomyces* sp. 6617-IAUFPe	Quinoxaline		Tumor cells (631)			
139	HADACIDIN	*P. frequentans*, *P. aurantio-violaceum*	Amino acid (3,4,8,632,633)	Bacteria, protozoa (634)	HAd-1, A-42, HS-1, KB (3,635)	Affected conversion of IMP to adenylosuccinic acid, incorporation of various precursors into nucleotides, competition with L-aspartate (636). See also (7,9,10,16,22,23).		0.3 ip

No.	Name	Source	Chemical group	Antimicrobial spectrum	Tumor activity	Mechanism/Remarks	LD50
140	HEDAMYCIN	*S. griseoruber* n.sp.		Bacteria, fungi	W-256, S-180 (2,24,638)	DNA (639), DNA polymerase *in vitro*, transcription by RNA polymerase (640). See also (9,11).	
141	HILAMYCIN A,B	*S. rochei*	(1)		Tumor cells (2,24,641)		
142	HYGROSCOPIN A,B,C	*S. hygroscopicus*	(1-3,5,6)	Fungi, viruses	YS (1,2,6)		8.75 ip
143	ILLUDIN M	*Clitocybe illudens*	Sesquiterpene (4,5,8)		Tumor cells (642). See also (22)		
144	ILLUDIN S	*Clitocybe illudens, Lampteromyces japonicus* (643)	Sesquiterpene (4,5,8,644)		ECA (644)	DNA, partly RNA synthesis (645).	
145	INOMYCIN	*S. griseus* var. *inomycini.*	Glutarimide	Fungi	ECA, S-180, S-37 (646)		390 iv
146	IODININ	*Waksmania aerata, Chromobact. iodinum*	Phenazine group (2,3,5)	Bacteria	Tumor cells (2)		
147	IYOMYCINS A,B₁,B₂,B₃,B₄,B₅	*S. phaeoverticil-latus*	Peptide group	Bacteria	ECA, S-180, YS, HeLa, AN-36 (1-3)	Cytological changes in the cells (647,648).	Complex 150 ip

(Continued)

303

TABLE I (Continued)

No.	Antibiotic	Produced by	Chemical nature	Biological effects			Footnotes or LD$_{50}$ (mg/kg)
				Effect on		Mechanism of action	
				Micro-organisms	Tumors		
148	JAWAHARENE	*Aspergillus niger*		Viruses	YS (649)	Respiration in tumor cells (650), inhibition of lipid synthesis (651).	
149	JULIMYCIN B-II	*S. shiodaensis*	Anthraquinone	Bacteria, viruses	ECA, AH-130 (1-3,21,24,652)	Affected the leukocytes (653). See also (21).	70 ip
150	KANCHANOMYCIN	*S. tumemacerans, Streptomyces* sp.	(1,2,654)	Bacteria, fungi	ECA, HeLa (1,2,24,655)	DNA and RNA synthesis, inter-action with polynucleotides (656), inhibition DNA-dependent DNA polymerase (657). See also (9-11).	
151	KIDAMYCIN	*S. phaeoverticil-latus* var. *takatsukiensis*		Bacteria, viruses	ECA, S-180, SN-36, FVL (658,659)	DNA, incorporation of thymidine into RNA, arginine into pro-teins (660).	200 iv
152	KIKUMYCIN A, B	*S. faeochromogenes*	(1,2,661)	Bacteria, viruses	HeLa, S-180 (2)		
153	KINAMYCIN	*S. murayamaensis*	Quinone (2)	Bacteria	ECA, S-180 (662)		
154	KINOMYCIN B	*Streptomyces* sp.			Tumor cells (24,663)		
155	KUNDRYMYCIN	*S. metachromogenes* (664)		Bacteria, fungi (2)	HeLa, W-256, P-388, L-1210, S-180 (665)		

No.	Name	Source	Nature (ref.)	Test systems	Activity	Remarks
156	KUNOMYCIN	*Streptomyces* sp.	(1)	HeLa (1,2,24)		
157	LABILOMYCIN	*S. albosporus* var. *labilomyceticus*	(1-3,666)	ECA, S-180, HeLa (667)		
158	LARGOMYCIN	*S. pluricolorescens*, NCRL 0367	Peptide (2,668)	ECA, S-180, SN-36, HeLa (2,669)	35 ip	
159	LATERIOMYCIN A,B,F	*S. griseoruber*	Indicator pH (2)	S-180, YS (2,670)	0.4 ip	
160	LENAMYCIN	*S. reticuli*, *S. chibaensis*	Acetylene (672,673)	HeLa, NF, EC (1-3,6,671)		
161	LENTINAN	*Lentinus edodes* Berk.	Polysaccharide	S-180 (674,675)		Induction of interferon and similar substances (676–678).
162	LEUCINE CEHYDROGENASE	*Bacillus sphaericus*		ECA (680)		
163	LEUCINOSTATIN	*P. lilacinum*	Peptide (679)	EC, HeLa (681)	1.6	
164	LIENOMYCIN	*A. diastatochromogenes* var. *lienomycini* (682,683)	Polyene group (684)	ECA, S-37, NK/Ly, L-5178 (685)	11.6 sc	
165	LIPIDS	*Crithidium oncopelti*		S-180 (686)		

(Continued)

305

TABLE I (Continued)

| No. | Antibiotic | Produced by | Chemical nature | Biological effects | | Mechanism of action | Footnotes or LD$_{50}$ (mg/kg) |
| | | | | Effect on | | | |
				Micro-organisms	Tumors		
166	LUCENSOMYCIN	*S. lucensis*	Polyene group (687)	Fungi, protozoa	ECA, NovH (688)	Incorporation of uridine and thymidine into RNA and DNA (689), interaction with the cell membrane (689).	44.6 iv
167	LUTEOMYCIN	*S. tanashiensis, S. flaveolus*	Indicator pH (1-3,5)	Bacteria, fungi	Tumor cells (1,690)		6.25 iv
168	LUTEOSKYRIN	*P. islandicum (691)*	Anthraquinone (4,5,21)	Protozoa	ECA, HeLa (692,695)	Complexes with DNA and RNA (696), affects RNA and protein synthesis (692,696), DNA repair (694), nucleotide chain elongation during transcription (693). Hepatotoxic, mycotoxin (2). See also (9,14,15).	6.55 iv Mutagen
169	LUTEOSTATIN	*Streptomyces* sp.		Bacteria	HeLa (1,2)		
170	LYMPHOMYCIN	*Streptomyces* sp. S-66	Peptide		S-180, SN-36, NQT-1 (2,697)		
171	MACROMOMYCIN	*S. macromomyceticus*	Peptide	Bacteria, fungi	S-180, L-1210, P-388, B-16 (2,24,702)	Inhibition of DNA synthesis, affected cell membrane (698-701). See also (19).	62 iv
172	MALEIMYCIN	*S. showdoensis*	Maleimide	Bacteria	L-1210 (703)		
173	MANNAN	*C. utilis, C. albicans*			S-180 (705)	Induction of interferone (704).	

306

No.	Name	Source	Chemical class	Spectrum	Tumor systems	Dose	Remarks
174	MARINAMYCIN	S. mariensis	Peptide (1-3)		ECA (6,706)		
175	MARINAMYCIN-like substance	A. candidus 10484			Tumor cells (2)		
176	MELANOMYCIN	S. melanogenes	Peptide	Bacteria, fungi, protozoa	ECA (1-3,5,6,707)		
177	MELINACIDINS I - IV	Acrostalagmus cinnabarinus var. melinacidicus	Dithiodiketo-piperazine	Bacteria, fungi	KB, L-1210 (708,709)		Synthesis of nicotinic acid and nicotin amide (710).
178	MEZZANOMYCIN	S. senocanescens	Indicator pH	Bacteria, fungi (1,2)	ECA (1,2,24,711)		
179	MINIATOMYCIN M$_2$,M$_3$	Streptomyces sp. 4360		Bacteria	W-256, YS (24,712)		
180	MINIMYCIN	S. hygroscopicus, Streptomyces sp. (713)	Oxazine nucleoside (714)	Bacteria	ECA, S-180 (713)	80 iv	
181	MINOMYCIN	S. minoensis	Anthracycline (1-3,21)	Bacteria, fungi	ECA, S-180 (1-3,715)	15 ip	Probably inhibition RNA synthesis (715). See also (7,22).

(Continued)

307

TABLE I *(Continued)*

No.	Antibiotic	Produced by	Chemical nature	Biological effects			Footnotes or LD_{50} (mg/kg)
				Effect on			
				Micro-organisms	Tumors	Mechanism of action	
182	MITHRAMYCIN	*S. athroolivaceus*, *S. ivernii*, *Streptomyces* sp., *S. argillaceus* (716,717)	Aureolic acid (1,2)	Bacteria	ECA, HeLa, S-180, Ca-755, HS-1, Hep-3 (1,2,718-720)	RNA synthesis (720-723). See also (7,9-11,17,19,20,22).	
183	MITIROMYCIN A	*S. verticillatus*	Mitomycin group (1-3,8,21,724)	Bacteria	Tumor cells (1,2)		
184	MITOCROMIN A,B	*S. viridochromogenes*	Anthracycline (2)	Bacteria	W-256, P-388 (2,24,725)		
185	MITOMALCIN	*S. malayensis*	Peptide (2)	Bacteria	L-1210, P-388, HeLa (2,24,726)		
186	MITOMYCIN A,B	*S. verticillatus*, *caespitosus* (727)	Mitomycin (1-3)	Bacteria	ECA, YS (1,2,6)		
187	MITOMYCIN C	*S. caespitosus* (728) (Biosynthesis) (751,752)	Mitomycin (1-3,5,8,21,753)	Bacteria, fungi, viruses	ECA, Ca-1025, RS, EC-771, C3HBA, AH-130, YS (1-3,6,8,729, 730)	Cross linking of DNA (731,732, 749), degradation of DNA (734), destruction of ribosomal RNA (745), inactivation of DNA binding (737), interaction with purine and pyrimidine bases (735). Affects DNA synthesis (733,736), RNA synthesis (755), lipid synthesis (754), protein synthesis (758),	Active also Mitomycin R, Mytomycin Y

308

No.	Name	Producer	Class	Active against	Activity	Notes	
188	MONAZOMYCIN	*S. mashiuensis*	(1-3)	Bacteria, fungi	BC-63, S-37, EC, (1-3,762)	chromosomal aberrations (742, 743,750), breakage (759,760), cell cycle (761), activity of DNase (746,747), thymidine kinase (756), synthesis of induced enzymes:penicillinase, β-galactosidase (757), activity of lysomal enzyme (741). Induction lysogenic phages (10,744). Immunosuppressant (740), teratogen (738), mutagen (739,748). See also (7,9-11,16-20,22).	5 iv
189	MONENSIN	*S. cinnamonensis*	Monensin group (2)	Bacteria, protozoa	Tumor cells (2)	See also (11).	
190	MONODEACETYL-ANGUIDIN	*F. concolor,* *F. diversisporium*	Trichothecane		Tumor cells	Interaction with lipids of the membrane (11). See also (11).	
191	MUTAMYCIN	*Streptomyces* sp.			Tumor cells (1,2)	Mitotic effect (763). See also (15).	
192	MUTOMYCIN	*A. atroolivaceus* var. *mutomycini*	(1-3,765)	Bacteria, viruses	ECA (6,764)	Inhibition of respiratory deficient staphylococci (2).	
193	MYCETIN D-17	*A. violaceus*	Anthracycline		S-180 (766)		

(Continued)

309

TABLE I *(Continued)*

No.	Antibiotic	Produced by	Chemical nature	Biological effects: Effect on Micro-organisms	Biological effects: Effect on Tumors	Biological effects: Mechanism of action	Footnotes or LD_{50} (mg/kg)
194	MYCOPHENOLIC ACID	*P. stoloniferum*, *P. brevicompactum Dierckx*	(3,5,8,767) (Analogs)(775)	Bacteria, fungi, viruses (3,768)	W-256, S-180, MH-134 (768-770)	Inhibition of DNA synthesis in intact cells (771), guanine nucleoside biosynthesis (773), IMP dehydrogenase and GMP synthetase (772). Affects interconversion of inosine, xanthine, and guanosine monophosphate (772). Immunosuppressant (774), mycotoxin (2). See also (10,11, 14,23).	550 iv
195	MYRORIDIN	*Myrothecium roridum*	Peptide	Fungi	HeLa (776)		250 ip
196	NAEMATOLIN	*Naematola fasciculare*			HeLa (777)		
197	NARANGOMYCIN	*S. lavenduli-griseus*		Bacteria, fungi, protozoa	S-180, Ca-755 (1,2,24)		
198	NEBULARINE	*Clitocybe nebularis*, *S. yokosukaensis* (2,778)	Purine nucleoside (3,5,6, 779)	Bacteria, viruses	S-180 (2,3,6,7)	Nucleoside synthesis and phosphoribosyl pyrophosphate (782), competitive inhibition of NAD (780), mitotic aberrations and breaking of chromosomes (781). See also (16).	100 sc

No.	Name	Source	Class	System	Mechanism	Dose
199	NEOCARZINO-STATIN	*S. carzinostaticus*, var.F-41	Peptide (1,2,783)	Bacteria	S-180, SN-36, HeLa (1,2,24,784, 789)	DNA synthesis (785-787,791), degradation of DNA (784,790, 792), induced scission of DNA (788). See also (10,11, 19,20).	50 ip
200	NEOCIDE	*Actinomyces* sp.	Peptide		Bacteria		
201	NEOPLURAMYCIN	*S. pluricolorescens*	(794,2)	Bacteria	S-45 (1,2,6,793)	Inhibition of nucleic acid synthesis (796).	12.5 iv
202	NETROPSIN	*S. ambofaciens*, *S. chromogenes*, *S. netropsis*, *Streptomyces* sp., *S. reticuli*	Oligopeptide (2)	Bacteria, fungi, protozoa (1,2)	L-1210, YS (2, 795)	Interaction with DNA (10,798), inhibition DNA, RNA and pro-tein synthesis (799).	
203	NIVALENOL	*F. nivale*	Trichotecane (4,14,15,800)	Bacteria	ECA, HeLa (801,802)	DNA and protein synthesis in tumor cells (802), mycotoxin (2).	
204	NOGALAMYCIN	*S. nogalater* var. *nogalater*	Anthracycline (1,2,21,24)	Bacteria	L-1210, KB (803)	DNA and RNA synthesis (804, 807,809), interaction with DNA (810,811), decay of RNA (806). Affects RNA polymerases (805, 807), chromosomal aberration (808). See also (7,9-11,17,18,22).	18 ip

(Continued)

311

TABLE I *(Continued)*

No.	Antibiotic	Produced by	Chemical nature	Biological effects: Effect on Micro-organisms	Biological effects: Effect on Tumors	Biological effects: Mechanism of action	Footnotes or LD_{50} (mg/kg)
205	NONACTIN	*S. tshushimaensis*, *S. viridochromogenes*, *S. werraensis*	Macrotetrolide (1-3,5,812,813)	Bacteria	Tumor cells (814)	Affected function of cytoplasmatic membrane (815,816). See also (7,9,12).	
206	OLEFICIN	*S. parvullus*	Polyene	Bacteria	YS (2,817)		40 iv
207	OLIVOMYCINS A,B,C,D	*A. olivoreticuli* (Biosynthesis) (829)	Aureolic acid (1-3,819)	Bacteria, viruses	S-180, lymphosarcoma (1-3,818)	Complexes with DNA (826). Affects RNA synthesis (820, 821,823), DNA synthesis (822), RNA and DNA polymerase (822, 824,825,828), elongation of RNA chain (824), thyrosine aminotransferase (827). See also (7,9-11,19,22).	10.6 iv
208	ONCOSTATIN B(x),C,K	*Streptomyces* sp.INA 16/58	Actinomycin (1-3)	Bacteria	Tumor cells (1,2)		
209	ORYZACHLORIN	*Aspergillus oryzae*	Dithiodiketopiperazine	Fungi	ECA (830)		
210	OSSAMYCIN	*S. hygroscopicus* var. *ossamyceticus*		Fungi, protozoa	HeLa, S-180, L-1210, KB (1,2,24,831)	Effect on ATPase (832). See also (11).	1.8 iv
211	OXAZINOMYCIN	*S. tanashiensis*	Oxazine nucleoside		ECA (833)		100 iv

No.	Name	Source organism			Test systems	Mode of action	
212	5-OXO-1H-PYRROLO[2,1-c][1,4]BENZODIAZEPIN-2-ACYLAMIDES	S. refuineus var. sermotolerans		Protozoa	Tumor cells (2)		
213	PACHYMAN	Poria cocos (834)	Polysaccharide		Tumor cells (835)		
214	PACTAMYCIN	S. pactum var. pactum	(1-3,836)	Bacteria	KB, L-1210, ME-1, Ca-755 (1-3,837)	Protein synthesis (838,842, 843). Affects transfer of aminoacyl-sRNA into ribosomes (839), interaction with ribosomal units (840,841,844), uridine kinase (845). See also (7,9,11,12,16,17,22).	15.6 iv
215	PANEPOXYDON PANEPOXYDION	Panus rudis Fr., P. conhatus			P-815 (846)		
216	PATHOCIDIN	S. albus var. pathocidicus, S. morookaensis	Purine (1-3)	Fungi	Ca-755, EO-771, C3H (1,2,847)	Incorporation into RNA in place of the corresponding normal purine nucleotides (848), breakage of heavy poly-ribosomes (849), genetic effect (850). See also (9,11,13,16,18,22).	Mutagen

(Continued)

313

TABLE I *(Continued)*

No.	Antibiotic	Produced by	Chemical nature	Effect on Micro-organisms	Effect on Tumors	Mechanism of action	Footnotes or LD$_{50}$ (mg/kg)
217	PATULIN	*Penicillium* sp., *Aspergillus* sp. (851)	Lactone (3-5,8,853)	Bacteria, fungi, protozoa	ECA (852)	Affects aerobic respiration in bacteria (853), cell division (855), chromosomal aberration (860), aldolase (856), lactic dehydrogenase (857,858), SH groups (854). Mutagen. Toxin (859). See also (7,14,15).	
218	PELIOMYCIN	*S. luteogriseus*	(1,2)	Bacteria, fungi	HeLa (2,861)	Inhibition of respiratory chain and energy transfer (11).	
219	PENICILLIC ACID	*Penicillium* sp., *Aspergillus* sp.	(3-5,8)	Fungi	ECA, SN-36, HeLa (862)	Inhibition of aldolase (863), lactic dehydrogenase (864), nuclei in cells (862), myco-toxin (2). See also (14).	
220	PEPTIMYCIN	*S. mauvecolor* (112)	Peptide (1-3)		ECA, S-180, Ca-755 (1-3)	(865)	25 iv
221	PERLIMYCIN	*S. chrysomallus*	(1,2,866)	Bacteria	ECA, S-180, Ca-755 (1,2,24)		
222	PHENOMYCIN	*S. fervens* var. *phenomyceticus*	Peptide		ECA, HeLa, S-180 (2,24,867)	Inhibition of protein synthesis and aminoacyl-sRNA in HeLa cells (868), blocks the initiation of globin synthesis (869).	

No.	Name	Source	Type	Test organisms	Activity	
223	PHLEOMYCIN A,B,C,D₁,D₂,E,F,G,H,I,J,K	*S. verticillus* (870)	Peptide (1-3,871)	Bacteria, fungi	ECA, S-180, YS Ca-755 (1-3,872)	Binding to DNA (873,874). Affects DNA (875), RNA synthesis (876), DNA breakdown (877-880), DNA polymerase (882), incorporation of thymidine triphosphate into DNA (881), replication of meningovirus (884), chromosomal aberrations (885,886), soluble cytoplasmatic proteins (883). See also (9-11,22).	Mutagen
224	PIGMENT	*S. padanus*, *S. xanthophaeus*		Bacteria	ECA (887)	
225	PILLAROMYCIN A,B,C	*S. flavovirens* (888,889)	Anthracycline (1-3,890)	Bacteria	Tumor cells (1-3,24,888)	
226	PLURALLIN	*S. pluricolorescens*	Glycoprotein (1,2)	Bacteria	ECA, HeLa (1,2,24)	Immunosuppressant (2).
227	PLURAMYCIN A,B	*S. pluricolorescens*	Indicator pH (1-3,5,6)	Bacteria, fungi	ECA, S-180, YS (1-3,6,891)	DNA synthesis lesser RNA synthesis (892,893), DNA-dependent DNA and RNA polymerase of *E. coli* (893,894). See also (7,9,11).
228	POINE	*F. poae*	(3,5,6,895)	Bacteria	ECA (3,6,896,897)	

(Continued)

315

TABLE I *(Continued)*

No.	Antibiotic	Produced by	Chemical nature	Biological effects		Mechanism of action	Footnotes or LD$_{50}$ (mg/kg)
				Effect on			
				Micro-organisms	Tumors		
229	POLYMYCIN	*Aspergillus* sp. 1787-9	Streptothricin (1-3)	Bacteria, viruses	ECA, S-180, NK/Ly (1-3,6,898,899)	Mitosis in HeLa cells (899)	50-60 iv
230	POLYSACCHA-RIDES	*Pseudomonas aeruginosa* (900), Basidiomycetes (901)			S-180 (902)	Enhanced the tumoral antibody response (903-905).	
231	PORFIROMYCIN	*S. verticillatus*, *S. ardus*	Mitosane (1-3)	Bacteria, viruses (906)	S-180, W-256, KB, HAd-1, L-1210, GC, MSL, ECA (1-3,908,907)	DNA synthesis (909), induction of deletion (910). See also (7,9,11,19,22).	
232	PORICIN	*Poria cortiocola*	Peptide (911)		S-180 (912)		
233	PRIMOCARCIN	*Nocardia fukayae*	Amino acid (1-3)	Bacteria	ECA, S-180, EO-771 (1-3,913)		
234	PRUNACETIN A	*S. griseus* var. *purpureus*, *S. californicus*	(2,24)	Bacteria	ECA, S-180 (2)	(914)	93.4 ip
235	PSICOFURANINE	*S. hygroscopicus* var. *decoicus*, *S. hygroscopicus* var. *angustmyceticus*, Streptomyces sp. 6A-704 (1-3,5,6, 8,915)	Purine	Bacteria (916)	W-256, JS, GC (1-3,6,917)	RNA and DNA synthesis (921), purine synthesis *de novo* by feedback mechanism (922). Affects incorporation of glycine into purine bases (918), XPM aminase (919,920), interaction of the biosynthesis of GPM from XPM (916,917). See also (7,9,11-13,16,22,23).	41 sc

						Mutagen
223	PHLEOMYCIN A,B,C,D$_1$,D$_2$,E, F,G,H,I,J,K	S. verticillus (870)	Peptide (1-3,871)	Bacteria, fungi	ECA, S-180, YS Ca-755 (1-3,872)	Binding to DNA (873,874). Affects DNA (875), RNA synthesis (876), DNA breakdown (877-880), DNA polymerase (882), incorporation of thymidine triphosphate into DNA (881), replication of meningovirus (884), chromosomal aberrations (885,886), soluble cytoplasmatic proteins (883). See also (9-11,22).
224	PIGMENT	S. padanus, S. xanthophaeus		Bacteria	ECA (887)	
225	PILLAROMYCIN A,B,C	S. flavovirens (888,889)	Anthracycline (1-3,890)	Bacteria	Tumor cells (1-3,24,888)	
226	PLURALLIN	S. pluricolorescens	Glycoprotein (1,2)	Bacteria	ECA, HeLa (1,2,24)	Immunosuppressant (2).
227	PLURAMYCIN A,B	S. pluricolorescens	Indicator pH (1-3,5,6)	Bacteria, fungi	ECA, S-180, YS (1-3,6,891)	DNA synthesis lesser RNA synthesis (892,893), DNA-dependent DNA and RNA polymerase of *E. coli* (893,894). See also (7,9,11).
228	POINE	F. poae	(3,5,6,895)	Bacteria	ECA (3,6,896,897)	

(Continued)

315

TABLE I *(Continued)*

No.	Antibiotic	Produced by	Chemical nature	Effect on Micro-organisms	Effect on Tumors	Mechanism of action	Footnotes or LD$_{50}$ (mg/kg)
					Biological effects		
229	POLYMYCIN	*Aspergillus* sp. 1787-9	Streptothricin (1-3)	Bacteria, viruses	ECA, S-180, NK/Ly (1-3,6,898,899)	Mitosis in HeLa cells (899)	50-60 iv
230	POLYSACCHA-RIDES	*Pseudomonas aeruginosa* (900), Basidiomycetes (901)			S-180 (902)	Enhanced the tumoral antibody response (903-905).	
231	PORFIROMYCIN	*S. verticillatus*, *S. ardus*	Mitosane (1-3)	Bacteria, viruses (906)	S-180, W-256, KB, HAd-1, L-1210, GC, MSL, ECA (1-3,908,907)	DNA synthesis (909), induction of deletion (910). See also (7,9,11,19,22).	
232	PORICIN	*Poria corticola*	Peptide (911)		S-180 (912)		
233	PRIMOCARCIN	*Nocardia fukayae*	Amino acid (1-3)	Bacteria	ECA, S-180, E0-771 (1-3,913)		
234	PRUNACETIN A	*S. griseus* var. *purpureus*, *S. californicus*	(2,24)	Bacteria	ECA, S-180 (2)	(914)	93.4 ip
235	PSICOFURANINE	*S. hygroscopicus* var. *decoicus*, *S. hygroscopicus* var. *angustmyceticus*, *Streptomyces* sp. 6A-704 (1-3,5,6, 8,915)	Purine	Bacteria (916)	W-256, JS, GC (1-3,6,917)	RNA and DNA synthesis (921), purine synthesis *de novo* by feedback mechanism (922). Affects incorporation of gly-cine into purine bases (918), XPM aminase (919,920), inter-action of the biosynthesis of GPM from XPM (916,917). See also (7,9,11-13,16,22,23).	41 sc

316

No.	Name	Source	Chemical class	Spectrum	Test systems	Remarks
236	PUROMYCIN	*S. alboniger*	3-Deoxypurine nucleoside (1-3,5,8,940)	Bacteria, protozoa	ECA (1-3,6)	Affects protein synthesis (923-927), DNA replication (932), transcription (931), ribosomes (928-930) accumulation of uridine into polynucleotides (933), amino acid transport (941), ornithine decarboxylase activity (935), phosphatidyl choline synthesis (936), decreased in oxygen consumption (934). Influence on locomotion and detachment of cells (939), vacuole formation and exocytosis in *T. pyriformis* (942). Inhibits antibody formation (937). See also (7,9,11-13, 16,22,23). 335 iv
237	PURPUROMYCIN (acetyl derivative)	*Actinoplanes ianthinnogenes* A1668		Bacteria	Leukemia (943)	
238	PYRAZOMYCIN	*S. candidus*	Pyrazole nucleotide (2,944)	Viruses	W-256, Ca-755, GL, X-5563, ROS (945)	Decarboxylation of orotidine monophosphate (945). See also (16).
239	PYRENOPHORIN	*Pyrenophora avenae*, *Stemphylium radicinum*	Lactone (3,4,948)	Fungi (946,947)	Tumor cells (948)	

(Continued)

TABLE I (Continued)

No.	Antibiotic	Produced by	Chemical nature	Biological effects			Footnotes or LD$_{50}$ (mg/kg)
				Effect on			
				Micro-organisms	Tumors	Mechanism of action	
240	5H-PYRROLO[2,1-c][1,4]BENZO-DIAZEPIN-5-ONE	S. achromogenes var. tomaymyceticus		Bacteria, fungi	Tumor cells (949)		
241	QUINOMYCIN B	Streptomyces sp.732, Streptomyces sp.1752	Quinoxaline (950)	Viruses, bacteria	HeLa (1,2,3)		0.054 iv
242	QUINOMYCIN C	Streptomyces sp.732	Quinoxaline (950)	Bacteria	ECA, EC (1,2,3,)	Incorporation of uracil into RNA (951). See also (9,11,18, 22).	0.025 ip
243	RACTINOMYCIN A,B	S. phaeochromogenes, Streptomyces sp.	Indicator pH (1-3,5)	Bacteria, fungi	ECA (1,2,6,952)		5-12.5 iv
244	RAROMYCIN	S. albochromogenes	(1,5)	Bacteria	ECA, S-180 (1,2,3)		200 iv
245	REQUINOMYCIN	S. filamentosus	Anthracycline (953)	Bacteria	YS	Transfer of R-factor and transport between E. coli strains (953-954).	
246	RETAMYCIN	Streptomyces sp., S. olidensis n.sp. (956)	Anthracycline	Bacteria	W-256, YS (955,956)		10.7 iv
247	REUMYCIN	Actinomyces sp. (957)	Pyrimidotriazine		ECA, S-180, LIO-1, W-256 (2,24)	RNA synthesis and transport of electrons (959).	

318

No.	Name	Source	Chemical class	Antimicrobial activity	Antitumor activity	Notes / Mechanism
248	RIFAMYCIN SV RIFAMPICIN	*S. mediterranei* (960)(Biosynthesis)(978)	(1-3) (Analogs)(982)	Bacteria, viruses (1,2,961,962)	W-256	RNA synthesis (971,979), protein synthesis (973,974). Affects DNA polymerase (964), RNA polymerase (968-970), RNA-directed DNA polymerase (963,965-967), AMP poly-nucleotidyl exotransferase (972), chromosome aberrations (976), transfer of R-factor (980), bacteriophage biogenesis (981). Antimitotic effect (975). Immunosuppressant (977). See also (7,9-11,22,23).
249	RORIDIN A	*Myrothecium roridum, Myrothecium verucaria*	Trichothecene (3,4,8,15)	Fungi	S-37, ECA, W-256, P-815 (3,15)	Inflammatory effect (2) — Active also: Roridins C,D,E,H
250	ROSEOLIC ACID	*Streptomyces* sp.	Indicator pH		Ca-755, HS-1 (1-3)	
251	ROSSIMYCIN	*S. chrysomallus*	Actinomycin		Tumor cells (1-3)	
252	RUBIFLAVIN	*Streptomyces* sp.SC 3728	Indicator pH	Bacteria, protozoa (984)	S-180, Ca-755 (1,2,24)	DNA synthesis lesser on RNA synthesis (983). See also (9,11). — 15 ip
253	RUBOMYCIN B	*S. coeruleorubidus*	Anthracycline (21,985)	Bacteria	ECA, S-37, L10-1 (1,2,986)	DNA and RNA synthesis (987). See also (9,19).

(Continued)

TABLE I *(Continued)*

No.	Antibiotic	Produced by	Chemical nature	Biological effects: Effect on Micro-organisms	Biological effects: Effect on Tumors	Biological effects: Mechanism of action	Footnotes or LD_{50} (mg/kg)
254	RUFOCHROMOMYCIN	*S. rufochromogenes*		Bacteria (1,2)	ECA, KB, S-180 (2,24,988)		2 iv
255	RUGULOSIN	*P. rugulosum* *P. bruneum*	Anthraquinone (3-5,8,21)	Fungi, protozoa, viruses (990)	ECA (989)		Toxin
256	RYEMYCIN A_1,A_2,B_1	*S. ryensis*	Anthracycline	Bacteria	Tumor cells (1,2)		
257	SANGIVAMYCIN	*S. rimosus*, *Streptomyces* sp.	Pyrrolopyrimidine nucleoside (1,8)	Bacteria, fungi	L-1210, S-180, HeLa, Ca-755 (2,24,991)	RNA synthesis (992,993). Affects ATP in various polymerization reactions (10). See also (7-11,13,16).	
258	SARKOMYCIN	*S. erythrochromogenes*	Carbocyclic compound (1-3,5,6)	Bacteria, fungi	ECA, YS (1,2,6,994)	DNA synthesis (995), incorporation of glycine and ^{32}P (996), inhibition of DNase (996). See also (7,11).	
259	SEPTACIDIN	*S. fimbriatus*	Adenine nucleoside. See also (14,16)	Fungi	S-180, Ca-755 (1-3)		
260	SHOWDOMYCIN	*S. showdoensis* (Biosynthesis) (999)	Maleimide nucleoside	Bacteria	ECA, AH-130 (1-3,1000)	DNA synthesis (1001), DNA-dependent RNA polymerase (1003). Affects interaction of Ac-phenyl-tRNA with ribosomal units (997), thymidilate synthetase (1004),	110 iv

No.	Name	Source	Class	Activity	Mechanism of action	
261	SIBIROMYCIN	*Streptosporangium sibiricum* (1006)	Aminoglycoside (1007,1008)	Bacteria (1009)	LIO-1, S-180, ECA, NK/Ly, S-37 (1010)	amino acid and sugar transport (998), reaction with SH-groups (1002,1005), induction of resistant mutants *E. coli* (1005). See also (9,11,13,16).	0.05 iv
262	SOEDOMYCIN	*S. hachioensis*			ECA, S-180 (24,1015)	DNA synthesis, induction of phage in lysogenic bacteria (1009,1011,1012). Affects mitotic activity (1013), activities of glutamate oxal-acetate transaminase and glutamate pyruvate transaminase (1014). See also (10,17,19).	
263	SPARSOMYCIN	*S. sparsogenes* var. *sparsogenes*	(1,3)	Bacteria, fungi	ECA, W-256, S-180, KB (2,1016)	Binding to ribosomal subunits (1017,1018). Affects initiation of polypeptide synthesis (1019), peptide-bound formation (1020,1021). See also (7,9,11,12-16,22).	2.4 ip
264	SPINAMYCIN	*S. albospinus*	Polyene hydrazide (1022)	Fungi	YS (2,24)		

(Continued)

TABLE I *(Continued)*

No.	Antibiotic	Produced by	Chemical nature	Micro-organisms	Effect on Tumors	Mechanism of action	Footnotes or LD$_{50}$ (mg/kg)
						Biological effects	
265	SPIRAMYCINS I,II,III	*S. ambofaciens*	Nonpolyene macrolide (1-3,5,8)	Bacteria	S-180, E-2, L-1210, Ca-755, MSL (1023)	Alteration of ribosomal RNA (1029). Affects ribosomes (1027), elongation of peptidyl chain (1024), peptidyl-tRNA and aminoacyl-tRNA (1025, 1026), protein synthesis by mitochondria (1028), morphological changes in *E. coli* (1030). See also (7,9,11,12, 16,22).	
266	SPORIDESMINS	*Pithomyces chartarum*	Dithiodiketo-piperazine (4,5)	Fungi	HeLa (15)	Effect on swelling and respiration of mitochondria (1031), hepatotoxic, mycotoxin (2). See also (15).	
267	STATOLON	*P. stoloniferum* (1032)	Polysaccharide	Viruses	S-180, RS, FVS (1033)		
268	STEFFIMYCIN	*S. steffiburgensis*	(2)	Bacteria, viruses	KB (2,1034)	RNA and protein synthesis complexes with tRNA, binding to RNA (1035,1036).	562 ip
269	STEFFIMYCIN B	*S. elgreteus*		Bacteria	L-1210 (1037)	Binding to double-stranded DNA (1038).	
270	STREPTOGAN	*S. streptogamesis*	Polypeptide	Bacteria	Tumor cells (2,24)		

271	STREPTONIGRIN	*S. floculus*, *Actinomyces* sp., *A. albus* var. *bruneomycini* (6,21)	(2,3,1040)	Bacteria, fungi	S-180, W-256, G-26, EO-771, ROS (1-3)	Interaction with DNA (1032). Affects DNA synthesis (1041), RNA synthesis (1042), cell division (1045), mitotic activity in fibroblasts (1048). Binding in nuclei (1046). Mutagen (1047). See also (7,9-11,19,20,22).
272	STREPTORUBIN A,B	*S. rubrireticuli* var. *pimprina* (1)	Indicator pH		Tumor cells (1,2)	
273	STREPTOVARI-CINS	*S. spectabilis*	Ansamycin group	Bacteria, fungi	HeLa	Inhibition of incorporation of nucleosides in HeLa (1050), reverse transcriptase of Raucher leukemia virus (1051,1052). (Appears to be identical with Rifamycins [1049].)
274	STREPTOVITA-CINS A,B,C$_1$,C$_2$,D,E	*S. griseus*	Glutarimide (5,6,1053)	Fungi, protozoa	ECA, L-1210 (1-3,1054)	Protein synthesis (1055), DNA synthesis (1056). Affects liver uridine kinase (1058), protects cells from the lethality inhibitors of DNA synthesis (1057). See also (7,9,11,22).

(Continued)

TABLE I (Continued)

No.	Antibiotic	Produced by	Chemical nature	Biological effects — Effect on: Micro-organisms	Tumors	Mechanism of action	Footnotes or LD$_{50}$ (mg/kg)
275	STREPTOZOTOCIN	*S. achromogenes* (1059,1060)	(1-3)	Bacteria	ECA, L-1210 (24,1061)	DNA synthesis, mononucleotides (1061,1062), DNA degradation (1063), NAD degradation (1064, 1065), Mutagen, carcinogen (1066,1067). Diabetogen (2). See also (19,22).	
276	SULFOCIDIN	*Streptomyces* sp.		Bacteria, fungi	Tumor cells (1-3,6)		
277	TENUAZONIC ACID	*Alternaria tenuis* (1068,1069)	Amino acid (4,5)		ECA, KB, A-42, FVL, Ca-1025 (1070)	Protein synthesis, blocked of peptide-bound formation (1071). Toxin (1069). See also (7,14,22).	
278	TERREIC ACID	*Aspergillus terreus*	Benzoquinone (3-5,21)	Bacteria	ECA, HeLa (1072)		120 ip
279	TOMAYAMYCIN	*S. achromogenes* var. *tomayamyceticus*	Pyrrolobenzo-diazepine	Bacteria	L-1210 (1073)	Complexes with DNA (1074)	3 ip
280	TOYOCAMYCIN	*S. toyocaensis*, *Streptomyces* sp.	Pyrrolopyrimidine nucleoside (3,5,8)	Fungi	HeLa, NF (1,2,6)	Inhibition of RNA synthesis (1075-1078), interferon inhibition (1079), conversion to Sangivamycin (1080). See also (7,9-11,16,18).	10 iv
281	TRICHOMYCIN	*S. hachioensis*	Polyene (1,3,5)	Fungi, protozoa	ECA, HeLa (1081)	Effect on morphology of ECA cells (1082). Cardiotonic (2).	2.2 iv

No.	Name	Source	Class	Test systems	Remarks	Dose
282	TRICHOTHECIN	*Trichothecium roseum* Link	Trichthecene (3-5,8)	Fungi, viruses (1083)	S-180, YS, KB (1084)	Inflammatory effect, toxin (2). See also (7,11,15)	300 iv
283	TRIENINE	*Streptomyces* sp.3725	Polyene	Bacteria, fungi	5-WM (2)		
284	TRIOSTIN C	*Streptomyces* sp.S-2-210, *S. aureus*	Quinoxaline (1,3)	Bacteria	HeLa, Ca-1025, ROS, Mecca (1-3,1085)		250 ip Active also: Triostin A,B
285	TUBERCIDIN	*S. tubercidicus*, *S. cuspidosporus*, *S. sparsogenes* (1086)	Purine (1,2,8)	Bacteria, fungi	ECA, S-180, KB, NF, JS (3,6,1087)	RNA, DNA, and protein synthesis (1088, 1089), feedback inhibition of purine synthesis (1090), interference with glucose utilization (1091). Substrate for terminal tRNA-C-C-A pyrophosphorylase (16). See also (7,9-11,13,16,18,22).	35 iv
286	TUOROMYCIN	*S. tuirus*, *S. collins* (1092, 1093)	Anthracycline	Bacteria, fungi	W-256 (2,24,1093)		
287	USSAMYCIN	*S. lavendulae* UV-9	Peptide (1094)	Bacteria	W-256 (1-3)		

(Continued)

TABLE I *(Continued)*

| No. | Antibiotic | Produced by | Chemical nature | Biological effects | | | Footnotes or LD$_{50}$ (mg/kg) |
| | | | | Effect on | | | |
				Micro-organisms	Tumors	Mechanism of action	
288	VARIAMYCIN	*A. olivovariabilis* (1095)	Aureolic acid (1096)	Bacteria	HeLa	Complexes with DNA (1098), inhibition RNA synthesis, RNA polymerase, mitose (1097).	
289	VARIOMYCIN	*Streptomyces* sp.			Tumor cells (1,2)		
290	VERMICULIN	*P. vermiculatum Dangeard* (1099)	Macrolide (1100)	Bacteria, protozoa	ECA, HeLa	RNA, DNA synthesis in ECA (1101), chromosomal aberation (1102).	420 ip
291	VERRUCARINS A,B,H,J,K	*M. roridum, M. verrucaria* (1103,1104)	Trichothecene (2,4,8)	Bacteria, fungi	ECA, S-37, W-256, JS, P-815 (3,1105)	Inhibition of peptide bound formation (1106). Inflammatory effect, toxin, mycotoxin (2).	
292	VERTICILLIN A,B	*Verticillium* sp.	Dithiodioxo-piperazine (1107)	Bacteria, fungi, viruses	ECA, HeLa (1108)	Toxin.	7.6 ip
293	VERTIMYCIN	*Streptomyces* sp.JA 4498	Cyclopentanone	Bacteria	ECA (1,2,5,1109)		
294	VIBRIOCIN	*Vibrio comma*			HeLa (1110)		
295	VIOLAMYCIN A,S	*S. violaceus* JA 6844 (1111)	Anthracycline (20)	Bacteria (20,1112)	S-180, L-1210 (20)		
296	VIROCYTIN	*S. fumigatus*			ECA, LIO-1, S-37, W-256, S-45(1113)		

326

No.	Name	Source	Class	Viruses	Bacteria	Tumor cells. (6)		Mechanism / Remarks
297	VIROLITIN	*C. albicans*						
298	XANTHOCILLIN- X METHYLETHER	*Dichotomyces albus* (1114,1115)	(3-5,8)		Bacteria	ECA, HeLa	40 ip	Protheosynthesis in NDV (1116)
299	ZEARALENONE	*Gibberella zeae*	Macrolide			ECA (1117)		Incorporation of precursors nucleic acids and protein synthesis in ECA (1118). Mycotoxin (2).
300	ZEDALAN	*S. achromogenes* var. *streptozoticus*	(2)			Tumor cells (2,1119)		
301	ZORBAMYCIN	*S. bikiniensis* var. *zorboensis*	Peptide			Tumor cells (1120)		Degradation of DNA and RNA in bacterial cells (1121), binding to DNA (1122).
302	A-114	*Streptomyces* sp.				ECA, HeLa (1123)		
303	A-216	*S. rubrireticuli*	Peptide (1,2)			ECA, S-180 (2,1123)		
304	A-280	*Streptomyces* sp. A-280	Peptide (1,2)		Bacteria	Tumor cells (1123)		

(Continued)

TABLE I *(Continued)*

No.	Antibiotic	Produced by	Chemical nature	Biological effects			Footnotes or LD$_{50}$ (mg/kg)
				Effect on		Mechanism of action	
				Micro-organisms	Tumors		
305	ARA-A	*S. antibioticus* (1124)	Arabinosyl nucleoside	Viruses (1125)	RS (16,1125)	DNA synthesis (16), adenine and glycine incorporation into DNA (1126), ribonucleotide reductase in NovH (1127), chromosome breakage (1128,1129). See also (16).	
306	BA-17039A	*S. longisporus*	Peptide		S-180, Ca-755 (1,2)		
307	BA-181314	*Streptomyces* sp.		Bacteria	HeLa, W-256 (1,2)		
308	BU-306	*Streptomyces* sp.	Actinomycin	Bacteria	Tumor cells (1-3,6,1130)		25 iv
309	C-1292	*Streptomyces* sp. ATCC 13748	Polysaccharide		Tumor cells (1,2)		
310	D-180	*Streptomyces* sp.			ECA, HeLa (1123)		
311	D-284	*Streptomyces* sp.			ECA, HeLa (1123)		
312	D-440	*Streptomyces* sp.			ECA, HeLa (1123)		
313	DCS	*S. nagasakiensis*	Amino acid		Tumor cells (1,2,1131)		

No.	Organism	Type	Antimicrobial	Tumor systems	Biological activity	Dose
314	DON *S. ambofaciens* *Streptomyces* sp.	Amino acid (3,6)	Bacteria	S-180, W-256, RS, L-1210, FJC (1,2,1132)	Purine synthesis (1133), effect on formylglycineamide synthetase (1134), glutaminase (1135), L-asparaginase (1136). Mutagen (1137). See also (7,10).	75 iv
315	E-73 *S. albus*, *Streptomyces* sp.	Glutarimide (3,5)	Fungi	S-180, Ca-755, HS-1 (2,6,1138)	Protein synthesis, incorporation of amino acids into tumor cells (1139).	0.7 ip
316	G-253 C,B *S. reticuli* var. *shimofusaensis*	Mitomycin (1140)	Bacteria, fungi	ECA, S-180 (1,2) (1141)		26 iv
317	HA-135 *Streptoverticillium sporiferum Thirum*		Fungi	S-180 (1142)		25 iv
318	M-259 *S. nigellus*	(1,2)	Bacteria	ECA, S-180, L-1210	Incorporation of guanine into RNA and DNA (1143)	1.47 ip
319	M-741 *Streptoverticillium septatum*	Indicator pH	Fungi	S-180 (1,2,1144)		
320	M-6672 *S. terminospiralis*			S-180 (1,1145)		
321	MT-10 *S. indicus Chakrabarty* sp.n	Actinomycin (1146)	Fungi, yeasts	Tumor cells		
322	N-294 *Aspergillus flaveolus*			NK/Ly, L-5178, S-37, ECA (1147)		

(Continued)

TABLE I *(Continued)*

No.	Antibiotic	Produced by	Chemical nature	Biological effects		Mechanism of action	Footnotes or LD$_{50}$ (mg/kg)
				Effect on			
				Micro-organisms	Tumors		
323	NSC A-649	*Streptomyces* sp. NSC A-649	Aureolic acid (1,2)	Bacteria	KB, EO-771, ROS, JS, W-256 (1,2,1148)		2 iv
324	0-2	*Streptomyces* sp. 0-2	Indicator pH	Bacteria	Tumor cells (1,2)		5.75 ip
325	0-5	*Streptomyces* sp.	Indicator pH	Bacteria	Tumor cells (1,2)		1.25 iv
326	OS-3256-B	*S. candidus* var. *azaticus*	Azaamino acid	Bacteria	Leukemia (1149)		
327	PBF	*Streptomyces* sp.	Anthracycline	Bacteria	Ca-755, S-180 (1150)		
328	PSX-1	*P. stipitatum Thom*		Fungi, protozoa	ECA, RS, HeLa, L-5178, L-1210	Nucleic acids and protein synthesis in ECA (1151,1152).	
329	P-42-13	*Aspergillus tumemacerans*		Bacteria, fungi	Tumor cells (2)		
330•	R-12	*Streptomyces* sp.			Tumor cells (1153)		
331	RP-9768	*S. livescens*	(2)	Bacteria	Tumor cells (2)		
332	RP-9865	*Streptomyces* sp. 31723		Bacteria	Tumor cells (2)		
333	RP-9971	*S. gascariensis*		Fungi	ECA (1,2)		

334	RP-13213	*S. coeruleorubidus*, *Streptomyces* sp.	Bacteria	Tumor cells (1,2)	
335	RP-13330	*S. coeruleorubidus*	Bacteria	Tumor cells (1,2)	
336	RP-16976	*S. livescens* (2)		Tumor cells (2)	
337	RP-17967	*S. roseopulatus*	Bacteria	Tumor cells (2)	
338	S-339-3	*Streptomyces* sp. No. S-339		Tumor cells (1,2,6)	Mitosis. 7.4 iv
339	SF-1306	*S. stachinatus* SF-1306	Fungi	HeLa (1154)	
340	SK-229 A,B	*S. griseus* var. 229 Aureolic acid	Bacteria	Tumor cells (1,2)	
341	SL-1846		Viruses	P-815 (1155)	
342	SL-2266	*Sordaria araneosa* Cain NRRL 3196	Fungi	P-815 (1156)	Mitosis.
343	SL-3364	*Sepedonium chrysospermum* NRRL 3489	Fungi	Tumor cells (1157)	
344	SL-3440	*Diheterospora chlamidosporia* NRRL 3472	Fungi	Tumor cells (1158)	

(Continued)

TABLE I (Continued)

No.	Antibiotic	Produced by	Chemical nature	Biological effects Effect on Micro-organisms	Tumors	Mechanism of action	Footnotes or LD$_{50}$ (mg/kg)
345	SQ-15859	*S. chrysomallus*	Cyclic polylactone	Bacteria	Tumor cells (1,2,3)		
346	SR-1768 A,B,C,D,E,F,G	*Streptomyces* sp. A-468	Aureolic acid	Bacteria	ECA, L-1210	Inhibited RNA synthesis (1159).	
347	T 12/3	*Streptomyces* sp. T 12/3		Bacteria	EC, LIO-1 (2,1160)		
348	TA-435 A	*Streptomyces* sp. 0-80	Indicator pH	Bacteria	HeLa, ECA, NF, YS, S-180 (1,2)		11 ip
349	TA-2590	*S. pluricolorescens*	Peptide	Bacteria	ECA (2)		
350	TS-885	*S. pluricolorescens* var. *yamashitaensis*	Glutarimide	Bacteria, fungi	Tumor cells (1161)		
351	U-20904	*Streptomyces* sp.	(2)	Fungi	KB (1,2)		56 ip
352	U-22956	*Streptoverticillium fervens*, *S. netropsorus*		Bacteria, fungi	KB (1,2)		
353	U-42126	*S. sviceus*	Chloroisoxazo-line amino acid (1164)	Fungi	L-1210, P-388 (1162,1163)	Glutamic antagonist in mam-malian cells (1165).	
354	U-43795	*S. sviceus*	Hydroxyderivate chloroisoxazo-line amino acid	Bacteria	L-1210, P388 (1166)		

355	WR-142	*S. olivaceus* WR-142		Bacteria	ECA, S-180		Inhibition of rRNA and protein synthesis (1167).
356	YA-56	*S. humidus* NRRL 0387	Peptide (1169)	Bacteria	ECA (1168)		
357	23-21	*Streptomyces* sp. 23-21	Actinomycin group		ECA (1,6,1170)		
358	173-t	*Pseudomonas aeruginosa* ATCC 9027			ECA, S-180, Ca-755 (1171)		
359	289 E	*S. phaeoverticillatus* var. *takatsukiensis*			Tumor cells (1,1172)		
360	289-F0	*S. phaeoverticillatus* var. *takatsukiensis*		Bacteria	Tumor cells (1173)		
361	323/58	*S. roseoflavus* var. *tauricus*, *Actinomyces* sp.		Bacteria	S-180, LIO-1, ECA (1,2,1174)	1 ip	
362	593 A	*S. griseoluteus*	Piperazine		Tumor cells (2,1175)		
363	661	*S. cinnamonensis*		Fungi	EC, S-180, L-5178		RNA, DNA, and protein synthesis (1176)

(Continued)

TABLE I (Continued)

No.	Antibiotic	Produced by	Chemical nature	Biological effects — Effect on — Micro-organisms	Tumors	Mechanism of action	Footnotes or LD_{50} (mg/kg)
364	757	*Streptomyces* sp. No. 757	Polyene	Fungi	Tumor cells (1177)		
365	1418-A$_1$	*Streptomyces* sp. No. 1418-A$_1$		Bacteria	ECA, HeLa (1,2)		2.5 iv
366	3002	*A. primicini*			Tumor cells (1177)		
367	3853	*A. luridus*			Tumor cells (1178)		
368	4035	*Micropolyspora*			ECA, S-37, L-5178, NK/Ly (1178)		
369	4418	*A. griseoplanus*			Tumor cells (2,1179)		
370	4695	(1,2,6)			Tumor cells		
371	5888 I,II	*A. galileus*			Tumor cells (2)		
372	6270	*A. flavochromogenes*	Quinoxaline	Bacteria	S-180, ECA (1-3,6)	RNA synthesis, RNA polymerase (1180-1182).	
373	6613	*A. daghestanicus*	Peptide	Bacteria	S-180, LIO-1 (1-3,1183)		

Acknowledgments

The assistance of Dr. Ladislav J. Hanka (Research Laboratories, The Upjohn Company, Kalamazoo, Michigan) with the final organization and language correction of the manuscript prior to publication is gratefully acknowledged. The authors are grateful to Dr. A. Fusková and Mrs. M. Adamková for assistance.

References

1. Umezawa, H., "Index of Antibiotics from Actinomycetes. Univ. of Tokyo Press, Tokyo, 1967.
2. Laskin, I. A., and Lechevalier, H. A., "Handbook of Microbiology," Vol. III. CRC Press, Cleveland, Ohio, 1973.
3. Korzybski, T., Kowszyk-Gindifer, Z., and Kurylowicz, W., "Antibiotics," Vols. 1 and 2. Academic Press, New York, 1967.
4. Turner, W. B., "Fungal Metabolites." Academic Press, New York, 1971.
5. Miller, M. W., "The Pfizer Handbook of Microbial Metabolites." McGraw-Hill, New York, 1961.
6. Shemyakin, M. M., Khokhlov, A. S., Kolosov, N. N., Bergelson, L. D., and Antonov, V. K., "Khimia Antibiotikov." Izd. Akad. Nauk USSR Moscow, 1961.
7. Gottlieb, D., and Shaw, P. D., eds., "Antibiotics," Vol. 1. Springer-Verlag, Berlin and New York, 1967.
8. Gottlieb, D., and Shaw, P. D., eds., "Antibiotics," Vol. 2. Springer-Verlag, Berlin and New York, 1967.
9. Gale, E. F., Cundliffe, E., Reynolds, P. E., Richmond, M. H., and Waring, M. J., "The Molecular Basis of Antibiotic Action." Wiley, New York, 1972.
10. Kersten, H., and Kersten, W., "Inhibitors of Nucleic Acid Synthesis. Biophysical and Biochemical Aspects." Springer-Verlag, Berlin and New York, 1974.
11. Hochster, R. M., Kates, M., and Quastel, J. H., eds., "Metabolic Inhibitors," Vol. 3. Academic Press, New York, 1973.
12. Hochster, R. M., Kates, M., and Quastel, J. H., eds., "Metabolic Inhibitors," Vol. 4. Academic Press, New York, 1973.
13. Roy-Burman, P., "Analogues of Nucleic Acid Compounds." Springer-Verlag, Berlin and New York, 1970.
14. Ciegler, A., Kadis, S., and Ajl, S. J., eds., "Microbial Toxins," Vol. VI. Academic Press, New York, 1971.
15. Kadis, S., Ciegler, A., and Ajl, S. J., eds., "Microbial Toxins," Vol. VII. Academic Press, New York, 1971.
16. Suhadolnik, R. J., "Nucleoside Antibodics." Wiley (Interscience), New York, 1970.
17. Hahn, F. E., "Progress in Molecular and Subcellular Biology," Vol. II. Springer-Verlag, Berlin and New York, 1971.
18. Balis, M. E., "Antagonists and Nucleic Acids." North-Holland Publ., Amsterdam, 1968.
19. Semonsky, M., Hejzlar, M., and Masák, S., eds., "Advances in Antimicrobial and Antineoplastic Chemotherapy, Proceedings of the 7th International Congress of Chemotherapy, Prague, 1971," Vol. 2. Avicenum, Prague, 1972.
20. Daikos, K. G., ed., Progress in Chemotherapy, Proceedings of the 8th International Congress of Chemotherapy," Vol. III. Hellenic Society for Chemotherapy, Athens, 1974.

21. Thomson, R. H., "Naturally Occurring Quinones," 2nd ed. Academic Press, New York, 1971.
22. Meek, E. S., "Antitumour and Antiviral Substances of Natural Origin." Springer-Verlag, Berlin and New York, 1970.
23. Franklin, T. J., and Snow, G. A., "Biochemistry of Antimicrobial Action." Chapman & Hall, London, 1971.
24. Strauss, D., Gegen Krebs wirksame Antibiotica aus Streptomyceten. Arzneim.-Forsch. 20, 301, 1970.
25. Waksman, S. A., "Actinomycin, Nature, Formation, and Activities." Wiley (Interscience), New York, 1968.
26. Pittillo, R. E., Fisher, M. W., McAlpine, R. J., Thompson, P. E., Ehrlich, J., Anderson, L. E., Fiskin, R. A., Galbraith, M., Kohberger, D. L., Manning, M. S., Reutner, T. F., Roll, P. R., and Weston, K., Antibiot. Annu. p. 497 (1959).
27. Antosz, F. J., Nelson, D. B., Herald, D. L., and Munk, M. E., J. Am. Chem. Soc. 92, 4933 (1970).
28. Nelson, D. B., and Munk, M. E., J. Org. Chem. 35, 3832 (1970).
29. Keele, B. B., Powell, H. L., Navia, J. M., and McGhee, J., Appl. Microbiol. 22, 957 (1971).
30. Burchenal, J. H., Holmberg, E. A. D., Reilly, H. S., Hemphill, S., and Reppert, J. A., Antibiot. Annu. p. 522 (1959).
31. Smithers, D., Bennett, L. L., and Struck, R. F., Mol. Pharmacol. 5, 433 (1969).
32. Sato, K., and Katagiri, K., Chemotherapy (Tokyo) 5, 182 (1957).
33. Smitz, H., Bradner, W. T., Gourewitch, A., Heinemann, B., Price, K. E., Lein, J., and Hooper, I. R., Cancer Res. 22, 163 (1962).
34. Bradner, W. T., and Sugiura, K., Cancer Res. 22, 167 (1962).
35. Ueda, M., Tanigawa, Y., Okami, Y., and Umezawa, H., J. Antibiot., Ser. A 7, 125 (1954).
36. Hori, M., Ito, E., Takeuchi, T., and Umezawa, H., J. Antibiot., Ser. A 16, 1 (1963).
37. Nitta, K., Takeuchi, T., Yamamoto, T., and Umezawa, H., J. Antibiot., Ser. A 8, 120 (1955).
38. Tereshin, I. M., Antibiotiki (Moscow) 14, 9 (1969).
39. Olzsewska, M. J., Cytobiologie 8, 371 (1974).
40. Wartell, R. M., Larson, J. E., and Wells, R. D., J. Biol. Chem. 250, 2698 (1975).
41. Meienhofer, J., and Atherton, E., Adv. Appl. Microbiol. 16, 203 (1973).
42. Perlman, K. L., Walker, J., and Perlman, D., J. Antibiot. 24, 135 (1971).
43. Rickards, R. W., Perlman, K. L., and Perlman, D. J. Antibiot. 26, 177 (1973).
44. Perlman, D., Cancer Chemother. Rep. 58, 93 (1974).
45. Orlova, T. I., Sorokina, N. V., Kuznetzov, V. D., and Silaev, A. B., Antibiotiki (Moscow) 18, 111 (1973).
46. Sorokina, N. V., Orlova, T. I., and Silaev, A. B., Antibiotiki (Moscow) 18, 595 (1973).
47. Vinogradova, K. A., Poltorak, V. A., Petrova, L. I., and Silaev, A. B., Antibiotiki (Moscow) 15, 1081 (1970).
48. Ponnuswany, P. K., McGuire, R. F., and Scheraga, H. A., Int. J. Pept. Protein Res. 5, 73 (1973).
49. Young, C. W., Am. J. Clin. Pathol. 52, 130 (1969).
50. Kessel, D., and Bosmann, B. H., Cancer Res. 30, 11 (1970).
51. Goldberg, I. H., and Friedman, P. A., Annu. Rev. Biochem. 40, 772 (1971).

52. Reich, E., and Golberg, I. H., *Prog., Nucleic Acid Res. Mol. Biol.* **3**, 183 (1964).
53. Reich, E., *Science* **143**, 684 (1964).
54. Sobell, H. M., *Fed. Proc., Fed. Am. Soc. Exp. Biol.* **30**, 1936 (1971).
55. Kirk, M., *Biochim. Biophys. Acta* **42**, 167 (1960).
56. Bickis, I. J., and Quastel, J. H., *Proc. Can. Food Biol. Soc.* **5**, 13 (1962).
57. Goldberg, I. H., and Rabinowitz, M., *Science* **136**, 315 (1962).
58. Hurwitz, J., Furth, J. J., Malamy, M., and Alexander, M., *Proc. Nat. Acad. Sci. U.S.A.* **48**, 1222 (1962).
59. Richardson, J. P., *J. Mol. Biol.* **21**, 83 (1966).
60. Perry, R. P., *Exp. Cell Res.* **29**, 400 (1963).
61. Penman, S., Vesco, C., and Penman, M., *J. Mol. Biol.* **34**, 49 (1968).
62. Laszlo, J., Miller, D. S., McCarty, K. S., and Hochstein, P., *Science* **151**, 1007 (1960).
63. Holden, J. T., and Utech, N. M., *Biochim. Biophys. Acta* **135**, 351 (1967).
64. Waring, M., *J. Mol. Biol.* **54**, 247 (1970).
65. Fletcher, R. D., Yayavasu, C., Yoo, S., and Albertson, J. N., *Antimicrob. Agents & Chemother.* **4**, 376 (1973).
66. Firtel, R. A., Baxter, L., and Lodish, H. F., *J. Mol. Biol.* **79**, 315 (1973).
67. Perry, R. P., and Kelley, D. E., *J. Cell Physiol.* **76**, 127 (1970).
68. Sawicki, S. G., and Godman, G. G., *J. Cell Biol.* **50**, 746 (1971).
69. Godard, C., Robin, J., and Boiron, M., *J. Gen. Virol.* **22**, 293 (1974).
70. Jordan, R. L., and Wilson, J. G., *Anat. Rec.* **168**, 549 (1970).
71. Küpper, H. A., McAllister, W. T., and Bautz, E. K. F., *Eur. J. Biochem.* **38**, 581 (1973).
72. Recher, L., Parry, N. T., Briggs, L. G., and Whitescaver, J., *Cancer Res.* **31**, 1915 (1971).
73. Flaconer, I. R., and Fidwer, T. J., *Biochim. Biophys. Acta* **218**, 508 (1970).
74. Weiss, L., and Huber, D., *J. Cell Sci.* **15**, 217 (1974).
75. Rutchenko, O. N., Likhacheva, N. A., Timakova, N. V., and Ilyashenko, B. N., *Vopr. Virusol.* **4**, 451 (1973).
76. Snyder, F. F., and Henderson, J. F., *Can. J. Biochem.* **52**, 263 (1974).
77. Calvin, H. I., and Bedford, J. M., *J. Reprod. Fertil.* **36**, 225 (1974).
78. Ho, Ch., Lipsich, L., Fischer, G., and Keller, S., *J. Cell Biol.* **63**, 140 (1974).
79. Conklin, K. A., Chou, S. C., Ramanathan, S., and Heu, P., *Pharmacology* **4**, 91 (1970).
80. Fok, J., and Waring, M., *Mol. Pharmacol.* **8**, 65 (1972).
81. Hyman, R. W., and Davidson, N., *Biochim. Biophys. Acta* **228**, 38 (1971).
82. Swindelhurst, M., *FEBS Lett.* **35**, 24 (1973).
83. Paston, I., and Friedman, R. M., *Science* **160**, 316 (1968).
84. Hamelin, R., Larsen, D. J., and Tavitian, A., *Eur. J. Biochem.* **35**, 350 (1973).
85. DePetrocellis, B., and Parisi, E., *Exp. Cell Res.* **82**, 351 (1973).
86. Dybing, E., *Biochem. Pharmacol.* **23**, 705 (1974).
87. Rhoads, R. E., McKnight, G. S., and Schimke, R. T., *J. Biol. Chem.* **248**, 2031 (1973).
88. Candela, J. L. R. and Garcia, M. C., *Life Sci. Part 2* **9**, 235 (1970).
89. Hazlett, L. D., and Ackermann, G. A., *J. Natl. Cancer Inst.* **47**, 1309 (1970).
90. Godman, G. C., Kemeklis, R. T. P., and King, M. E., *Exp. Cell Res.* **77**, 159 (1973).
91. Goldblatt, P.-J., and Sullivan, R. J., *Cancer Res.* **30**, 1349 (1970).
92. Heenen, M., Premount, A. M., and Galand, P., *Cancer Res.* **33**, 2624 (1973).
93. Hyman, R. W., and Davidson, N., *J. Mol. Biol.* **50**, 421 (1970).

94. Bal, A. K., (1970). Z. *Pflanzenphysiol.* **63**, 261 (1970).
95. Doida, Y., and Okada, S. *Cell Tissue Kinet.* **5**, 15 (1972).
96. Williams, J. G., and McPherson, I. A., *J. Cell Biol.* **67**, 148 (1973).
97. Ostertag, W., and Haake, J., Z. *Vererbungsl.* **98**, 299 (1966).
98. Rocchi-Brasiello, A., and DiCastro, M., *Caryologia* **27**, 339 (1974).
99. Kang, Y. S., and Yang, J. S., *Seoul Natl. Univ. Fac. Pap., Biol. Agric. Ser. 3*, 27 (1974).
100. Pater, M. M., and Mak, S., *Nature (London)* **250**, 786 (1974).
101. Ficek, S., and Wieckowski, S., *Acta Soc. Bot. Pol.* **43**, 251 (1974).
102. Kawamata, J., and Imanashi, M., *Nature (London)* **187**, 1112 (1960).
103. Apple, M. A., *J. Clin. Pharmacol.* **15**, 29 (1975).
104. Khokhlov, A. S., Cherkes, B. Z., Reshetov, P. D., Smirnova, G. M., Sorokina, I. B., Prokopheva, T. A., Koloditskaya, T. A., and Smirnov, V. V., *Prog. Antimicrob. Anticancer Chemother., Proc. Int. Congr. Chemother., 6th, 1969* Vol. I, p. 99 (1969).
105. Boyko, V. I., and Koroleva, V. G., *Antibiotiki (Moscow)* **6**, 1058 (1961).
106. Arcamone, F., Franceschi, G., Tanco, S., and Selva, A., *Tetrahedron Lett.* p. 1007 (1969).
107. Schwartz, H. S., and Grindey, G. B., *Cancer Res.* **33**, 1837 (1973).
108. Kim, S. H., and Kim, J. H., *Cancer Res.* **32**, 323 (1972).
109. Kitaura, K., Imai, R., Shihara, Y., Yamai, H., and Takahari, H., *J. Antibiot.* **25**, 509 (1972).
110. Müller, W. E. G., Yamazaki, Z., and Forster, W., *Klin. Wochenschr.* **50**, 790 (1972).
111. Meriwether, W. D., and Bachur, N. R., *Cancer Res.* **32**, 1137 (1972).
112. Silvestrini, R., and Lenaz, L., *European Association for Cancer Research Symp. Mech. Action Cytotoxic Agents, 1972* Abstr. Book, p. 60 (1972).
113. Barranco, S. C., Gerner, E. W., Burk, K. H., and Humphrey, R. M., *Cancer Res.* **33**, 11 (1973).
114. Goodmann, M. F., Bessman, M. J., and Bachur, N. R., *Proc. Natl. Acad. Sci. U.S.A.* **71**, 1193 (1974).
115 Čihák, A., Veselý, J., and Harrap, K. R., *Biochem. Pharmacol.* **23**, 1087 (1974).
116. Bhuyan, B. K., and Fraser, T. J., *Cancer Chemother. Rep., Part 1* **58**, 149 (1974).
117. Egorin, M. J., Hildebrand, R. C., Cimino, E. F., and Bachur, N. R., *Cancer Res.* **34**, 2243 (1974).
118. Atassi, G., Tagnon, H. J., Bournonville, F., and Wynands, M., *Eur. J. Cancer* **10**, 399 (1974).
119. Gosalvez, M., Blanco, M., Hunter, J., Miko, M., and Chance, B., *Eur. J. Cancer* **10**, 567 (1974).
120. Gericke, G. D., and Chandra, P., Z. *Krebsforsch. Klin. Onkol.* **79**, 277 (1973).
121. Massimo, L., Dagna-Bricarelli, F., and Eossati-Gugliemoni, A., *Rev. Eur. Etud. Clin. Biol.* **15**, 793 (1970).
122. Israel, M., Modest, E. J., and Frei, E., *Cancer Res.* **35**, 1365 (1975).
123. Henry, W., *Cancer Chemother. Rep., Part 2* **4**, 5 (1974).
124. Hurwitz, E., Levy, R., Maron, R., Wilchek, M., Arnon, R., and Sela, M., *Cancer Res.* **35**, 1175 (1975).
125. Zunino, F., Gambetta, R., and DiMarco, A., *Biochem. Pharmacol.* **24**, 309 (1975).
126. Schwartz, H. S., *Res. Commun. Chem. Pathol. Pharmacol.* **10**, 51 (1975).

127. Price, P. J., Suk, W. A., Skeen, F. C., Chirigos, M. A., and Huebner, R. J., *Science* **187**, 1200 (1975).
128. Shin, C. N., and Marth, E. H., *Biochim. Biophys. Acta* **338**, 286 (1974).
129. Dolimpio, D., Jacobsen, S., and Legator, M., *Proc. Soc. Exp. Biol. Med.* **127**, 559 (1968).
130. Green, S., *Nature (London)* **220**, 931 (1968).
131. Butler, W. H., and Neal, G. E., *Cancer Res.* **33**, 2878 (1973).
132. Sarasin, A., and Moulé, Y., *FEBS Lett.* **29**, 329 (1973).
133. Garner, R. C., *Chem.Biol. Interact.* **6**, 125 (1973).
134. Williams, D. Y., Clark, R. P., and Rabin, B. R., *Br. J. Cancer* **27**, 283 (1973).
135. Thaxton, J. P., Tung, M. T., and Hamilton, P. B., *Poultry Sci.* **53**, 721 (1974).
136. Moulé, Y., *Cancer Res.* **33**, 514 (1973).
137. Gurtoo, H. L., *Biochem. Biophys. Res. Commun.* **50**, 649 (1973).
138. Wan Bang Lo and Black, H. S., *Experientia* **28**, 1278 (1972).
139. Wogan, G. N., *Methods Cancer Res.* **7**, 309 (1973).
140. Garvican, L., Casone, F., and Rees, K. R., *Chem.-Biol. Interact.* **7**, 39 (1973).
141. Alexandrov, K., and Frayssinet, C., *Cancer Res.* **34**, 3289 (1974).
142. Akinrimisi, E. O., Beneck, B. J., and Seifart, K. H., *Eur. J. Biochem.* **42**, 333 (1974).
143. Edwards, G. S., and Allfrey, V. G., *Biochim. Biophys. Acta* **299**, 354 (1973).
144. Grisan, E. V., *Appl. Microbiol.* **25**, 342 (1973).
145. DiPaolo, J. A., Ellis, J., and Erwin, H., *Nature (London)* **215**, 638 (1967).
146. Legator, M., *Bacteriol. Rev.* **30**, 471 (1966).
147. Detroy, R. W., and Freer, S. N., *Dev. Ind. Microbiol.* **15**, 124 (1974).
148. U.S. Patent 3, 676, 490 (1972).
149. Pitillo, R. F., and Wodley, C., *Antimicrob. Agents & Chemother.* **3**, 739 (1973).
150. Gale, G. R., and Smidt, G. B., *Biochem. Pharmacol.* **17**, 363 (1968).
151. Gale, G. R., Osttander, E. W., and Atkins, L. M., *Biochem. Pharmacol.* **17**, 1823 (1968).
152. DeVoe, T. C., Rigler, N. E., Shay, A. J., Martin, J. H., Boyd, T. C., Backus, E. J., Mowat, J. H., and Bohonos, N., *Antibiot. Annu.* p. 730 (1957).
153. Hata, T., Umezawa, I., Iway, Y., Katagiri, M., Awaya, Y., Komiyama, K., Oiwo, R., and Atsumi, K., *J. Antibiot.* **26**, 181 (1973).
154. Barg, W., Bogiano, W., Sloane, N., and DeRenzo, E. C., *Fed. Proc., Fed. Am. Soc. Exp. Biol.* **16**, 150 (1957).
155. Rosenfeld, G. S., Rostovtseva, L. I., Baikina, V. M., Trakhtenberg, D. M., and Khokhlov, A. S., *Antibiotiki (Moscow)* **8**, 201 (1963).
156. Fukushima, K., Yazawa, K., and Arai, T., *J. Antibiot.* **26**, 175 (1973).
157. Olson, B. H., and Gorner, G. L., *Appl. Microbiol.* **13**, 314 (1965).
158. Olson, B. H., Jennins, B. C., Roga, V., Junek, A. J., and Schurmans, D. M., *Appl. Microbiol.* **13**, 322 (1965).
159. Hinuma, Y., Kuroya, M., Yajima, T., Ishibara, K., Hamada, S., Watanabe, K., and Kikuchi, K., *J. Antibiot. Ser. A* **8**, 148 (1955).
160. Hanesian, S., and Haskell, T., *Tetrahedron Lett.* p. 2451 (1964).
161. Burchenal, J. H., Yuceoglu, M., Dagg, M. Y., and Stock, C. C., *Proc. Soc. Exp. Biol. Med.* **86**, 891 (1954).
162. Pestka, J., *Arch. Biochem. Biophys.* **136**, 89 (1970).
163. Bloch, T. D., and Coutsogergopoulos, C., *Biochemistry* **5**, 3345 (1966).
164. Bloch, T. D., *J. Bacteriol.* **85**, 527 (1963).
165. Chassy, B. M., and Suhadolnik, R. J., *Biochim. Biophys. Acta* **182**, 316 (1969).
166. Guarino, A. J., and Kredien, N. M., *Biochim. Biophys. Acta* **68**, 317 (1963).

167. Suhadolnik, R. J., Chassy, B. M., and Waller, G. R., *Biochim. Biophys. Acta* **179**, 258 (1969).
168. Bloch, A., and Nichol, C. A., *Antimicrob. Agents Chemother.* p. 530 (1964).
169. Truman, J. T., and Klenow, H., *Mol. Pharmacol.* **4**, 77 (1968).
170. Truman, J. T., and Fredriksen, S., *Biochim. Biophys. Acta* **182**, 36 (1969).
171. Suzuki, T., and Nakanishi, T., Japan Kokai 74 110, 891 (1974); *C. A.* **82**, 137759 (1975).
172. Strong, R. C., Bodanszky, A. A., and Bodanszky, M., *Antimicrob. Agents & Chemother.* **10**, 42 (1970).
173. Bodanszky, M., Sigler, G. F., and Bodanszky, A. A., *J. Am. Chem. Soc.* **95**, 2352 (1973).
174. Anand, S. R., *Hind. Antibiot. Bull.* **6**, 57 (1963).
175. U. S. Patent 3,126,317 (1964).
176. Stähelin, H., Kalberer-Rüsch, M. E., Singer, E., and Kazary, S., *Arzneim.-Forsch.* **18**, 989 (1968).
177. Sigg, H.P., Mauli, R., Flura, E., and Hauser, D., *Helv. Chim. Acta* **48**, 962 (1965).
178. French Patent 1,372,122 (1964); C. A. **62**, 5856 (1965).
179. Ewans, R. C., "The Technology of the Tetracyclines." Quadrangle Press, New York, 1968.
180. Fuska, J., Podojil, M., Fuskova, A., and Vaněk, Z., *Experientia* **30**, 403 (1974).
181. Grollman, A., *J. Biol. Chem.* **242**, 3226 (1967).
182. Neth, R., Monro, R. E., Heller, G., Battaner, E., and Vasquez, D., *FEBS Lett.* **6**, 198 (1970).
183. Barbacid, M., and Vasquez, D., *J. Mol. Biol.* **84**, 603 (1974).
184. Carrasco, L., and Vasquez, D., *J. Antibiot.* **25**, 732 (1972).
185. Jimenez, A., Sanchez, L., and Vasquez, D., *Biochim. Biophys. Acta* **383**, 427 (1975).
186. Hurley, L. H., and Zmizjewski, M., *J. Chem. Soc., Chem. Commun.* p. 337 (1974).
187. Leimgruber, W., and Czajkowski, R. C., *J. Am. Chem. Soc.* **90**, 5641 (1968).
188. Grunberg, E., Prince, H. N., Titsworth, E., Beskid, G., and Tendler, M. D., *Chemotherapia* **11**, 249 (1966).
189. Kann, L. E., and Kohn, K. W., *J. Cell. Physiol.* **79**, 331 (1972).
190. Kohn, K. W., and Spears, C. L., *J. Mol. Biol.* **51**, 551(1970).
191. Glaubiger, D., Kohn, K. W., and Charney, E., *Biochim. Biophys. Acta* **361**, 303 (1974).
192. Kohn, K. W., Glaubiger, D., and Spears, C. L., *Biochim. Biophys. Acta* **361**, 288 (1974).
193. Harris, C., Grady, H., and Svoboda, D., *Cancer Res.* **28**, 97 (1968).
194. Horwitz, S. B., *Prog. Mol. Subcell. Biol.* **2**, 40 (1971).
195. Cargill, C., Bachman, E., and Zbinden, G., *J. Natl. Cancer Inst.* **53**, 481 (1974).
196. Bates, H. M., Kuenzing, W., and Weston, W. B., *Cancer Res.* **29**, 2195 (1969).
197. Adamson, R. H., Hart, L. G., DeVita, V. T., and Oliviero, V. Z., *Cancer Res.* **28**, 343 (1968).
198. Sigdestad, C. P., and Okunewick, J. P., *Chemotherapy (Basel)* **18**, 1 (1973).
199. Ayukawa, S., Takeuchi, T., Sezaki, M., Hara, T., and Umezawa, H., *J. Antibiot.* **21**, 350 (1968).
200. Nagatsu, T., Ayukawa, S., and Umezawa, H., *J. Antibiot.* **21**, 354 (1968).
201. Nozaki, M., Okuno, S., and Fujisawa, H., *Biochem. Biophys. Res. Commun.* **44**, 1109 (1971).

202. Mizuno, K., Ando, T., and Abe, J., *J. Antibiot.* **23**, 493 (1970).
203. Kusaka, T., Yamamoto, H., Shibata, M., Muroi, M., Kishi, T., and Mizuno, K., *J. Antibiot.* **21**, 255 (1968).
204. Bennet, L. L., Allan, P. W., and Hill, D. L., *Mol. Pharmacol.* **4**, 208 (1968).
205. Tamura, G., Suzuki, S., Takatsuki, S., Ando, K., and Arima, K., *J. Antibiot.* **21**, 539 (1968).
206. Minato, H., Katayama, T., Hayakawa, S., and Katagiri, K., *J. Antibiot.* **25**, 315 (1972).
207. Aldridge, D. C., Borrow, A., Foster, R. G., Large, M. S., Spencer, H., and Turner, W. B., *J. Chem. Soc., Perkin Trans. 1* p. 2136 (1972).
208. Sasaki, K., Hosokawa, T., Nawata, Y., and Ando, K., *Agric. Biol. Chem.* **38**, 1463 (1974).
209. Hayakawa, S., Minato, H., and Katagiri, K., *J. Antibiot.* **24**, 653 (1971).
210. Mashburn, L. T., and Wriston, J. C., *Arch. Biochem. Biophys.* **105**, 451 (1964).
211. Igarashi, S., Imada, A., Nakahama, K., Matsumoto, T., and Ootsu, K., *Experientia* **30**, 814 (1974).
212. Kafkewitz, D., and Goodman, D., *Appl. Microbiol.* **27**, 206 (1974).
213. Arima, K., Tamura, G., and Sakamoto, M., Japanese Patent 71 42592; *C. A.* **76**, 84543 (1972).
214. Kato, F., Katsumoto, R., Tatayama, K., and Kawagata, T., German Patent 2,126,181 (1971); *C. A.* **76**, 44692 (1972).
215. Petterson, R. E., and Ciegler, A., *Appl. Microbiol.* **17**, 929 (1969).
216. Broome, J. D., *J. Exp. Med.* **118**, 99 (1963).
217. Boeck, L. D., and Ho, P. P. K., *Can. J. Microbiol.* **19**, 1251 (1973).
218. Arens, A., Rauenbusch, E., Iridon, E., Wagner, O., Bauer, K., and Kaufmann, W., *Hoppe Seyler's Z. Physiol. Chem.* **315**, 197 (1970).
219. Maral, R., and Werner, G. H., *Nature (London) New Biol.* **232**, 187 (1971).
220. King, O. Y., Wilbur, J. R., Munford, D. H., and Sutow, W. W., *Cancer* **33**, 611 (1974).
221. Sandberg, J., and Goldin, A., *Cancer Chemother. Rep., Part 1* **55**, 233 (1971).
222. Mashburn, L. T., *Cancer* **28**, 1321 (1971).
223. Neish, W. J. P., and Smith, J. A., *Z. Krebsforsch.* **79**, 78 (1973).
224. Kessel, D., *Biochim. Biophys. Acta* **240**, 554 (1971).
225. Stevens, J., Mashburn, L. T., and Hollander, V. P., *Biochim. Biophys. Acta* **186**, 332 (1969).
226. Seerer, S., and Weser, U., *Nature (London)* **225**, 652 (1970).
227. Ehrman, M., Cedar, H., and Schwartz, J. H., *J. Biol. Chem.* **246**, 88 (1971).
228. Bosmann, H. B., and Kessel, D., *Biochim. Biophys. Acta* **338**, 280 (1974).
229. Han, T., and Ohnuma, T., *Immunology* **26**, 169 (1974).
230. Jaffe, N., Tragis, D., Das, L., Moloney, W. C., Hann, H. W., Kim, B. S., and Nair, R., *Cancer Res.* **31**, 942 (1971).
231. Hersh, E. M., *Transplantation* **12**, 368 (1971).
232. Liu, S. F., and Zajic, J. E., *Appl. Microbiol.* **25**, 92 (1973).
233. Kvasnikov, E. I., Nagornaya, S. S., and Shchelokova, I. F., *Mikrobiol. Zh.* **36**, 638 (1974).
234. Ozolins, R., Garkavaya, L. F., Grivina, P., and Kraizman, Z., *Fermentatsiya* p. 22 (1974).
235. Mashburn, L. T., and Landin, L. M., *Cancer Res.* **34**, 313 (1974).
236. Nikolaev, A. Ya., Kozlov, E. A., Sokolov, N. N., Kondrateva, N. A., and Dobrynin, Ya, V., *Vopr. Med. Khim.* **20**, 272 (1974).

237. Ashworth, L. A. E., and McLennan, A. P., *Cancer Res.* 34, 1353 (1974).
238. Nahas, A., and Capizzi, R. L., *Cancer Res.* 34, 2689 (1974).
239. Shifrin, S., Parrott, C. L., and Lubrosky, S. W., *J. Biol. Chem.* 249, 1335 (1974).
240. Wriston, J. C., Jr., and Yellin, T. O., *Adv. Enzymol.* 39, 185 (1973).
241. Grundmann, E., and Oettgen, H. F., "Experimental and Clinical Effects of L-Asparaginase." Springer-Verlag, Berlin and New York, (1970).
242. Argoudelis, A. D., and Zieserl, J. F., *Tetrahedron Lett.* p. 69 (1969).
243. Owen, F., and Bhuyan, B. K., *Antimicrob. Agents Chemother.* p. 804 (1965).
244. Kuznetsova, B. S., Orlova, T. I., and Silaev, A. B., *Antibiotiki (Moscow)* 17, 322 (1971).
245. Planelles, J. J., Solovyeva, J. V., Belova, Z. N., Silaev, A. B., Ebert, M. K., Gracheva, N. P., Kharitonova, A. M., Gosheva, A. E., and Akopyants, S. S., *Acta Unio Int. Cancrum* 20, 297 (1964).
246. Sazykin, J. O., and Borisova, G. N., *Fed. Proc., Fed. Am. Soc. Exp. Biol.* 23, Trans. Suppl., 380 (1964).
247. Deev, A. I., Dobretsov, G. E., and Petrov, V. A., *Antibiotiki (Moscow)* 19, 584 (1974).
248. Kravchenko, L. S., and Tereshin, I. M., *Antibiotiki (Moscow)* 19, 675 (1974).
249. Matsuura, S., and Katagiri, K., *J. Antibiot., Ser. A* 14, 353 (1961).
250. Katagiri, K., and Sugiura, K., *J. Antibiot., Ser. A* 15, 233 (1962).
251. Haňka, L. J., Evans, J. S., Mason, D. J., and Dietz, A., *Antimicrob. Agents Chemother.* p. 619 (1966).
252. Shugar, D., *FEBS Lett., Suppl.* 40, 48 (1974).
253. Levitan, I. B., and Webb, T. E., *Biochim. Biophys. Acta* 182, 491 (1969).
254. Li, L. H., Olin, E. J., Buskirk, H. H., and Reineke, L. M., *Cancer Res.* 30, 2760 (1970).
255. Zain, B. S., Adams, R. L. P., and Imrie, R. C., *Cancer Res.* 33, 40 (1973).
256. Reichmann, M., Karlan, D., and Penman, S., *Biochim. Biophys. Acta* 299, 173 (1973).
257. Li, L. H., Olin, E. J., Fraser, T. J., and Bhuyan, B. K., *Cancer Res.* 30, 2770 (1970).
258. Heby, O., and Russel, D. H., *Cancer Res.* 33, 159 (1973).
259. Fučik, V., Zadaražil, S., Šormova, Z., and Šorm, F., *Collect. Czech. Chem. Commun.* 30, 2883 (1965).
260. Weiss, J. W., and Pitot, H. C., *Arch, Biochem. Biophys.* 160, 119 (1974).
261. Čihak, A., Veselý, J., and Harrap, K. R., *Biochem. Pharmacol.* 23, 1087 (1974).
262. Lee, T., Karon, M., and Momparler, R. L., *Cancer Res.* 34, 2482 (1974).
263. Raška, K., Jurovčik, M., Fučik, M., Tykva, R., Šormova, Z., and Šorm, F., *Collect. Czech. Chem. Commun.* 31, 2809 (1966).
264. Čihǎk, A., Narurkar, L. M., and Pittot, H. C., *Collect. Czech. Chem. Commun.* 38, 948 (1973).
265. Reichman, M., and Penman, S., *Biochim. Biophys. Acta* 324, 282 (1973).
266. Shutt, R. H., and Krueger, R. G., *J. Immunol.* 108, 819 (1972).
267. Pittot, H. C., *Biochemistry* 14, 316 (1975).
268. Weiss, J. W., and Pittot, H. C., *Arch. Biochem. Biophys.* 165, 588 (1974).
269. Stock, C. C., Reilly, H. C., Buckely, S. M., Clarke, D. A., and Rhoads, C. P., *Nature (London)* 173, 171 (1954).
270. Johnson, I. S., Baker, L. A., and Wright, H. F., *Ann. N.Y. Acad. Sci.* 76, 861 (1958).

271. Hartmann, S. C., Leveberg, B., and Buchanan, J. M., *J. Am. Chem. Soc.*, **77**, 501 (1955).
272. Fernandes, J. F., LePage, G. A., and Lindner, A., *Cancer Res.* **16**, 154 (1966).
273. Levenberg, B., Melnick, I., and Buchanan, J. M., *J. Biol. Chem.* **225**, 163 (1957).
274. Connors, T. A., *Biochem. Soc. Trans.* **1**, 1049 (1973).
275. David, I. B., French, T. C., and Buchanan, J. M., *J. Biol. Chem.* **238**, 2178 (1963).
276. Chu, S. Y., and Henderson, J. F., *Biochem. Pharmacol.* **21**, 401 (1972).
277. Deery, D. J., and Taylor, K. W., *Biochem. J.* **134**, 557 (1973).
278. Zampieri, A., and Grenberg, J., *Genetics* **57**, 41 (1967).
279. Gauze, G. F., Maksimova, T. S., Olkhovatova, O. L., Terekhova, L. P., Kochetkova, G. V., and Ilchenko, G. B., *Antibiotiki* (*Moscow*) **17**, 8 (1972).
280. Kudinova, M. K., Borisova, V. M., Petukhova, N. M., and Brazhnikova, M. G., *Antibiotiki* (*Moscow*) **17**, 689 (1972).
281. Shorin, V. A., Averbukh, L. A., Bazhanov, V. S., Rossolimo, O. K., Lepeshkina, G. N., and Grinstain, A. M., *Antibiotiki* (*Moscow*) **17**, 432 (1972).
282. Bazhanov, V. S., *Antibiotiki* (*Moscow*) **19**, 802 (1974).
283. Goldberg, L. E., Filiposyants, T. S., and Stepanov, E. S., *Antibiotiki* (*Moscow*) **17**, 616 (1972).
284. Shapovalova, S. P., Malkov, I. V., and Artemova, L. K., *Antibiotiki* (*Moscow*) **19**, 128 (1974).
285. Balan, J., Fuska, J., Kuhr, I., and Kuhrova, V., *Folia Microbiol.* (*Prague*) **15**, 479 (1970).
286. Cornforth, J. W., Ryback, G., Robinson, P. M., and Park, D., *J. Chem. Soc.* C p. 2786 (1971).
287. Kjaer, D., Kjaer, A., Pedersen, C., Bu'Lock, J. D., and Smith, J. R., *J. Chem. Soc.* C p. 2792 (1971).
288. Fuska, J., Nemec, P., Veselý, P., Ujházy, V., and Horáková, K., in *Prog. Chemother., Proc. Int. Congr. Chemother., 8th, 1973.* Vol. III, p. 835 (1974).
289. Henderson, J. F., personal communication (1974).
290. Fuska, J., Proksa, B., and Fusková, A., *Neoplasma* **22**, 335 (1975).
291. Seto, H., *Agric. Biol. Chem.* **37**, 24 (1973).
292. Yonehara, H., and Otake, N., *Tetrahedron Lett.* p. 3785 (1966).
293. Tanaka, N., Sakagami, Y., Nishimura, T., Yamaki, H., and Umezawa, H., *J. Antibiot., Ser. A* **14**, 121 (1961).
294. Lichtenthaler, F. W., and Trummlitz, G., *FEBS Lett.*, **38**, 237 (1974).
295. Yamaguchi, H., Yamamoto, C., and Tanaka, N., *Biochemistry* (*Tokyo*) **57**, 667 (1965).
296. Pestka, S., *Proc. Natl. Acad. Sci. U.S.A.* **64**, 709 (1969).
297. Černá, J., Rychlík, I., and Lichtenthaler, F. W., *FEBS Lett.* **30**, 147 (1973).
298. Timberlake, W. E., and Griffin, D. H., *Biochim. Biophys. Acta* **353**, 248 (1974).
299. Batra, P. P., Harbin, T. L., Howes, C. D., and Bernstein, S. C., *J. Biol. Chem.* **246**, 7125 (1971).
300. Kinoshita, M., Wada, M., Aburagi, S., and Umezawa, H., *J. Antibiot.* **24**, 724 (1971).
301. Umezawa, H., Maeda, K., Takeuchi, T., and Okami, Y., *J. Antibiot.* **19**, 200 (1966).
302. Umezawa, H., *Euchem. Conf. Antibiot. 1972*, p. 1.
303. Umezawa, H., *Biomedicine* **18**, 459 (1973).
304. Fuji, A., Takita, T., Shimada, U., and Umezawa, H., *J. Antibiot.* **27**, 73 (1974).

344 J. FUSKA AND B. PROKSA

305. Takita, T., Fuji, A., Fukuoka, T., and Umezawa, H., *J. Antibiot.* **26**, 252 (1973).
306. Umezawa, H., Takahashi, Y., Fuji, A., Saino, T., Shirai, T., and Takita, T., *J. Antibiot.* **26**, 117 (1973).
307. Fuji, A., Takita, T., Maeda, K., and Umezawa, H., *J. Antibiot.* **26**, 396 (1973).
308. Fuji, A., Takita, T., Maeda, K., and Umezawa, H., *J. Antibiot.* **26**, 398 (1973).
309. Nakayama, Y., Kunishima, M., Omoto, S., Takita, T., and Umezawa, H., *J. Antibiot.* **26**, 400 (1973).
310. Blum, R. H., Carter, S. K., and Agre, K., *Cancer* **31**, 903 (1973).
311. Urano, M., Fukuda, N., and Koike, S., *Cancer Res.* **33**, 2894 (1973).
312. Michaels, L., Grey, P. A., and Rowson, K. E. K., *J. Pathol.* **109**, 315 (1973).
313. Suzuki, H., Nagai, H., Akutsu, E., Yamaki, H., Tanaka, M., and Umezawa, H., *J. Antibiot.* **23**, 473 (1970).
314. Fujiwara, Y., and Kondo, T., *Biochem. Pharmacol.* **22**, 323 (1973).
315. Shirakawa, I., Azegami, M., Ishi, S., and Umezawa, H., *J. Antibiot.* **24**, 761 (1971).
316. Umezawa, H., Asakura, H., Oda, K., and Hori, S., *J. Antibiot.* **26**, 521 (1973).
317. Krueger, W. C., Pschigoda, L. M., and Reusser, P., *J. Antibiot.* **26**, 424 (1973).
318. Haidle, C. W., Kuo, M. T., and Weiss, K. K., *Biochem. Pharmacol.* **21**, 3305 (1972).
319. Sartiano, G. P., Winkelstein, A., Linch, W., and Boggs, S. S., *J. Antibiot.* **26**, 437 (1973).
320. Umezawa, H., *Fed. Proc., Fed. Am. Soc. Exp. Biol.* **33**, 2296 (1974).
321. Yamazaki, Z. I., Mueller, W. E. G., and Zahn, R. K., *Biochim. Biophys. Acta* **308**, 412 (1973).
322. Watanabe, M., Takabe, Y., Katsumato, K., Terasima, T., and Umezawa, H., *J. Antibiot.* **26**, 417 (1973).
323. Nagai, K., Yamaki, H., Suzuki, H., Tanaka, N., and Umezawa, H., *Biochim. Biophys. Acta* **179**, 165 (1969).
324. Yamaki, H., Suzuki, H., Nagai, K., Tanaka, N., and Umezawa, H., *J. Antibiot.* **24**, 178 (1971).
325. Mueller, W. E. G., Yamazaki, Z., Zöllner, J. E., and Zahn, R. K., *FEBS Lett.* **31**, 217 (1973)
326. Miyaki, M., Ono, T., and Umezawa, H., *J. Antibiot.* **24**, 587 (1971).
327. Mueller, W. E. G., Yamazaki, Z., and Zahn, R. K., *Biochem. Biophys. Res. Commun.* **46**, 1667 (1972).
328. Tanaka, N., *J. Antibiot.* **23**, 523 (1970).
329. Onishi, T., Shimada, K., and Takagi, Y., *Biochim. Biophys. Acta* **312**, 248 (1973).
330. Barranco, S. C., Luce, J. K., Romsdahl, M. M., and Humphrey, R. M., *Cancer Res.* **33**, 882 (1973).
331. Watanabe, M., Tanabe, Y., Katsumata, T., and Terashima, T., *Cancer Res.* **34**, 878 (1974).
332. Krishan, A., *Cancer Res.* **33**, 777 (1973).
333. Paika, K. D., and Krishan, A., *Cancer Res.* **33**, 961 (1973).
334. Hittelman, W. N., and Rao, P. N., *Cancer Res.* **34**, 3433 (1974).
335. Tamura, H., Sugiyama, Y., and Sugahara, T., *Gann* **65**, 103 (1974).
336. Reiter, H., Milewskiy, M., and Kelley, P., *J. Bacteriol.* **111**, 586 (1972).
337. Mueller, V. E. G., Yamazaki, Z., Breter, H., and Zahn, R. K., *Eur. J. Biochem.* **31**, 518 (1973).
338. Yamaki, H., Tanaka, N., and Umezawa, H., *J. Antibiot.* **22**, 315 (1969).
339. Dlugi, A. M., Robie, K. M., and Mitchell, M. S., *Cancer Res.* **34**, 2504 (1974).

340. Promchainaut, C., *Mutat. Res.* **28**, 107 (1975).

341. Miyaki, M., Kitayama, T., and Ono, T., *J. Antibiot.* **27**, 647 (1974).

342. Hittelman, W. N., and Rao, P. N., *Cancer Res.* **34**, 3433 (1974).

343. Takeshita, M., Horwitz, S. B., and Grollman, A. P., *Virology* **60**, 455 (1974).

344. Desai, L. S., Krishan, A., and Foley, G. E., *Cancer* **34**, 1873 (1974).

345. Hütter, R., Porrala, K., Zachau, H. G., and Zähner, H., *Biochem. Z.* **344**, 190 (1966).

346. Poralla, K., and Zähner, H., *Arch. Microbiol.* **61**, 143 (1968).

347. Mass, G., Poralla, K., and Zähner, H., *Biochem. Biophys. Res. Commun.* **34**, 84 (1969).

348. Nass, G., and Hasenbank, R., *Mol. Gen. Genet.* **108**, 28 (1970).

349. Hirakawa, T., Morinaga, H., and Watanabe, K., *Agric. Biol. Chem.* **38**, 85 (1974).

350. Japanese Patent 70/05,036 (1970). *C.A.* **72**, 120087 (1970).

351. Noda, T., Take, T. Watanabe, T., and Abe, Y., *Tetrahedron* **26**, 1339 (1970).

352. Arsenault, G. P., Althaus, J. R., and Divekar, P. V., *Chem. Commun.* p. 1414 (1969)

353. McCurry, M. P., Jr., and Abe, K., *J. Am. Chem. Soc.* **95**, 17 (1973).

354. Japanese Patent 73/56,984 (1973); *C.A.* **79**, 135295 (1973).

355. Mizuno, K., Tsujino, M., Takeda, M., Hayashi, M., Atsumi, K., Asano, K., and Matsuda, T., *J. Antibiot.* **27**, 775 (1974).

356. Betina, V., Fuska, J., Kjaer, A., Kutkova, M., Nemec, P., and Shapiro, R. H., *J. Antibiot., Ser. A* **19**, 115 (1966).

357. Singleton, V. L., Bohonos, N., and Ulstrup, A. J., *Nature (London)* **181**, 1072 (1958).

358. Suzuki, Y., Tanaka., H., Aoki, H., and Tamura, T., *Agric. Biol. Chem.* **34**, 395 (1970).

359. Sigg, H. P., *Helv. Chim. Acta* **47**, 1401 (1964).

360. Betina, V., *Neoplasma* **16**, 23 (1969).

361. Cross, B. E., and Hendley, P., *J. Chem. Soc., Chem. Commun.* p. 124 (1975).

362. Hayashi, T., Takatsuki, A., and Tamura, G., *J. Antibiot.* **27**, 65 (1974).

363. Japanese Patent 396 (1955); *C.A.* **77**, 12409 (1956).

364. Rolland, J. F., Chmieliewicz, Z. F., Weiner, B. A., Cross, M. A., Boning, O. P., Luck-Rosen, J. V., Bardos, T. J., Reilly, H. C., Sugiura, K., and Stock, C. C., *Science* **132**, 1897 (1960).

365. Sterneberg, S. S., Philips, F. S., Cronin, A. C., Sodergren, I. E., and Vidal, P. M., *Cancer Res.* **23**, 1036 (1963).

366. Vavra, J. J., and Dietz, A., *Antimicrob. Agents & Chemother.* p. 75 (1964).

367. Bergy, M. L., and Herr, R. R., *Antimicrob. Agents Chemother.* p. 80 (1964).

368. Akita, E., Okami, Y., Suzuki, M., Maeda, K., Takeuchi, T., and Umezawa, H., *J. Antibiot.* **15**, 130 (1962).

369. Hosoya, S., *Chemotherapy (Tokyo)* **3**, 128 (1955).

370. Gauze, G. F., Sveshnikova, M. A., Ukholina, R. S., Gavrilina, G. V., Filicheva, V. A., and Gladkikh, F. G., *Antibiotiki (Moscow)* **18**, 675 (1973).

371. Brazhnikova, M. G., Zbarsky, V. G., Ponomarenko, V. J., and Potapova, N. P., *J. Antibiot.* **27**, 254 (1974).

372. Guaze, G. F., Brazhnikova, M. G., and Shorin, V. A., *Cancer Chemother. Rep., Part 1* **58**, 255 (1974).

373. Dudnik, V., Yu., Ostanina, L. N., Kozmyan, L. I., and Gauze, G. G., *Antibiotiki (Moscow)* **19**, 514 (1974).

374. Shapovalova, S. P., *Antibiotiki (Moscow)* **19**, 260 (1974).

375. Shapovalova, S. P., Malkov, I. V., and Artemova, L. K., *Antibiotiki* (*Moscow*) **19**, 128 (1974).
376. Onda, M., Konda, Y., Omura, S., and Hata, T., *Chem. Pharm. Bull.* **19**, 2013 (1971).
377. Stoll, B. A., *Cancer* **13**, 439 (1960).
378. Tanko, B., *Biochem. J.* **84**, 92P (1962).
379. Terawaki, A., and Greenberg, J., *Biochim. Biophys. Acta* **119**, 59 (1966).
380. Kumagai, K., Miyazaki, K., Rikimaru, M., and Ishida, N., *J. Antibiot., Ser. A* **15**, 154 (1962).
381. Kumagai, K., *J. Antibiot., Ser. A* **15**, 53 (1962).
382. Harada, Y., Nara, T., and Okamoto, F., *J. Antibiot., Ser. A* **9**, 6 (1956).
383. Marumo, S., Sasaki, K., Ohkuma, K., Anzai, K., and Suzuki, S., *Agric. Biol. Chem.* **32**, 209 (1968).
384. Itakura, C., Sega, T., Suzuki, S., and Sumiki, Y., *J. Antibiot., Ser. A* **16**, 231 (1963).
385. Hauser, D., Weber, H. P., and Sigg, H. P., *Helv. Chim. Acta* **53**, 1061 (1970).
386. Hauser, D., Loosli, H. R., and Niklaus, P., *Helv. Chim. Acta* **55**, 2182 (1972).
387. Safe, S., and Taylor, A., *J. Chem. Soc., Perkin Trans. 1* p. 472 (1972).
388. Trown, P. W., *Biochem. Biophys. Res. Commun.* **33**, 402 (1968).
389. Close, A., and Huguenin, R., *Helv. Chim. Acta* **57**, 533 (1974).
390. Stähelin, H., and Tripmacher, A., *Eur. J. Cancer* **10**, 801 (1974).
391. Kaziwara, K., Watanabe, J., Komeda, T., and Usui, T., *Cancer. Chemother. Rep.* **13**, 99 (1961).
392. Wakisaka, G., Uchino, H., Nakamura, T., Sotobayashi, H., Shirakawa, S., Adachi, A., and Sakurai, M., *Nature* (*London*) **198**, 385 (1963).
393. Kuwano, M., Kamiya, T., Endo, H., and Komiyama, S., *Antimicrob. Agents & Chemother.* **3**, 580 (1974).
394. Kersten, W., *Nucleic Acid Norm. Cancer Tissues, Proc. Symp., 1967* GANN Monogr. No. 6, p. 65 (1968).
395. Nayak, R., Sirsi, M., and Podder, S. K., *Biochim. Biophys. Acta* **378**, 195 (1975).
396. Honikel, K. O., and Santo, R. E., *Biochim. Biophys. Acta* **269**, 354 (1972).
397. Fok, J., and Waring, M., *Mol. Pharmacol.* **8**, 65 (1972).
398. Kamiyama, M., *J. Biochem.* (*Tokyo*) **63**, 566 (1968).
399. Koschel, K., Hartman, G., Kersten, W., and Kersten, H., *Biochem. Z.* **344**, 76 (1966).
400. Yano, M., Kusakari, T., and Miura, Y., *J. Biochem.* (*Tokyo*) **53**, 461 (1963).
401. Behr, W., Honikel, K., and Hartman, G., *Eur. J. Biochem.* **9**, 82 (1969).
402. Hartman, G., Behr, W., Bock, L., Honikel, K., Lill, H., Lill, U., and Sippel, A., *Zentralbl. Bakteriol., Parasitenkd., Infektionskr. Hyg., Abt. I: Orig.* **212**, 224 (1970).
403. Kida, M., Ujihara, M., Ohmura, E., and Kaziwara, K., *J. Biochem.* (*Tokyo*) **59**, 353 (1966).
404. Kuwata, T., and Sekita, S., *Arch. Gesamte Virusforsch.* **37**, 386 (1972).
405. Ayukawa, S., Hamade, M., Kojiri, K., Takeuchi, T., Hara, T., Nagatsu, T., and Umezawa, H., *J. Antibiot., Ser. A* **22**, 303 (1969).
406. Maevskij, M. M., Urazova, A. P., Romanenko, E. A., Molkov, J. N., Bondareva, A. S., Timofeevskaya, E. A., Vyazova, O. I., Mazaeva, V. G., and Talyzina, V. A., *Antibiotiki* (*Moscow*) **9**, 33 (1964).
407. Richle, W., Winkler, E. K., Hawley, D. M., Dobler, M., and Keller-Schierlein, W., *Helv. Chim. Acta* **55**, 467 (1972).

408. Calendi, E., DiMarco, A., Reggiani, M., Scarpinato, B., and Valentini, L., *Biochim. Biophys. Acta* **103**, 25 (1965).
409. Kersten, W., and Kersten, H., *Biochem. Z.* **341**, 174 (1965).
410. Apple, M. A., and Haskell, C. M., *Physiol. Chem. Phys.* **3**, 307 (1971).
411. Apple, M. A., *J. Clin. Pharmacol.* **15**, 29 (1975).
412. Richle, W., Winkler, E. K., Hawley, D. M., Dobler, M., and Keller-Schierlein, W., *Helv. Chim. Acta* **55**, 467 (1972).
413. Bhuyan, B. K., and Dietz, A., *Antimicrob. Agents Chemother.* p. 426 (1967).
414. Reusser, F., and Bhuyan, B. K., *J. Bacteriol.* **94**, 576 (1967).
415. ElKhadem, H. S., and ElAshry, E. S. H., *Carbohydr. Res.* **32**, 339 (1974).
416. Ikehara, M., and Tada, H., *Chem. Pharm. Bull.* **15**, 94 (1967).
417. Jagger, D. V., Kredich, N..M., and Guarino, A. J., *Cancer Res.* **21**, 216 (1961).
418. Siev, M., Weinberg, R., and Penman, S:, *J. Cell Biol.* **41**, 510 (1969).
419. Rottman, F., and Guarino, A. J., *Biochim. Biophys. Acta* **80**, 640 (1964).
420: Shiguera, H. T., and Gordon, C. N., *J. Biol. Chem.* **240**, 806 (1965).
421. Klenow, H., *Biochim. Biophys. Acta* **76**, 354 (1963).
422. Klenow, H., and Overgaard-Hansen, K., *Biochim. Biophys. Acta* **80**, 500 (1964).
423. Anderson, J. M., and Roth, R. M., *Biochim. Biophys. Acta* **335**, 285 (1974).
424. Podobed, O. V., Leitim, V. L., Brykina, E. V., Mantjeva, V. L., and Lerman, M. I., *Mol. Biol.* (*Moscow*) **7**, 343 (1973).
425. Butcher, F. R., Bushmell, D. E., Becker, Y. E., and Potter, V. R., *Exp. Cell Res.* **74**, 115 (1972).
426. Latorre, J., and Perry, R. P., *Biochim. Biophys. Acta* **335**, 93 (1973).
427. Dietz, J. L., *Chromosoma* **42**, 345 (1973).
428. Fouguet, H., Wick, R., Boehme, R., Sauer, H. W., and Scheller, K., *Arch. Biochem. Biophys.* **168**, 273 (1975).
429. Richardson, L. S., Ting, R. C., Gallo, R. C., and Wu, A. M., *Int. J. Cancer* **15**, 451 (1975).
430. Takeuchi, T., Iinuma, H., Iwanaga, J., Takahashi, S., Takita, T., and Umezawa, H., *J. Antiobiot.* **22**, 215 (1969).
431. Gláz, E. T., Czanyi, E., and Gyimesi, J., *Nature* (*London*) **212**, 617 (1966).
432. Uraguchi, K., *Int. Encycl. Pharmacol. Toxicol.* Sect. 71, Part V. (1971).
433. Ueno, Y., Kaneko, M., Tatsuno, T., Ueno, I., and Uraguchi, K., *Seikagaku* **35**, 224 (1963).
434. Yamazoe, S., Nakao, M., Hayashi, K., Nagano, K., Motegi, T., Uesugi, S., and Kanoh, T., *J. Med. Soc. Gunma Univ.* **12**, 73 (1963).
435. White, F. R., *Cancer Chemother. Rep.* **5**, 48 (1959).
436. Farber, J. L., and Farmar, R., *Biochem. Biophys. Res. Commun.* **51**, 626 (1973).
437. Luettge, U., Lauchli, A., Ball, E., and Pitman, M. G., *Experientia* **30**, 470 (1974).
438. Timberlake, W. E., and Griffin, D. H., *Biochim. Biophys. Acta* **353**, 248 (1974).
439. Fujiwara, Y., *Cancer Res.* **32**, 2089 (1972).
440. Highfield, D. P., and Dewey, W. C., *Exp. Cell Res.* **75**, 1972 (1972).
441. Werry, P. A. T. J., and Wanka, F., *Biochim. Biophys. Acta* **287**, 232 (1972).
442. Cikes, M., and Klein, G., *J. Natl. Cancer Inst.* **48**, 509 (1972).
443. Malhotra, S. S., Solomons, T., and Spencer, M., *Planta* **14**, 169 (1973).
444. Craig, E. A., and Raskas, H. J., *J. Virol.* **14**, 26 (1974).
445. Cowtan, E. R., Yoda, B., and Israels, L. G., *Arch. Biochem. Biophys.* **155**, 194 (1973).

446. Yomain-Braun, M., Garber, A. Y., Farber, E., and Hamson, R. W., *J. Biol. Chem.* **248**, 1536 (1973).
447. Weiss, L., and Huber, D., *J. Cell Sci.* **15**, 217 (1974).
448. Čihák, A., Veselý, J., and Harrap, K. R., *Biochem. Pharmacol.* **23**, 1087 (1974).
449. Young, C. W., and Dowling, M. D., Jr., *Cancer Res.* **35**, 1218 (1975).
450. Dybing, E., *Biochim. Biophys. Acta* **373**, 100 (1974).
451. Greenberg, J. S., and Aaronson, S. A., *J. Virol.* **15**, 64 (1975).
452. Mitchell, J. L. A., and Sedory, M. J., *FEBS Lett.* **49**, 120 (1974).
453. Ricketts, T. R., and Rappitt, A. F., *Arch. Microbiol.* **102**, 1 (1975).
454. Japanese Patent 70/18,278 (1970); *C.A.* **73**, 86553 (1970).
455. Rothweiler, W., and Tamm, C., *Helv. Chim. Acta* **53**, 696 (1970).
456. Graf, W., Robert, J. L., Vederas, J. C., Tamm, C., Solomon, P. H., Miura, I., and Nakanishi, K., *Helv. Chim. Acta* **57**, 1801 (1974).
457. Katagiri, K., and Matsuura, S., *J. Antibiot.* **24**, 722 (1971).
458. Nishikawa, T., and Marks, R., *Br. J. Dermatol.* **88**, 469 (1973).
459. Ehtensen, R. D., *Proc. Soc. Exp. Biol. Med.* **136**, 1256 (1971).
460. Zigmond, S. H., and Hirsch, J. G., *Exp. Cell Res.* **73**, 383 (1972).
461. Axline, S. G., and Reaven, E. P., *J. Cell Biol.* **62**, 647 (1974).
462. Hoffmann, E. K., Rasmussen, L., and Zeuthen, E., *J. Cell Sci.* **15**, 403 (1974).
463. Kletzien, R. F., and Perdue, J. F., *J. Biol. Chem.* **248**, 711 (1973).
464. Bloch, R., *Biochemistry* **12**, 4799 (1975).
465. Taverna, R. D., and Langdon, R. G., *Biochim. Biophys. Acta* **323**, 207 (1973).
466. Mizel, S. B., *Nature (London) New Biol.*, **243**, 125 (1973).
467. Westermark, B., *Exp. Cell Res.* **82**, 341 (1973).
468. Lin, S., Lin, D. C., Spudlich, A., and Kun, E., *FEBS Lett.* **37**, 241 (1973).
469. Ono, M., and Hozumi, M., *Biochem. Biophys. Res. Commun.* **53**, 342 (1973).
470. Zurier, R. B., Hoffstein, S., and Weissmann, G., *Proc. Natl. Acad. Sci. U.S.A.* **70**, 884 (1973).
471. Lin, S., Santi, D. V., and Spudich, J. A., *J. Biol. Chem.* **249**, 2268 (1974).
472. Bos, C. J., and Emmelot, P., *Chem.-Biol. Interact.* **8**, 349 (1974).
473. Weiss, L., *Exp. Cell Res.* **74**, 21 (1972).
474. Boyde, A., Bailey, E., and Veselý, P., *Scanning Electron Microsc.* **1–4**, 597 (1974).
475. Gerschenbaum, M. R., Shay, J. W., and Porter, K. R., *Scanning Electron Microsc.* **1–4**, 589 (1974).
476. O'Neil, F. J., *J. Natl. Cancer Inst.* **52**, 653 (1974).
477. Strah, S. K., and Holzer, H., *J. Cell Biol.* **63**, 337 (1974).
478. Davies, P., Allison, A. C., and Haswell, A. P., *Biochem. J.* **134**, 33 (1973).
479. Plagemann, P. G. W., and Estensen, R. D., *J. Cell Biol.* **55**, 179 (1972).
480. Estensen, R. D., and Plagemann, P. G. W., *Proc. Natl. Acad. Sci. U.S.A.* **69**, 1430 (1972).
481. Partrige, I., Jones, G. E., and Gillet, R., *Nature (London)* **253**, 632 (1975).
482. Hirano, A., and Kurimura, T., *Exp. Cell Res.* **89**, 111 (1974).
483. Mayhew, E., Poste, G., Cowden, M., Tolson, N., and Maslow, D., *J. Cell. Physiol.* **84**, 373 (1974).
484. Dombach, M., and Weber, W., *Comp. Biochem. Physiol. C* **50**, 49 (1975).
485. Minato, H., and Matsumoto, M., *J. Chem Soc.* C p. 38 (1970).
486. Lebet, C. R., and Tamm, C., *Helv. Chim. Acta* **57**, 1785 (1974).
487. Godman, G. C., Miranda, A. F., Deitch, A. D., Tanenbaum, S. W., and Cottral, G., *J. Cell Biol.* **64**, 644 (1975).
488. Tanaka, N., Sagami, Y., Nishimura, T., Yamaki, H., and Umezawa, H., *J. Antibiot., Ser. A* **14**, 121 (1961).

489. Berdy, J., Zsadani, J., Halasz, M., Horvath, I., and Magyar, K., *J. Antibiot.* 24, 209 (1971).
490. DiMarco, A., Casazza, A. M., Guillani, F., and Seranzo, C., *EACR Symp. Mech. Action Cytostatic Agents. 1972*, Abstr. Book, p. 29 (1972).
491. Jones, B., *Cancer Res.* 31, 84 (1971).
492. Weill, M., and Glidewell, O. L., *Cancer Res.* 33, 921 (1973).
493. Rusconi, A., and DiMarco, A., *Cancer Res.* 29, 1507 (1969).
494. Danø, K., Frederiksen, S., and Hellung-Larsen, P., *Cancer Res.* 32, 1307 (1972).
495. Crook, L., Rees, K. R., and Cohen, A., *Biochem. Pharmacol.* 21, 281 (1972).
496. Silvestrini, R., Lenaz, L., DiFronzo, G., and Sanfilippo, O., *Cancer Res.* 33, 2954 (1973).
497. DiMarco, A., Zunino, F., Silvestrini, R., Gambarucci, C., and Gambetta, R. A., *Biochem. Pharmacol.* 20, 1328 (1971).
498. Zunino, F., Gambetta, R. A., DiMarco, A., and Zaccara, A., *Biochim. Biophys. Acta* 277, 489 (1972).
499. Doskočil, J., and Fric, I., *FEBS Lett.* 37, 58 (1973).
500. Dall 'Acqua, F., Terbojevich, M., Marcian, S., Vedaldi, D., and Rodigniero, G., *Farmaco, Ed. Sci.* 29, 682 (1974).
501. Goodman, M. F., Bessman, M. J., and Bachur, N. R., *Proc. Natl. Acad. Sci. U.S.A.* 71, 1193 (1974).
502. Müller, W. E. G., Obermeier, J., Totsuoka, A., and Zahn, R. K., *Nucleic Acids Res.* 1, 63 (1974).
503. Linstead, E. D., Rhodes, P. M., and Wilkie, D., *Biochim. Biophys. Acta* 312, 323 (1973).
504. Čihák, A., Veselý, J., and Harrap, K. R., *Biochem. Pharmacol.* 23, 1087 (1974).
505. Tormensen, T., and Friesen, J. D., *Mol. Genet.* 124, 174 (1973).
506. Vig, B. K., Samuels, L. D., and Kontras, S. B., *Chromosoma* 29, 62 (1970).
507. Jensen, M. K., and Philip, P., *Mutat. Res.* 12, 91 (1971).
508. Bemfong, A. M., *Can. J. Genet. Cytol.* 15, 587 (1973).
509. Vig, B. K., Kontras, S. B., and Aubele, A., *Mutat. Res.* 7, 91 (1969).
510. Linden, W. A., Brisch, H., Canstein, L., König, K., and Canstein, M., *Eur. J. Cancer* 10, 647 (1974).
511. Crook, L. E., Rees, K. R., and Cohen, A., *Chem.-Biol. Interact.* 4, 343 (1972).
512. Schwartz, H. S., and Grindey, G. B., *Cancer Res.* 33, 1837 (1973).
513. Zunino, F., DiMarco, A., Zaccara, A., and Luoni, G., *Chem.-Biol. Interact.* 9, 25 (1974).
514. Zunino, F., Gambetta, R., DiMarco, A., Zaccara, A., and Luoni, G., *Cancer Res.* 35, 754 (1975).
515. Mizuno, N. S., Zakis, B., and Decker, W. R., *Cancer Res.* 34, 1542 (1975).
516. Linden, W. A., Baisch, H., von Canstein, L., Koenig, K., and von Canstein, M., *Eur. J. Cancer* 10, 647 (1974).
517. Zunino, F., Gambetta, R., Zaccara, A., and Luoni, G., *Cancer Res.* 35, 754 (1975).
518. Cargill, C., Bachmann, E., and Zbinden, G., *J. Natl. Cancer Inst.* 53, 481 (1974).
519. Tanaka, N., Nishimura, T., Yamaguchi, H., and Umezawa, H., *J. Antibiot., Ser. A* 14, 98 (1961).
520. Miyairi, N., Tanaka, N., and Umezawa, H., *J. Antibiot., Ser. A* 14, 119 (1961).
521. Tanaka, N., *J. Antibiot., Ser. A* 16, 163 (1963).

522. Bloch, A., and Nichol, C. A., *Fed. Proc., Fed. Am. Soc. Exp. Biol.* **23**, 324 (1964).
523. U.S. Patent 3,326,761 (1967).
524. DeVault, R. L., Schmitz, H., and Hooper, I. R., *Antimicrob. Agents Chemother.* p. 796 (1965).
525. Myiazaki, Y., Yoshida, H., Hidaka, T., Takeuchi, S., and Yonehara, H., *J. Antibiot., Ser.* A **22**, 393 (1969).
526. Netherlands Patent 6,516,312 (1966); *C.A.* **65**, 19271 (1966).
527. Betina, V., Baráthova, E., and Nemec, P., *J. Antibiot.* **22**, 129 (1969).
528. Japanese Patent 70/26,715 (1970); *C.A.* **74**, 2673 (1971).
529. Aoki, H., Miyairi, M., Ajisaka, M., and Sasaki, H., *J. Antibiot.* **22**, 201 (1969).
530. Takeuchi, T., Takahashi, S., Iinuma, H., and Umezawa, H., *J. Antibiot.* **24**, 631 (1971).
531. Kunimoto, T., Hori, M., and Umezawa, H., *Biochim. Biophys. Acta* **298**, 513 (1973).
532. Ishizuka, M., Iinuma, H., Takeuchi, Z., and Umezawa, H., *J. Antibiot.* **25**, 320 (1972).
533. Chandra, P., Zimmer, C., and Thrumm, H., *FEBS Lett.* **7**, 90 (1970).
534. Zimmer, C., and Luck, G., *FEBS Lett.* **10**, 339 (1970).
535. Krey, A. K., Allison, R. G., and Hahn, F. E., *FEBS Lett.* **29**, 59 (1973).
536. Müller, W. E. G., Obermeier, J., Maidhof, A., and Zaan, R. K., *Chem.-Biol. Interact.* **8**, 183 (1974).
537. Rossi, G., and Zaalberg, O. B., *J. Immunol.* **113**, 424 (1974).
538. Krey, A. K., and Hahn, H., *FEBS Lett.* **10**, 175 (1970).
539. Becker, Y., and Weinberg, A., *Ind. J. Med. Sci.* **8**, 75 (1972).
540. Puschendorf, B., Petersen, E., Wolf, H., Werchan, H., and Grunicke, H., *Biochem. Biophys. Res. Commun.* **43**, 617 (1971).
541. Kotler, M., and Becker, Y., *Nature (London), New Biol.* **234**, 212 (1971).
542. DeRadult, Y., and Werner, G. H., *Prog. Antimicrob. Anticancer Chemother., Proc. Int. Congr. Chemother., 6th, 1969* Vol. 2, p. 14 (1970).
543. Zimmer, C., Puschendorf, B., Grunicke, P., and Venner, H., *Eur. J. Biochem.* **21**, 269 (1971).
544. Zimmer, C., Reinert, K. E., Luck, G., Wähnert, U., Löber, G., and Thrum, H., *J. Mol. Biol.* **58**, 329 (1971).
545. Küpper, H. A., McAllister, N. T., and Bautz, E. K. F., *Eur. J. Biochem.* **38**, 581 (1973).
546. Haupt, J., Zimmer, C., and Thrumm, H., *Ergeb. Exp. Med.* **9**, 368 (1972).
547. Rao, K. V., Brooks, S. C., Kugelman, M., and Romano, A. A., *Antibiot. Annu.* p. 943 (1960).
548. Golden, A., Venditti, J. M., and Kline, J., *Cancer Res.* **22**, 748 (1962).
549. Brockman, R. W., *Adv. Cancer Res.* **7**, 129 (1963).
550. Rosenberg, S., and Calabresi, P., *Nature (London)* **199**, 1101 (1963).
551. Ansfield, F. J., *Cancer Chemother. Rep.* **46**, 37 (1970).
552. Brockman, R. W., Pitillo, F. F., Shadix, S., and Hill, D. L., *Antimicrob. Agents Chemother.* p. 56 (1969).
553. Hersch, E. M., and Brown, B. W., *Cancer Res.* **31**, 834 (1971).
554. Shapovalova, S. P., *Antibiotiki* **18**, 635 (1973).
555. Pittillo, R. F., Wooley, C., and Brockman, R. W., *Cancer Chemother. Rep.* **55**, 47 (1971).
556. German Patent 2,202,690 (1972). *C.A.* **77**, 164320 (1972).

557. Kuhr, I., Fuska, J., Sedmera, P., Podojil, M., Vokoun, J., and Vaněk, Z., J. Antibiot. 26, 535 (1973).
558. Fuska, J., Kuhr, I., Nemec, P., and Fusková, A., J. Antibiot. 27, 123 (1974).
559. Loshkareva, N. P., Gauze, G. F., and Zbarski, I. B., Biokhimiya 34, 500 (1969).
560. Gauze, G. G., Loshkareva, N. P., and Zbarski, I. B., Mol. Biol. (Moscow) 3, 566 (1969).
561. Sato, K., Niinomi, Y., Katagiri, K., Matsukage, A., and Minagawa, T., Biochim. Biophys. Acta 174, 230 (1969).
562. Altman, S., and Lerman, L. S., J. Mol. Biol. 50, 263 (1970).
563. Czech Patent 140, 970 (1970).
564. Fuska, J., Podojil, M., Fusková, A., Vaněk, Z., and Vokoun, J., EACR Symp. Mech. Action Cytostatic Agents, 1972 Abstr. Book, p. 24 (1972).
565. Kudinova, M. K., Kovsharova, I. N., Proshlyakova, V. V., Prozorovskaya, N. A., and Brazhnikova, M. G., Antibiotiki (Moscow) 10, 488 (1965).
566. Mizuno, S., Nitta, K., and Umezawa, H., J. Biochem. (Tokyo) 61, 373 (1967).
567. Mizuno, S., Nitta, K., and Umezawa, H., J. Antibiot., Ser. A 19, 97 (1966).
568. French Patent 1,545,790 (1969); C. A. 71, 122302 (1969).
569. Montagni-Marelli, A., Testorelli, G., and Valenti, L., Arch. Ital. Patol. Clin. Tumori 11, 301 (1968).
570. Kato, A., Ando, K., Tamura, G., and Arima, K., Cancer Res. 31, 501 (1971).
571. Fuska, J., Kuhr, I., and Koman, V., Folia Microbiol. (Prague) 19, 301 (1974).
572. Kato, A., Ando, K., Kimura, T., Suzuki, S., Tamura, G., and Arima, K., Prog. Antimicrob. Anticancer Chemother., Proc. Int. Congr. Chemother., 6th, 1969 Part II, p. 137 (1970).
573. Kudinova, M. K., Kovsharova, I. N., Proshlyakova, V. V., Prozorovskaya, N. A., and Brazhnikova, M. G., Antibiotiki (Moscow) 10, 488 (1965).
574. Mondovi, B., Strom, R., Argo, A. F., Caiafa, P., DeSole, P., Botzi, A., Rotilio, G., and Tanelli, A. R., Cancer Res. 31, 505 (1971).
575. Schroeder, F., Holland, J. F., and Bieber, L. L., Biochemistry 12, 4785 (1973).
576. Bittman, R., Chen, W. C., and Anderson, O. R., Biochemistry 13, 1364 (1974).
577. DeKruyff, B., Gerritsen, W. J., Oerlemans, A., Demel, R. A., and Van Deenen, L. L. M., Biochim. Biophys. Acta 339, 30 (1974).
578. Mondovi, B., Strom, R., Rotilio, G., Finazzi, A. A., Cavalieri, R., and Rossi, F. A., Eur. J. Cancer 5, 129 (1969).
579. Mondovi, B., Finazzi, A. A., Rotilio, G., Strom, R., Moricca, G., and Rossi, F. A., Eur. J. Cancer 5, 137 (1969).
580. Norman, A. W., Demel, R. A., DeKruyff, B., Guerts van Kessel, W. S. M., and Van Deenen, L. L M., Biochim. Biophys. Acta 290, 1 (1972).
581. Drabikowski, W., Lagwinska, E., and Sarzala, M. G., Biochim. Biophys. Acta 291, 61 (1973).
582. Shroeder, F., and Bieber, L. L., Insect Biochem. 5, 201 (1975).
583. Komatsu, N., Terakawa, H., Nakanishi, K., and Watanabe, Y., J. Antibiot., Ser. A 16, 139 (1963).
584. Japanese Patent 73/10,554 (1973); C. A. 80, 35824 (1974).
585. Aizawa, S., Hidaka, T., Otake, M., Yonehara, H., Isono, K., Igarashi, M., and Suzuki, S., Agric. Biol. Chem. 29, 375 (1965).
586. Takeuchi, T., Iwanaga, J., Aoyagi, T., and Umezawa, H., J. Antibiot., Ser. A 19, 286 (1966).
587. Hori, M., Wakashiro, T., Ito, E., Sawa, T., Takeuchi, T., and Umezawa, H., J. Antibiot., Ser. A 21, 264 (1968).

588. Tanami, Y., Yamada, Y., Suzuki, K., and Tazaki, T., *Uirusu* **19**, 121 (1969).
589. Henderson, J. F., Patterson, A. R. P., Caldwell, I. C., and Hori, M., *Cancer Res.* **27**, 715 (1967).
590. Warnick, C. T., Muzik, H., and Patterson, A. R. P., *Cancer Res.* **32**, 2017 (1972).
591. Darlix, J. L., Fromageot, P., and Reich, E., *Biochemistry* **10**, 1525 (1971).
592. Howard, E., *J. Cell Biol.* **63**, 145 (1974).
593. Ward, D. C., and Reich, E., *Annu. Rep. Med. Chem.* p. 272 (1969).
594. Nishimura, C., and Tsukeda, H., *Prog. Antimicrob. Anticancer Chemother. Proc. Int. Congr. Chemother., 6th, 1969* Part II, p. 20 (1970).
595. Bossa, R., Dibini, F., Galatulas, I., and Taniguchi, M., *Arch. Ital. Patol. Clin. Tumori* **13**, 135 (1970).
596. Muyama, A., Hata, K., and Tamura, S., *Agric. Biol. Chem.* **33**, 1599 (1969).
597. Fuska, J., Nemec, P., Kuhr, I., and Fusková, A., *Adv. Antimicrob. Antineoplast. Chemother., Proc. Int. Congr. Chemother., 7th, 1971* Vol. II, p. 97 (1972).
598. Corey, E. Y., and Snider, B. B., *J. Am. Chem. Soc.* **97**, 2549 (1972).
599. DiPaolo, M. C., Tarbell, D. S., and Moore, G. E., *Antibiot. Annu.* p. 541 (1960).
600. Tursunkodzhaev, N. B., *Antibiotiki (Moscow)* **10**, 316 (1965).
601. Jaronski, S. T., *J. Antibiot., Ser. A* **25**, 327 (1972).
602. Ando, K., Suzuki, S., Saeki, T., Tamura, G., and Arima, K., *J. Antibiot., Ser. A* **22**, 189 (1969).
603. Ueno, Y., Ueno, I., Tatsuno, T., Ohkubo, K., and Tsunoda, H., *Experientia* **25**, 1062 (1969).
604. Arai, T., and Ito, T., *Prog. Antimicrob. Anticancer Chemother., Proc. Int. Congr. Chemother., 6th, 1969* Part I, p. 87 (1970).
605. DeBoer, C., Muelman, P. A., Wnuk, R. K., and Peterson, D. H., *J. Antibiot., Ser. A* **23**, 442 (1970).
606. Miller, P. A., Trown, P. W., Fulmor, W., Morton, G. O., and Karliner, J., *Biochem. Biophys. Res. Commun.* **33**, 219 (1968).
607. Reilly, H. C., Stock, C. C., and Buckley, S. M., *Cancer Res.* **13**, 684 (1953).
608. Hadler, H. I., Hadler, M. R., and Daniel, B. G., *J. Antibiot.* **26**, 30 and 36 (1973).
609. Maxia, L., LaColla, P., Spano, P. F., and Loddo, B., *Prog. Antimicrob. Anticancer Chemother., Proc Int. Congr. Chemother., 6th, 1969* Part II, p. 37 (1970).
610. Trown, P. W., and Bilello, J. A., *Antimicrob. Agents & Chemother.* **2**, 261 (1972).
611. Miller, P. A., Milstrey, K. P., and Trown, P. W., *Science* **159**, 431 (1968).
612. Miller, P. A., Linsay, H. L., Cormier, M., Mayberry, B. R., and Trown, P. W., *Ann. N.Y. Acad. Sci.* **173**, 151 (1970).
613. Krasilnikov, M. A., and Kuimova, T. F., *Antibiotiki (Moscow)* **12**, 1059 (1966).
614. Alekhina, R. P., Belousova, A. K., and Remesova, M. I., *Biokhimiya* **37**, 813 (1972).
615. Sugiura, M., Ito, H., Fujii, K., Sumida, K., Kunihisa, M. and Nishigata, Y., Japan Kokai 7435, 588 (1974); *C.A.* **81**, 103234 (1974).
616. Sakai, S., Takada, S., Kamasuka, T., Monoki, Y., and Saguyama, J., *Cancer Chemother. Abstr.* **9**, 68 (1968).
617. Ito, H., Fujii, K., Naruse, S., and Miyazaki, T., *Mie Med. J.* **23**, 117 (1973); *C.A.* **82**, 114727 (1975).

618. Soda, K., Misono, H., Mori, K., and Sakato, H., *Biochem. Biophys. Res. Commun.* **44**, 931 (1971).
619. Oki, T., Shirai, M., Ohshima, M., Yamamoto, T., and Soda, K., *FEBS Lett.* **33**, 286 (1973).
620. Roberts, J., Holcenberg, J. S., and Dolowy, W. C., *Nature (London)* **227**, 1137 (1970).
621. Ikeuchi, T., Kitame, F., Kikuchi, M., and Ishida, N., *J. Antibiot.* **25**, 548 (1972).
622. Pestka, S., *Arch. Biochem. Biophys.* **136**, 89 (1970).
623. Barbacid, M., and Vasquez, D., *Eur. J. Biochem.* **44**, 445 (1974).
624. Čihak, A., Veselý, J., and Harrap, K. R., *Biochem. Pharmacol.* **23**, 1087 (1974).
625. Chang, C. J., Floss, H. G., Soong, P., and Gang, C. T., *J. Antibiot.* **28**, 156 (1975).
626. Hungarian Patent 157,600 (1970). *C.A.* **73**, 119214 (1970).
627. Ogata, Y., *J. Antibiot., Ser. A* **12**, 133 (1959).
628. Nakamura, S., Maeda, K., and Umezawa, H., *J. Antibiot., Ser. A* **17**, 33 (1964).
629. Gachon, P., Kergomard, A., Vechambre, H., Esteve, C., and Staron, T., *J. Chem. Soc.* p. 1421 (1970).
630. French Patent 2,091,913 (1972); *C.A.* **77**, 124 772 (1972).
631. Fernandes de Albuquerque, M. M., Goncalves de Lima, O., Magali de Araucho, J., Coelho, J. S., Lyra, F. D., Cavalcanti, M., and Lima de Oliviera, L., *Rev. Inst. Antibiot., Univ. Fed. Pernambuco, Recife* **3**, 8 (1968).
632. Kaczka, E. A., Gitterman, C. O., Dulaney, E. L., and Folkers, K., *Biochemistry* **1**, 340 (1962).
633. Fritz, H. P., and von Stetten, O., *Z. Naturforsch., Teil B* **27**, 1457 (1972).
634. Mego, J. L., *Biochim. Biophys. Acta* **79**, 221 (1964).
635. Neuman, R. E., and Tytell, A. A., *Proc. Soc. Exp. Biol. Med.* **112**, 57 (1963).
636. Shigeura, H. T., and Gordon, C. M., *J. Biol. Chem.* **237**, 1932 (1962).
637. Fairbanks, M. B., and Kollar, E. J., *Teratology* **9**, 169 (1974).
638. Smitz, H., Crook, K. E., and Bush, J. A., *Antimicrob. Agents Chemother.* p. 613 (1966).
639. White, H. L., and White, J. R., *Biochemistry* **8**, 1030 (1969).
640. Joel, P. B., and Goldberg, I. H., *Biochim. Biophys. Acta* **224**, 361 (1970).
641. Japanese Patent 63/2,796 (1963); *C.A.* **59**, 5741 (1963).
642. McMorris, T. C., and Anchel, M., *J. Am. Chem. Soc.* **87**, 1594 (1965).
643. Shirahama, H., Fukuoka, Y., and Matsumoto, T., *Nippon Kagaku Zasshi* **83**, 1289 (1962).
644. McMorris, T. C., and Anchel, M., *J. Am. Chem. Soc.* **85**, 831 (1963).
645. Walser, J., and Heinstein, P. F., *Antimicrob. Agents & Chemother.* **3**, 357 (1973).
646. Tyc, M., *Arch. Immunol. Ther. Exp.* **18**, 129 (1970).
647. Hoshino, M., Umezawa, H., Mimura, Y., and Hata, T., *J. Antibiot., Ser. A* **20**, 30 (1967).
648. Umezawa, I., Kanda, N., and Hata, T., *J. Antibiot., Ser. A* **21**, 30 (1968).
649. Ranadive, K. J., Gothoskar, S. V., Bapat, C., and Cotinho, W. G., *Indian J. Med.* **54**, 229 (1966).
650. Chaudhury, S., and Roy, D. K., *Indian J. Cancer* **8**, 54 (1971).
651. Chaudhury, S., and Roy, D. K., *Indian J. Cancer* **11**, 285 (1974).

652. Matsuura, S., Shiratori, G., Harada, Y., and Katagiri, K., *J. Antibiot., Ser. A* **20**, 282 (1967).
653. Katagiri, K., Nishiyama, S., and Sato, K., *Prog. Antimicrob. Anticancer Chemother., Proc. Int. Congr. Chemother., 6th, 1969* Part II, p. 11 (1970).
654. Fukushima, K., Ishiwata, K., Kuroda, S., and Arai, T., *J. Antibiot.* **26**, 65 (1973).
655. Kuimova, T. F., Fukushima, K., Kuroda, S., and Arai, T., *J. Antibiot.* **24**, 69 (1971).
656. Friedman, P. A., Joel, P. B., and Goldberg, I. H., *Biochemistry* **8**, 1535 (1969).
657. Joel, P. B., Friedman, P. A., and Goldberg, I. H., *Biochemistry* **9**, 4421 (1970).
658. Kanda, N., *J. Antibiot.* **24**, 595 (1971).
659. Kanda, N., Kono, M., and Asano, K., *J. Antibiot.* **25**, 553 (1972).
660. Umezawa, I., Komiyama, K., Takeshima, H., Hata, K., Kono, M., and Kanda, N., *J. Antibiot.* **26**, 669 (1973).
661. Takaishi, T., Suzuki, M., and Tatematsu, A., *Org. Mass Spectrom.* **9**, 635 (1974).
662. Hata, T., Omura, S., Iwai, T., Nakagawa, A., Otami, M., Ito, C., and Matsuya, T., *J. Antibiot.* **24**, 353 (1971).
663. Japanese Patent 7398 (1964); *C.A.* **62**, 11116 (1965).
664. Bush, J. A., Cassidy, C. S., Crook, K. E., and German, L. B., *J. Antibiot.* **24**, 143 (1971).
665. Bradner, W. ., and Nettleton, D. E., *J. Antibiot.* **24**, 149 (1971).
666. Akita, E., Maeda, K., and Umezawa, H., *J. Antibiot., Ser. A* **17**, 200 (1964).
667. Ishizuka, M., Takeuchi, T., Nitta, K., Hori, M., and Umezawa, H., *J. Antibiot., Ser. A* **17**, 124 (1964).
668. Yamaguchi, T., Kashida, T., Nawa, K., Yajita, T., Miyagishima, T., Ito; H., Okuda, T., Ishida, N., and Kumagai, K., *J. Antibiot.* **23**, 373 (1970).
669. Yamaguchi, T., Sato, M., Omura, Y., Arai, Y., Enomoto, K., Ishida, M., and Kumagai, K., *J. Antibiot.* **23**, 382 (1970).
670. Highashide, E., Hasegawa, T., and Shibata, M., *J. Antibiot.* **22**, 409 (1969).
671. Suzuki, S., Nakamura, C., Ohkuma, K., and Tomiyama, Y., *J. Antibiot., Ser. A* **11**, 81 (1958).
672. Suzuki, S., and Ohkuma, K., *J. Antibiot., Ser. A* **11**, 84 (1958).
673. Jones, R. E. A., Keeping, J. W., Pellatt, M. G., and Haller, V., *J. Chem. Soc., Perkin Trans. 1* p. 148 (1973).
674. Ikekawa, T., Nakanishi, M., Uehara, N., Chihara, G., and Fukuoka, F., *Gann* **59**, 155 (1968).
675. Chihara, G., Hamuro, J., Maeda, Y., Arai, Y., and Fukuoka, F., *Cancer Res.* **30**, 2776 (1970).
676. Maeda, Y., and Chihara, G., *Int. J. Cancer* **11**, 153 (1973).
677. Maeda, Y., Hamura, J., Yamada, Y., Ishimura, K., and Chihara, G., *Immunopotentiation, Ciba Found. Symp., 1973* p. 259 (1973).
678. Maeda, Y., Hamuro, J., and Chihara, G., *Int. J. Cancer* **8**, 41 (1971).
679. Soda, K., Misono, H., Mori, K., and Sakato, H., *Biochem. Biophys. Res. Commun.* **44**, 931 (1971).
680. Oki, T., Shirai, M., Ohshima, M., Yamamoto, T., and Soda, K., *FEBS Lett.* **33**, 286 (1973).
681. Arai, T., Nikami, M., Fukushima, K., Utsumi, T., and Yazawa, K., *J. Antibiot.* **26**, 157 (1973).
682. Gauze, G. F., Maximova, T. S., Olkhovatova, O. L., Kudrina, E. S., Ilchenko,

G. B., Kochetkova, G. V., and Volkova, L. J., *Antibiotiki* (*Moscow*) **16**, 387 (1971).

683. Sveshnikova, M. A., Maximova, T. S., Olkhovatova, O. L., Tulyakova, T. V., Lavrova, M. F., and Grishin, I. A., *Antibiotiki* (*Moscow*) **18**, 99 (1973).

684. Brazhnikova, M. G., Kudinova, M. K., Lavrova, M. F., Borisova, V. N., Kruglyak, E. B., Kovsharova, I. N., and Proshlyakova, V. V., *Antibiotiki* (*Moscow*) **16**, 483 (1971).

685. Shorin, A. V., Rossolimo, O. K., Bazhanov, V. S., Averbukh, L. A., Kryatkova, G. A., and Lepeshkina, G. N., *Antibiotiki* (*Moscow*) **16**, 708 (1971).

686. Sukhareva-Nemakova, N. N., Kozlova, J. P., Zeleneva, P. M., and Silaev, A. B., *Antibiotiki* (*Moscow*) **16**, 596 (1971).

687. Chong, C. N., and Rickards, R. W., *Tetrahedron Lett.* p. 5053 (1972).

688. Mondovi, B., Strom, R., Agro, A. F., Caiafa, P., DeSole, P., Bozzi, A., Rotilio, G., and Tanellii A. R., *Cancer Res.* **31**, 505 (1971).

689. Strom, R., Bozzi, A., Scioscia Santoro, A., Crifo, C., Mondovi, B., and Rossi Fanelli, A., *Cancer Res.* **32**, 868 (1972).

690. Hata, T., Higuchi, T., Sano, Y., and Sawachi, K., *J. Antibiot.*, *Ser. A* **3**, 313 (1950).

691. Ueno, Y., and Ishikawa, I., *Appl.. Microbiol.* **18**, 406 (1969).

692. Schachtshabel, D. O., Ziliken, F., Shito, M., and Foley, G. E., *Exp. Cell Res.* **57**, 19 (1969).

693. Ruet, A., Sentenac, A., Simon, E. J., Bouhet, J. C., and Fromageot, P., *Biochemistry* **12**, 2318 (1973).

694. Mouton, R. F., and Fromageot, P., *FEBS Lett.* **15**, 45 (1971).

695. Ueno, Y., Platel, A., and Fromageot, P., *Biochim. Biophys. Acta* **134**, 27 (1967).

696. Goldberg, I. H., and Friedman, P. A., *Annu. Rev. Biochem.* **40**, 775 (1971).

697. Ishida, N., Suzuki, F., Maeda, H., Ozo, K., and Kumagai, K., *J. Antibiot.* **22**, 218 (1969).

698. Kunimoto, T., Hori, M., and Umezawa, H., *Cancer Res.* **32**, 1251 (1972).

699. Kunimoto, T., Hori, M., and Umezawa, H., *J. Antibiot.* **24**, 203 (1971).

700. Lippman, M. M., *Cancer Chemother. Rep.* **58**, 181 (1974).

701. Lippman, M. M., *In Vitro* **9**, 370 (1974).

702. Lippman, M. M., Laster, W. R., Abbot, B. J., Venditti, J., and Baratta, N., *Cancer Res.* **35**, 939 (1975).

703. Elstner, F. A., Carnes, D. M., Suhadolnik, R. J., Kreitshman, G. P., Schweizer, M. P., and Robins, R. K., *Biochemistry* **12**, 4992 (1973).

704. Kumano, N., *Sci. Rep. Res. Inst., Tohoku Univ., Ser. C* **19**, 89 (1972).

705. Kumano, N., Kurita, K., and Oka, S., *Gann* **64**, 529 (1973).

706. Soeda, M., *Cancer Chemother. Rep.* **18**, 9 (1962).

707. Takita, H., Takaoka, M., and Hata, T., *J. Antibiot.*, *Ser. A* **13**, 172 (1960).

708. Argoudelis, A. D., and Reusser, F., *J. Antibiot.* **24**, 383 (1971).

709. Argoudelis, A. D., *J. Antibiot.* **25**, 171 (1972).

710. Reusser, F., *J. Bacteriol.* **96**, 1285 (1968).

711. Japanese Patent 64/7,397 (1964); *C.A.* **62**, 11115 (1965).

712. DeAlbuquerque, M., *Rev. Inst. Antibiot., Univ. Fed. Pernambuco, Recife* **6**, 35 (1966).

713. Kusakabe, Y., Nagatsu, J., Shibuya, M., Kawaguchi, O., Hirose, C., and Shirato, S., *J. Antibiot.* **25**, 44 (1972).

714. Sasaki, K., Kusakabe, Y., and Esumi, S., *J. Antibiot.* **25**, 151 (1972).

715. Nishimura, H., Sasaki, K., Mayama, M., Shimaoka, N., Tawara, K., Okamoto, S., and Nakajima, K., *J. Antibiot., Ser. A* 13, 327 (1960).

716. Torbokhina, L. I., Bobikov, E. V., Sazykin, Yu.O., Lokshin, G. B., Zhadanovich, Yu.V., and Kuzkov, A. D., *Antibioitiki (Moscow)* 18, 7 (1973).

717. U.S. Patent 3,646,194 (1972); *C.A.* 76, 152038 (1972).

718. Kennedy, B. J., Yabro, J. W., Kickertz, V., and Wolheim, M. S., *Cancer Res.* 28, 91 (1968).

719. Kofman, S., Perlia, S. P., and Economou, S. G., *Cancer* 31, 889 (1973).

720. Northrop, G., Taylor, S. G., and Northrop, R. L., *Cancer Res.* 29, 1916 (1969).

721. Kenedy, B. J., *Am. J. Med.* 49, 494 (1970).

722. Fok, J., and Waring, M., *Mol. Pharmacol.* 8, 65 (1972).

723. Yabro, J. W., *Proc. Chemother. Conf. Mithramycin (Mithracin) Dev. Appl. Symp. Ther. Testicular Neoplasms, 1970* p. 8 (1972); *C.A.* 79, 100 331 (1973).

724. Morton, G. O., Van Lear, G. E., and Fulmor, W., *J. Am. Chem. Soc.* 92, 2588 (1970).

725. Liu, W., Cullen, W. P., and Rao, K. V., *J. Antibiot.* 22, 608 (1969).

726. Morrison, R. K., Brown, D. E., and Timmens, E. K., *Cancer Chemother. Rep.* 54, 217 (1970).

727. Hata, T., Sano, Y., Sugawara, R., Matsumae, A., Kanamuri, K., Shima, T., and Hoshi, T., *J. Antibiot., Ser. A* 9, 141 (1956).

728. Sugawara, R., and Hata, T., *J. Antibiot., Ser. A* 9, 147 (1956).

729. Sugiura, K., *Cancer Chemother. Rep.* 13, 51 (1961).

730. Usubuchi, L., Oboshi, S., Tsushida, R., and Tanabe, H., *Gann* 49, 209 (1958).

731. Iyer, U. N., and Szybalski, W., *Proc. Soc. Natl. Acad. Sci. U.S.A.* 50, 355 (1963).

732. Szybalski, W., and Iyer, U. N., *Fed. Proc., Fed. Am. Soc. Exp. Biol.* 23, 946 (1964).

733. Schwartz, H. S., Sternberg, S. S., and Philips, F. S., *Cancer Res.* 23, 1125 (1963).

734. Shadkin, A. J., Reich, E., Franklin, R. H., and Tatum, L. L., *Biochim. Biophys. Acta* 55, 277 (1962).

735. Kodama, M., *Biochem. J.* 61, 162 (1967).

736. Dyachenko, M. S., and Nosach, L. M., *Mikrobiol. Zh. (Kiev)* 33, 604 (1971).

737. Shinagawa, H., and Itoh, T., *Mol. Gen. Genet.* 126, 103 (1973).

738. Lai, Y.-Y., and Wei, H.-H., *Sheng Wu K'o Hsueh* 1, 37 (1972); *C.A.* 81, 85976 (1974).

739. Murayama, I., and Otsuji, N., *Mutat. Res.* 18, 117 (1973).

740. Majevski, J. A., and Sole, B. T., *Antimicrob. Agents & Chemother.* 4, 495 (1973).

741. Pokrovski, A. A., Kravchenko, L. V., and Tuteljan, V. A., *Biokhimiya* 36, 690 (1971).

742. Shah, V. C., Rao, S. R. V., and Arora, O. P., *Indian J. Exp. Biol.* 10, 431 (1972).

743. Adler, I. D., *Mutat. Res.* 23, 369 (1974).

744. Pricer, W. E., Jr., and Weisbach, A., *Biochem. Biophys. Res. Commun.* 14, 91 (1964).

745. Kate, N., Kobayashi, K., and Mizuno, D., *J. Biochem. (Tokyo)* 67, 175 (1970).

746. Fujimura, S., Makino, I., and Hayashi, P., *Jpn. J. Bacteriol.* 25, 316 (1970).

747. Kersten, H., Shneider, B., Leopold, G., and Kersten, W., *Biochim. Biophys. Acta* 108, 619 (1965).

748. Tereshin, I. M., *Antibiotiki (Moscow)* 14, 796 (1969).

749. Kageyama, M., Hasegawa, M., Inagaki, A., and Egami, F., *J. Biochem.* (*Tokyo*) **67**, 549 (1970).

750. Bourgeois, C. A., *Chromosoma* **48**, 203 (1974).

751. Hornemann, U., Kehrer, J. P., Nuñez, C. S., Ranieri, R. L., and Ho, Y. K., *Dev. Ind. Microbiol.* **15**, 82 (1974).

752. Hornemann, U., Kehrer, J. P., Nuñez, C. S., and Ranieri, R. L., *J. Am. Chem. Soc.* **96**, 320 (1974).

753. Remers, W. A., and Schepman, C. S., *J. Med. Chem.* **17**, 729 (1974).

754. Chaudhury, S., and Roy, D. K., *Indian J. Cancer* **11**, 285 (1974).

755. Kersten, H., and Kersten, W., in "Inhibitor Tools in Cell Research" (T. Büchner and H. Sies, eds.), p. 11. Springer-Verlag, Berlin and New York, (1969).

756. Ross, V. C., and Solymosi, I., *Fed. Proc., Fed. Am. Soc. Exp. Biol.* **26**, 291 (1967).

757. Basu, S. K., Chakrabarty, A. M., and Roy, S. C., *Biochim. Biophys. Acta* **108**, 713 (1965).

758. Lerman, M. I., and Benyumovich, M. S., *Nature* (*London*) **206**, 1231 (1965).

759. Sinkus, A. G., *Tsitologiya* **11**, 933 (1969).

760. Arora, O. P., Shah, V. C., and Rao, S. R. V., *Exp. Cell Res.* **56**, 443 (1969).

761. Diordievic, B., and Kim, J. H., *J. Cell Biol.* **38**, 477 (1968).

762. Nishimura, T., and Tanaka, N., *J. Antibiot., Ser. A* **16**, 179 (1963).

763. German Patent 1,233,527, (1967); *C.A.* **66**, 84744 (1967).

764. Gauze, G. F., Maksimova, T. S., Popova, O. L., Brazhnikova, N. G., Uspenskaya, T. A., and Rossolimo, O. K., *Antibiotiki* (*Moscow*) **4**, 20 (1959).

765. Gurevich, A. I., Kiseleva, O. A., Kolosov, M. N., Kuznetsov, V. D., Onoprienko, V. B., and Rosynov, B. V., *Antibiotiki* (*Moscow*) **12**, 880 (1967).

766. Kasyan, A. I., *Antibiotiki* (*Kiev*) p. 86 (1965).

767. Canonica, L., Rindone, B., Santanello, E., and Solastico, C., *Tetrahedron* **28**, 4395 (1972).

768. Noto, T., Sawada, M., Ando, K., and Koyama, K., *J. Antibiot.* **22**, 165 (1969).

769. Carter, S. B., Franklin, T. J., Jones, F. D., Leonard, B. J., Niells, S. D., Turner, R. W., and Turner, W. B., *Nature* (*London*) **223**, 848 (1969).

770. Sweeney, N. J., Gerzon, K., Harris, P. M., Holmes, R. E., Poore, G. A., and Williams, R. H., *Cancer Res.* **32**, 1795 (1972).

771. Franklin, T. J., and Cook, J. N., *Biochem. J.* **113**, 515 (1969).

772. Sweeney, M. J., Hoffman, D. H., and Esterman, M. A., *Cancer Res.* **32**, 1803 (1972).

773. Franklin, T. J., and Cook, J. M., *Biochem. Pharmacol.* **20**, 1334 (1971).

774. Mutsui, A., Suzuki, S., Koyama, K., and Akiba, T., *Prog. Antimicrob. Anticancer Chemother., Proc. Int. Congr. Chemother., 6th, 1969* Part II, p. 130 (1970).

775. Beister, J. A., and Hillery, S. S., *J. Pharm. Sci.* **64**, 84 (1975).

776. Japanese Patent 70/12276 (1970); *C.A.* **73**, 54611 (1970).

777. Japanese Patent 70/16795 (1970); *C.A.* **73**, 86573 (1970).

778. Brown, G. W., and Weliky, V. S., *J. Biol. Chem.* **204**, 1019 (1953).

779. Kapuler, A., Ward, D. C., Mendelsohn, N., Klett, H., and Acs, G., *Virology* **37**, 701 (1969).

780. Brink, N. G., *Acta Chem. Scand.* **7**, 1081 (1953).

781. Löfgren, N., Lüning, B., and Hedstrom, H., *Acta Chem. Scand.* **79**, 3252 (1957).

782. Smith, M. C., Snyder, F. F., Fontenelle, L. S., and Henderson, J. F., *Biochem. Pharmacol.* 23, 2023 (1974).
783. Maeda, H., *J. Antibiot.* 27, 303 (1974).
784. Shaeppi, U., Menninger, F., Fleischman, R. W., Bogden, A. E., Schein, P. S., and Cooney, D. A., *Cancer Chemother. Rep.* 5, 43 (1974).
785. Ono, Y., Watanabe, Y., and Ishida, N., *Biochim. Biophys. Acta* 119, 46 (1966).
786. Sawada, H., Tatsumi, K., Sasada, M., Shirakawa, S., Nakamura, T., and Wakisaka, G., *Cancer Res.* 34, 3341 (1974).
787. Tsuruo, T., Sato, H., and Ukita, T., *J. Antibiot.* 24, 423 (1971).
788. Beerman, T. A., and Goldberg, I. H., *Biochem. Biophys. Res. Commun.* 59, 1254 (1974).
789. Sakamoto, K., Endo, N., and Sakka, M., *Eur. J. Cancer* 9, 725 (1973).
790. Ohtsuki, K., and Ishida, N., *J. Antibiot.* 28, 143 (1975).
791. Sawada, H., Tatsumi, K., Sasada, M., Shirakawa, S., Nakamura, T., and Wakisaka, G., *Cancer Res.* 34, 3341 (1974).
792. Tatsurui, K., Nakamura, T., and Wakaisaka, G., *Gann* 65, 459 (1974).
793. Derkach, E. M., *Antibiotiki (Moscow)* 11, 619 (1966).
794. Kondo, S., Wakashiro, T., Hamada, M., Maeda, K., Takeuchi, T., and Umezawa, H., *J. Antibiot.* 23, 354 (1970).
795. Hisamatsu, T., and Koeda, T., *J. Antibiot.* 24, 200 (1971).
796. Tsukuda, I., Hamada, M., and Umezawa, H., *J. Antibiot.* 24, 189 (1971).
797. Zygmunt, W. A., *Biochem. Biophys. Res. Commun.* 6, 324 (1961).
798. Zimmer, Ch., Reinert, K. E., Luck, G., Wähnert, U., Löber, G., and Thrum, H., *J. Mol. Biol.* 58, 329 (1971).
799. Haupt, J., Zimmer, Ch., and Thrum, H., *Ergeb. Exp. Med.* 9, 368 (1972).
800. Tatsuno, T., Saito, M., Enomoto, M., and Tsunoda, H., *Chem. Pharm. Bull.* 16, 2519 (1968).
801. Ueno, Y., and Fukushima, K., *Experientia* 24, 1032 (1968).
802. Tatsuno, T., *Cancer Res.* 28, 2393 (1968).
803. Bhuyan, B. K., and Reusser, F., *Cancer Res.* 30, 984 (1970).
804. Bhuyan, B. K., and Smith, C. G., *Proc. Natl. Acad. Sci. U.S.A.* 54, 566 (1965).
805. Ward, D., Reich, E., and Goldberg, I. H., *Science* 149, 1259 (1965).
806. Fok, J., and Waring, M., *Mol. Pharmacol.* 8, 65 (1972).
807. DiMarco, M., Zunino, F., Silvestrini, R., Gambarucci, C., and Gambetta, R. A., *Biochem. Pharmacol.* 20, 1323 (1971).
808. Bemfong, M. A., *Mutat. Res.* 21, 323 (1973).
809. Ellen, K. A. O., and Rhode, S. L., *Biochim. Biophys. Acta* 209, 415 (1970).
810. Neogy, R. K., Chowdhury, K., and Thakurta, G. G., *Biochim. Biophys. Acta* 299, 241 (1973).
811. DasGokul, C., DasGupta, S., and DasGupta, N. N., *Biochim. Biophys. Acta* 353, 274 (1974).
812. Frolova, V. I., Bozynov, B. V., and Kuzovkov, A. D., *Antibiotiki (Moscow)* 18, 777 (1973).
813. Keller-Schierlein, W., *Fortschr. Chem. Org. Natursst.* 26, 161 (1968).
814. Mayers, E., Pansy, F. E., Perlman, D., Smith, D. A., and Weisenborn, F. I., *J. Antibiot.* 18 ,128 (1965).
815. Henderson, P. J. F., McGivan, J. D., and Chappell, J. D., *Biochem. J.* 111, 521 (1969).
816. Tostenson, D. C., Andreoli, T. E., Tieffenberg, M., and Cook, P., *J. Gen. Physiol.* 51, 373 (1968).

817. Gyimesi, J., Ott, I., Horvath, I., Koczka, I., and Magyar, K., *J. Antibiot.* **24**, 277 (1971).
818. Karpov, V. L., Romanova, L. G., and Kiseleva, V. I., *Antibiotiki (Moscow)* **19**, 394 (1974).
819. Berlin, J. A., Kolosov, M. N., and Piotrovich, L. A., *Khim. Prir. Soedin.* **4**, 519, 526, 535, 537, and 542 (1972).
820. Zalmanson, E. S., Zelenin, A. V., Kafiani, K. A., Lobareva, S. L., Lyapunova, E. A., and Timofeeva, M. J., *Antibiotiki (Moscow)* **10**, 613 (1965).
821. Loshkareva, N. P., Gauze, G. F., and Zbarsky, I. B., *Biokhimiya* **34**, 500 (1969).
822. Gauze, G. G., Loshkareva, N. P., and Dudnik, Yu. V., *Antibiotiki (Moscow)* **10**, 307 (1965).
823. Gauze, G. G., Loshkareva, N. P., and Zbarsky, I. B., *Mol. Biol. (Moscow)* **3**, 566 (1969).
824. Gauze, G. G., Loshkareva, N. P., and Zbarsky, I. B., *Biochim. Biophys. Acta* **166**, 752 (1968).
825. Gauze, G. F., *Prog. Antimicrob. Anticancer Chemother., Proc. Int. Congr. Chemother., 6th, 1969* Part II, p. 428 (1970).
826. Ward, D., Reich, E., and Goldberg, I. H., *Science* **149**, 1259 (1965).
827. Popov, P. G., Kavrakirova, S. V., and Malev, A. C., *Biochem. Pharmacol.* **22**, 1526 (1973).
828. Gauze, G. G., Loshkareva, N. P., and Zbarsky, I. B., *Mol. Biol. (Moscow)* **3**, 566 (1969).
829. Karpov, V. L., Romanova, L. G., and Kiseleva, V. I., *Antibiotiki (Moscow)* **19**, 394 (1974).
830. Kato, A., Saeki, T., Suzuki, S., Ando, K., Tamura, G., and Arima, K., *J. Antibiot.* **22**, 322 (1969).
831. Schmitz, H., Jubinski, S. D., Hooper, I. R., Crook, E. K., Price, K. E., and Leim, J., *J. Antibiot.* **18**, 82 (1965).
832. Walter, P., Lardy, H. A., and Johnson, D., *J. Biol. Chem.* **242**, 5014 (1967).
833. Haneishi, T., Okazaki, T., Hata, T., Tamura, K., Nomura, M., Naito, A., Seki, I., and Arai, M., *J. Antibiot.* **24**, 797 (1971).
834. Warsi, S. A., and Whelan, W. J., *Chem. Ind. (London)* p. 1573 (1957).
835. Chihara, G., Hamuro, J., Maeda, Y., Arai, Y., and Fumiko, F., *Nature (London)* **225**, 943 (1970).
836. Wiley, P. F., Jahnke, H. K., MacKellar, F., Kelly, R. B., and Argoudelis, A. D., *J. Org. Chem.* **35**, 1420 (1970).
837. White, F. R., *Cancer Chemother. Rep.* **24**, 75 (1962).
838. Ayuso, M., and Goldberg, I. H., *Biochim. Biophys. Acta* **294**, 118 (1973).
839. Bhuyan, B. K., *Biochem. Pharmacol.* **16**, 1411 (1967).
840. Stewart, M. L., and Goldberg, I. H., *Biochim. Biophys. Acta* **294**, 123 (1973).
841. Cheng, C. P., Stewart, M. L., and Gupta, N. K., *Biochem. Biophys. Res. Commun.* **54**, 1092 (1973).
842. Goldberg, I. H., *Cancer Chemother. Rep., Part 1* **58**, 479 (1974).
843. Kresse, H., and Buddecke, E., *Hoppe Seyler's Z. Physiol. Chem.* **349**, 1507 (1968).
844. Tai, P. C., Wallace, B. J., and Davis, B. D., *Biochemistry* **12**, 616 (1973).
845. Čihak, A., Veselý, J., and Harrap, K. R., *Biochem. Pharmacol.* **23**, 1087 (1974).
846. Kis, Z., Close, A., Sigg, H. P., Hruban, L., and Snatzke, G., *Helv. Chim Acta* **53**, 1577 (1970).

847. Kidder, G. W., Dewey, V. C., Parks, R. E., and Woddside, G. L., *Science* **109**, 511 (1949).
848. Levin, D. H., *J. Biol. Chem.* **238**, 1098 (1963).
849. Webb, T. E., *Biochim. Biophys. Acta* **138**, 307 (1967).
850. Bull, A. T., and Faulknar, B. M., *Nature (London)* **203**, 506 (1964).
851. Florey, H. W., Chain, N. G., Heatley, N., Jennings, A. G., Sanders, E. O., and Florey, M. E., "Antibiotics," Vol. II. University Press, London, 1949.
852. Vollmar, H., *Z. Hyg. Infectionskr.* **127**, 316 (1947).
853. Chain, E., Florey, H. W., and Jennings, M. A., *Br. J. Exp. Pathol.* **23**, 202 (1942).
854. Dickens, F., and Jones, H. E. H., *Br. J. Cancer* **15**, 85 (1961).
855. Ellis, J. R., and McCalla, T. M., *Appl. Microbiol.* **25**, 562 (1973).
856. Ashoor, S. H., and Chu, F. S., *Food Cosmet. Toxicol.* **11**, 995 (1973).
857. Ashoor, S. H., and Chu, F. S., *Food Cosmet. Toxicol.* **11**, 617 (1973).
858. Sommer, N. F., Buchanan, J. R., and Fortlage, R. J., *Appl. Microbiol.* **28**, 589 (1974).
859. Mayer, V. W., and Legator, M. S., *J. Agric. Food Chem.* **17**, 454 (1969).
860. Reis, J., *Cytologia* **39**, 703 (1975).
861. Price, K. E., Schlein, A., Bradner, W. T., and Lein, J., *Antimicrob. Agents Chemother.* p. 95 (1962/1963).
862. Suzuki, S., Kimura, P., Saito, F., and Ando, K., *Agric. Biol. Chem.* **35**, 287 (1971).
863. Ashoor, S. H., and Chu, F. S., *Food Cosmet. Toxicol.* **11**, 995 (1973).
864. Ashoor, S. H., and Chu, F. S., *Food Cosmet. Toxicol.* **11**, 617 (1973).
865. Murase, H., Hikii, T., Nitta, K., Okami, Y., Takeuchi, T., and Umezawa, H., *J. Antibiot., Ser. A* **14**, 113 (1961).
866. Dutcher, J. D., *Antimicrob. Agents Chemother.* p. 173 (1961/1962).
867. Nishimura, T., *J. Antibiot.* **21**, 106 (1968).
868. Nishimura, T., *J. Antibiot.* **21**, 110 (1968).
869. Yamaki, H., Nishimura, T., Kubota, K., Kinoshita, T., and Tanaka, N., *Biochem. Biophys. Res. Commun.* **59**, 482 (1974).
870. Maeda, K., Kosaka, H., Yagishita, K., and Umezawa, H., *J. Antibiot., Ser. A* **9**, 82 (1956).
871. Makita, T., Muraoka, Y., Yoshioka, T., Fuji, A., Maeda, K., and Umezawa, H., *J. Antibiot.* **25**, 735 (1972).
872. Ishizuka, M., Takayama, H., Takeuchi, T., and Umezawa, H., *J. Antibiot., Ser. A* **19**, 260 (1966).
873. Pietsch, P., *Biotechnol. Bioeng.* **15**, 1039 (1973).
874. Krueger, W. C., Pschigoda, L. M., and Reusser, F., *J. Antibiot.* **26**, 424 (1973).
875. Shove, S. J., and Rauth, A. M., *Cancer Res.* **31**, 1422 (1971).
876. Watanabe, M., and August, J. T., *J. Mol. Biol.* **33**, 21 (1968).
877. Grigg, G. W., *Mol. Gen. Genet.* **107**, 162 (1970).
878. Sleigh, M. J., and Grigg, G. W., *FEBS Lett.* **39**, 35 (1974).
879. Farell, L. L., and Reiter, H., *Antimicrob. Agents & Chemother.* **4**, 320 (1973).
880. Reiter, H., Milevsky, M., and Kelley, P., *J. Bacteriol.* **111**, 586 (1973).
881. Friedman, R. M., Stern, R., and Rose, J. A., *J. Natl. Cancer Inst.* **55**, 693 (1974).
882. Falashi, A., and Kornberg, A., *Fed. Proc., Fed. Am. Soc. Exp. Biol.* **23**, 940 (1964).
883. Wheatley, D. N., *Chem.-Biol. Interact.* **9**, 187 (1974).
884. Korbecki, M., *Arch. Gesamte Virusforsch.* **40**, 265 (1973).

885. Jacobs, N. F., Neu, R. L., and Gardner, L. I., *Mutat. Res.* **7**, 251 (1969).
886. Mattingly, E., *Mutat. Res.* **4**, 51 (1967).
887. Baldacci, E., Farina, G., Piazena, E., and Fabbria, G., *G. Microbiol.* **16**, 9 (1968).
888. Shibata, M., Asai, M., Mizuno, K., Miyake, A., and Tatsuoka, S., *Proc. Jpn. Acad.* **40**, 296 (1964).
889. Japanese Patent 66/7839 (1966); *Chem. Abstr.* **65**, 10432 (1966).
890. Asai, M., Mizuta, E., Mizuno, K., and Tatsuoka, S., *Chem. Pharm. Bull.* **18**, 1720 (1970).
891. Nishibori, A., *J. Antibiot., Ser. A* **10**, 213 (1957).
892. Nagai, K., Yamaki, H., Tanaka, N., and Umezawa, H., *J. Biochem. (Tokyo)* **62**, 321 (1967).
893. Tanaka, N., Nagai, H., Yamaguchi, H., and Umezawa, H., *Biochem. Biophys. Res. Commun.* **21**, 328 (1965).
894. Tanaka, N., *J. Antibiot.* **23**, 523 (1970).
895. Elpidina, O. L., *Antibiotiki* **4**, 46 (1959).
896. Zaretskaya, I. I., Krasilina, A. Y., Terentieva, E. I., and Koretskaya, T. I., *Vopr. Onkol.* **7**, 68 (1961).
897. Elpidina, O. L., Dunaeva, R. D., Samoylova, E. R., and Semenov, V., *Kazan. Med. Zh.* **6**, 37 (1966).
898. Navashin, S. M., Fomina, I. P., and Koroleva, V. G., *Antibiotiki (Moscow)* **6**, 912 (1961).
899. Terentieva, T. G., Fomina, I. P., and Navashin, S. M., *Antibiotiki (Moscow)* **15**, 442 (1970).
900. Hoshi, A., Kanzawa, T., Kuretani, K., Homma, J. Y., and Abe, C., *Gann* **64**, 523 (1973).
901. Yoshioka, Y., Sano, T., and Ikekawa, T., *Chem. Pharm. Bull.* **21**, 1772 (1973).
902. Ikekawa, T., Yoshioka, Y., Emori, M., Sano, T., and Fukuoka, F., *Cancer Chemother. Rep.* **57**, 85 (1973).
903. Haruse, S., Takeda, S., Ito, M., Fujii, K., Terada, Y., Shimura, K., Sugiura, M., and Miyazaki, T., *Mie Med. J.* **23**, 207 (1974); *C.A.* **82**, 25861 (1975).
904. Hamuro, J., and Chihara, G., *Nature (London)* **245**, 40 (1973).
905. Okuda, T., Yoshioka, Y., Ikekawa, T., Chihara, G., and Nishioka, K., *Nature (London), New Biol.* **238**, 59 (1972).
906. Sidewall, R. W., Dixon, G. J., and Schabel, W., *Prog. Antimicrob. Anticancer Chemother., Proc. Int. Congr. Chemother., 6th, 1969* Part II, p. 26 (1970).
907. Izbicki, R., Ai-Saprat, M., Reed, M. L., Waughn, C. B., and Vaitkevicius, V. K., *Cancer Chemother. Rep.* **56**, 615 (1972).
908. Evans, J. S., Mussar, E. A., and Gray, J. E., *Antibiot. Chemother. (Washington, D.C.)* **11**, 445 (1961).
909. Smith-Kielland, I., *Biochim. Biophys. Acta* **114**, 254 (1966).
910. Iyer, V. N., and Szybalski, W., *Science* **145**, 55 (1964).
911. Schillings, R. T., and Ruelius, H. W., *Arch. Biochem. Biophys.* **127**, 672 (1968).
912. Ruelius, H. W., Janssen, F. W., Kerwin, R. M., Goodwin, C. W., and Schillings, T. R., *Arch. Biochem. Biophys.* **125**, 126 (1968).
913. Itakura, Ch., Sega, T., Isono, K., and Suzuki, S., *J. Antibiot., Ser. A* **15**, 250 (1962).
914. Arai, T., Kushikata, S., Takamiya, K., Yanagisavo, F., and Koyama, T., *J. Antibiot., Ser. A* **20**, 334 (1967).
915. Yünsten, H., Yonehara, H., and Ui, H., *J. Antibiot., Ser. A* **7**, 113 (1954).

916. Haňka, L. J., Burch, M. R., and Sokolski, W. T., *Antibiot. Chemother. (Basel)* 9, 419 (1959).
917. Evans, J. S., and Gray, J. E., *Antibiot. Chemother. (Basel)* 9, 675 (1959).
918. Haňka, L. J., and Burch, M. R., *Antibiot. Chemother. (Basel)* 10, 484 (1960).
919. Kuramitsu, H., and Moyed, H. S., *J. Biol. Chem.* 241, 1596 (1966).
920. Zyk, N., Citri, N., and Moyed, H. S., *Biochemistry* 8, 2787 (1969).
921. Beppu, T., Nose, M., and Arima, K., *Agric. Biol. Chem.* 32, 197 (1968).
922. Henderson, J. F., *Biochem. Pharmacol.* 12, 551 (1967).
923. Pestka, S., *Arch. Biochem. Biophys.* 136, 80 (1970).
924. Nichols, D. M., and Cohen, A., *Can. J. Biochem.* 48, 858 (1970).
925. Grollman, A. P., and Huang, M. T., *Symp. Soc. Pharmacol. Exp. Th., 1972* p. 1673 (1972).
926. Hirashima, A., Childs, G., and Inouye, M., *J. Mol. Biol.* 79, 373 (1973).
927. Schwartz, J. L., Katagiri, M., Omura, S., and Tischler, M., *J. Antibiot.* 27, 379 (1974).
928. Harris, R., Hanlon, J., and Symon, R., *Proc. Biochem. Soc.* 3, 39 (1970).
929. Azzan, M. E., and Algranati, I. D., *Proc. Natl. Acad. Sci. U.S.A.* 70, 1866 (1973).
930. Vince, R., and Daluge, S., *J. Med. Chem.* 17, 578 (1974).
931. Cremer, K., Imamoto, F., and Schlessinger, D., *Mol. Gen. Genet.* 130, 183 (1974).
932. Hori, T., and Lark, K. G., *J. Mol. Biol.* 77, 391 (1973).
933. Timberlake, W. E., and Griffin, D. H., *Biochim. Biophys. Acta* 353, 248 (1974).
934. Bilder, G. E., and Dencala, W. D., *Science* 185, 1060 (1974).
935. Beck, W., Bellantone, R. A., and Camellakis, E. S., *Nature (London)* 241, 775 (1973).
936. O'Doherty, P. J. A., and Kuksis, A., *Can. J. Biochem.* 52, 1705 (1974).
937. Maraldy, N. M., Biagini, G., Simoni, P., Barlieri, M., Mariani, M., and Bersani, F., *J. Ultrastruct. Res.* 44, 265 (1973).
938. Meller, K., Mestres, P., Breipohl, V., and Waelsh, M., *Cell Tissue Res.* 148, 227 (1974).
939. Weiss, L., and Huber, D., *J. Cell Sci.* 15, 217 (1974).
940. Egges, S. H., Biedron, S. I., and Hawtrey, A. O., *Tetrahedron Lett.* p. 3271 (1966).
941. Phang, J. M., Valle, D. L., Fischer, R. L., and Granger, A., *Am. J. Physiol.* 228, 23 (1975).
942. Ricketts, T. R., and Rappitt, A. F., *Arch. Microbiol.* 102, 1 (1975).
943. German Patent 2,412,890 (1974); *C.A.* 82, 110310 (1975).
944. Gerzon, K., Williams, R. H., Hoehn, M., Gorman, M., and DeLong, D. C., *Int. Congr. Heterocycl. Chem., 2nd, 1969* Abstr. Book C-30 (1969).
945. Sweeney, M. J., Davis, P. A., Gutowski, G. E., Hamill, R. L., Hoffman, D. H., and Poore, G. A., *Cancer Res.* 33, 2619 (1973).
946. Ishibashi, K., *J. Antibiot., Ser. A* 15, 166 (1960).
947. Ishibashi, K., *J. Agric. Chem. Soc. Jpn.* 35, 257 (1961).
948. Nozoe, S., Hirai, K., Tsuda, K., Ishibashi, K., Shirasaka, M., and Grove, J. F., *Tetrahedron Lett.* p. 4675 (1965).
949. U.S. Patent 3,692,777 (1972).
950. Otsuka, H., and Shoji, J., *J. Antibiot., Ser. A* 19, 128 (1966).
951. Sato, K., Yoshida, T., and Katagiri, K., *J. Antibiot., Ser. A* 20, 188 (1967).

952. Utahara, R., Oyagi, H., Yagishita, K., Okami, K., and Umezawa, H., *J. Antibiot., Ser. A* **18**, 132 (1955).
953. Hori, M., Takemoto, K., Takeuchi, T., Kondo, S., Hamada, M., Okazaki, T., Okami, Y., and Umezawa, H., *J. Antibiot.* **25**, 3 (1972).
954. Hori, M., Takemoto, K., Watanabe, J., Umezawa, M., and Umezawa, H., *J. Antibiot.* **25**, 629 (1972).
955. Cotias, C. T., Santana, C. F., Pinto, K. V., Filho, A. M., Lacerda, A., and Moreira, C. T., *Rev. Inst. Antibiot., Univ. Fed. Pernambuco, Recife* **11**, 51 (1971).
956. DeLaMonache, F., Marini Betollo, G. B., and DeAlbuquerque, L., *Ann. Ist. Super. Sanita* **6**, 537 (1970).
957. Stegelman, L. A., *Antibiotiki (Moscow)* **15**, 1021 (1970).
958. Esipov, S. E., Kolosov, M. N., and Saburova, L. A., *J. Antibiot.* **26**, 537 (1973).
959. Navashin, S. M., *Antibiotiki (Moscow)* **12**, 892 (1967).
960. Karlsson, A., Sartori, G., and White, R., *Eur. J. Biochem.* **47**, 251 (1974).
961. Yang, S. S., Herrera, F. M., Smith, R. G., Reitz, M. S., Lancini, G., Ting, R. C., and Gallo, R. C., *J. Natl. Cancer Inst.* **49**, 7 (1972).
962. Adamson, R. H., *Arch. Int. Pharmacodyn. Ther.* **192**, 61 (1971).
963. Hartman, G., Honikel, K. O., Knuesel, F., and Nuesh, J., *Biochim. Biophys. Acta* **145**, 843 (1967).
964. Waalkes, T. P., Sanders, K., Smith, R. G., and Adamson, R. H., *Cancer Res.* **34**, 385 (1974).
965. Wu, A. M., and Gallo, R. C., *Biochim. Biophys. Acta* **340**, 419 (1974).
966. Horszewicz, J. S., and Carter, W. A., *Antimicrob. Agents & Chemother.*, **5**, 196 (1974).
967. Ischler, A. N., Joss, V. R., Thompson, F. M., and Calvin, M. J., *J. Med. Chem.* **16**, 1071 (1973).
968. Kerrich-Santo, R. E., and Hartman, G. R., *Eur. J. Biochem.* **43**, 521 (1974).
969. Tsai, N., and Saunders, G. F., *Proc. Natl. Acad. Sci. U.S.A.* **70**, 2072 (1973).
970. Liu, U. I., and Hartman, G. R., *Eur. J. Biochem.* **38**, 336 (1973).
971. Srb, V., Puža, V., Spurná, V., and Keprtová, J., *Experientia* **30**, 484 (1974).
972. Mans, R. J., *FEBS Lett.* **32**, 245 (1973).
973. Weinberg, A., and Becker, J., *Isr. J. Med. Sci.* **7**, 1084 (1971).
974. Hirashima, A., Childs, G., and Inouye, M., *J. Mol. Biol.* **79**, 373 (1973).
975. Paunescu, E., Vasilescu, F., and Danalache-Dumitrescu, M., *Rev. Roum. Med. Interne* **10**, 193 (1973).
976. Klen, R., Skalská, H., Srb, V., and Heger, J., *Folia Biol. (Prague)* **19**, 354 (1973).
977. Tierry, R. C., *Eur. J. Immunol.* **3**, 320 (1973).
978. White, R. J., Martinelli, E., and Lancini, G., *Proc. Natl. Acad. Sci. U.S.A.* **71**, 3260 (1974).
979. Tittawella, I. P. B., and Hayward, R. S., *Mol. Gen. Genet.* **134**, 181 (1974).
980. Mandi, Y., and Belady, I., *Acta Microbiol. Acad. Sci. Hung.* **21**, 385 (1974).
981. Riveros-Moreno, V., *FEBS Lett.* **51**, 249 (1975).
982. Brufani, M., Cerrini, S., Fedeli, W., and Vaciago, A., *J. Mol. Biol.* **87**, 409 (1974).
983. Aszalos, A., Jelinek, M., and Berk, B., *Antimicrob. Agents Chemother.* p. 68 (1964).
984. Ebringer, L., *Neoplasma* **19**, 579 (1974).

985. Brazhnikova, M. G., Konstantinova, N. V., Mezentsev, A. S., and Pomaskova, V. A., *Antibiotiki (Moscow)* 13, 781 (1968).
986. Shorin, V. A., Rossolimo, O. K., and Sokolov, I. K., *Antibiotiki (Moscow)* 14, 249 (1969).
987. Lapshinskaya, O. A., *Antibiotiki (Moscow)* 15, 494 (1970).
988. Maral, P., and Ganter, P., *Bull. Cancer* 53, 201 (1966).
989. Ueno, Y., Ueno, I., Sato, N., Itoi, Y., Saito, N., Enomoto, M., and Tsunoda, H., *Jpn. J. Exp. Med.* 41, 177 (1971).
990. Nakamura, S., Nii, F., Inoue, S., Nakanishi, I., and Shimizu, M., *Jpn. J. Microbiol.* 18, 1 (1974); *Microbiol. Abstr.* 10A, 1991 (1975).
991. Rao, K. V., *J. Med. Chem.* 11, 939 (1968).
992. Suhadolnik, R. J., Uematsu, T., Uematsu, H., and Wilson, R. G., *J. Biol. Chem.* 243, 2761 (1968).
993. Uematsu, T., and Suhadolnik, R. J., *Arch. Biochem. Biophys.* 162, 614 (1974).
994. Takeuchi, T., *J. Antibiot., Ser. A* 7, 37 (1954).
995. Sung, S. C., and Quastel, J. H., *Cancer Res.* 23, 1549 (1963).
996. Bickis, I. J., Creaser, E. H., Quastel, J. H., and Shofield, P. G., *Nature (London)* 180, 1109 (1957).
997. Jimenez, A., Monro, R. E., and Vasquez, D., *FEBS Lett.* 7, 103 (1970).
998. Roy-Burman, S., Huang, Y. H., and Visser, D. W., *Biochem. Biophys. Res. Commun.* 42, 445 (1971).
999. Elstner, E. F., and Suhadolnik, R. J., *Biochemistry* 11, 2578 (1972).
1000. Matsuura, S., Shiratori, O., and Katagiri, K., *J. Antibiot., Ser. A* 17, 234 (1964).
1001. Komatsu, Y., and Tanaka, K., *Agric. Biol. Chem.* 34, 891 (1970).
1002. Watanabe, S., *J. Antibiot.* 23, 313 (1970).
1003. Maryanka, D., and Johnston, I. R. J., *FEBS Lett.* 7, 125 (1970).
1004. Kalman, T. I., *Biochem. Biophys. Res. Commun.* 49, 1007 (1972).
1005. Titani, Y., and Tsuruta, Y., *J. Antibiot.* 27, 956 (1974).
1006. Gauze, G. F., Preobrazhenskaya, T. P., Ivanitskaya, L. P., and Sveshnikova, M. A., *Antibiotiki (Moscow)* 14, 963 (1969).
1007. Brazhnikova, M. G., Konstantinova, N. V., and Mezentsev, A. S., *J. Antibiot.* 25, 668 (1972).
1008. Rubasheva, L. M., Mezentsev, A. S., Vlasov, T. F., and Anisimova, O. S., *Antibiotiki (Moscow)* 18, 216 (1973).
1009. Gauze, G. G., Dudnik, Yu. B., and Dolgilevitch, S. M., *Antibiotiki (Moscow)* 17, 413 (1972).
1010. Shorin, V. A., *Adv. Antimicrob. Antineoplast. Chemother., Proc. Int. Congr. Chemother., 7th, 1971* Part II, p. 89 (1972).
1011. Gauze, G. F., and Dudnik, Yu. V., *Adv. Antimicrob. Antineoplast. Chemother., Proc. Int. Congr. Chemother., 7th, 1971* Part II, p. 87 (1972).
1012. Dudnik, Yu. V., and Netyksa, E. M., *Antibiotiki (Moscow)* 17, 44 (1972).
1013. Polikarpova, S. I., *Dokl. Akad. Nauk SSSR* 209, 469 (1973).
1014. Shepalevtseva, N. G., *Antibiotiki (Moscow)* 20, 145 (1975).
1015. Soeda, M., Kitahara, T., and Homma, S., *Prog. Antimicrob. Anticancer Chemother., Proc. Int. Congr. Chemother., 6th, 1969* Part II, p. 1018 (1970).
1016. Owen, S. P., Dietz, A., and Camiener, G. W., *Antimicrob. Agents Chemother.* p. 772 (1962).
1017. Hirashima, A., Child, G., and Inouye, M., *J. Mol. Biol.* 79, 373 (1973).
1018. Goldberg, I. H., *Cancer Chemother. Rep. Part II* 58, 479 (1974).

1019. Goldberg, I. H., Stewart, M. L., Ayuso, M., and Kappen, L. S., *Fed. Proc., Fed. Am. Soc. Exp. Biol.* 32, 1688 (1973).
1020. Busiello, E., and DiGirolamo, M., *Biochim. Biophys. Acta* 312, 581 (1973).
1021. Pestka, S., *Arch. Biochem. Biophys.* 136, 89 (1970).
1022. Naganawa, H., Takita, T., Maeda, K., and Umezawa, H., *J. Antibiot.* 21, 241 (1968).
1023. Back, M., Shields, R. P., and Munson, A. E., *Antibiot. Chemother. (Basel)* 11, 652 (1961).
1024. Shimizu, M., Sayto, T., Hashimoto, H., and Mitsuhashi, S., *J. Antibiot.* 23, 63 (1970).
1025. Černá J., and Rychlík, I., *Biochim. Biophys. Acta* 157, 436 (1968).
1026. Mao, J. C. H., and Wiegand, R. G., *Biochim. Biophys. Acta* 157, 404 (1968).
1027. Neth, R., Monro, R. E., Heller, G., Battaner, E., and Vasquez, D., *FEBS Lett.* 6, 198 (1970).
1028. DeVries, H., Andersen, A. J., and Croon, A. M., *Biochim. Biophys. Acta* 331, 264 (1973).
1029. Lai, C. Y., Weisblum, B., Fahnestock, S. R., and Nomura, M., *J. Mol. Biol.* 74, 67 (1973).
1030. Nakao, M., and Nakazawa, S., *J. Antibiot.* 27, 970 (1974).
1031. Midleton, M. C., *Biochem. Pharmacol.* 23, 801 and 811 (1974).
1032. Rhim, J. S., and Huebner, R. J., *Antimicrob. Agents Chemother.* p. 177 (1968).
1033. Rhim, J. S., Levy, H. B., Baron, S., and Huebner, R. J., *Proc. Soc. Exp. Biol. Med.* 136, 524 (1974).
1034. Bergy, M. E., and Reusser, F., *Experientia* 23, 254 (1967).
1035. Reusser, F., *Biochem. Pharmacol.* 17, 2001 (1968).
1036. Reusser, F., *Biochem. Pharmacol.* 18, 287 (1968).
1037. Brodansky, T. F., and Reusser, F., *J. Antibiot.* 27, 809 (1974).
1038. Reusser, F., *Biochim. Biophys. Acta* 383, 266 (1975).
1039. Dudnik, Yu. V., Gauze, G. G., Kaprov, V. L., Kozmyan, L. I., and Padron, E., *Antibiotiki (Moscow)* 18, 968 (1973).
1040. Rao, K. V., Biemann, K., and Woodward, R. B., *J. Am. Chem. Soc.* 85, 2532 (1963).
1041. Mikhaylov, V. S., and Gauze, G. G., *Mol. Biol. (Moscow)* 8, 108 (1974).
1042. Woods, V. A., Massicot, J. G., Webb, J. A., and Chirigos, M. A., *In Vitro* 9, 24 (1974).
1043. Chirigos, M. A., Pearson, J. W., Papas, T. S., Woods, W. A., Woods, N. B., and Spahn, G., *Cancer Chemother Rep.* 57, 305 (1973).
1044. Mizuno, N. S., and Gilboe, D. P., *Biochim. Biophys. Acta* 224, 319 (1970).
1045. Mizuno, N. S., *Biochim. Biophys. Acta* 108, 394 (1965).
1046. Padron, E., Kaprov, V. L., Gauze, G. G., and Dudnik, Yu, V., *Antibiotiki (Moscow)* 19, 387 (1974).
1047. Lapshinskaya, O. A., *Antibiotiki (Moscow)* 15, 494 (1970).
1048. Sinkus, A. H., *Genetika* 5, 149 (1969).
1049. Yamazaki, H., Mizuno, S., Nitta, K., Utahara, R., and Umezawa, H., *J. Antibiot.* 21, 63 (1968).
1050. Tan, K. B., and McAuslan, B. R., *Biochem. Biophys. Res. Commun.* 42, 230 (1971).
1051. Rinehart, K. L., Jr., Antosz, F. J., Sasaki, K., Martin, P. K., Maheshavari, M. L., Reusser, F., Li, L. H., Moran, D., and Wiley, P. F., *Biochemistry* 13, 861 (1974).
1052. Carter, W. A., Borden, E. C., Brockman, W. W., Byrd, D., Ligon, W.,

Antosz, E. J., and Rinehart, K. L., Jr., *Collect. Pap. Annu. Symp. Fund., Am. Cancer Res.* **25**, 303 (1974).

1053. Johnson, F., *Fortschr. Chem. Org. Naturst.* **29**, 140 (1971).
1054. Smith, C. G., Lummis, W. L., and Grady, J. E., *Cancer Res.* **19**, 847 (1959).
1055. Bennet, L. L., Ward, V. L., and Brockman, E. W., *Biochim. Biophys. Acta* **103**, 478 (1965).
1056. Bennet, L. L., Smithers, D., and Ward, C. T., *Biochim. Biophys. Acta* **87**, 60 (1964).
1057. Bhuyan, B. K., and Fraser, T. J., *Cancer Res.* **34**, 778 (1974).
1058. Čihák, A., Veselý, J., and Harrap, K. R., *Biochem. Pharmacol.* **23**, 1087 (1974).
1059. Vavra, J. J., DeBoer, C., Dietz, A., Haňka, L. J., and Sokolski, W. T., *Antibiot. Annu.* p. 230 (1960).
1060. Herr, R. R., Jahnke, H. K., and Argoudelis, A. D., *J. Am. Chem. Soc.* **89**, 4808 (1967).
1061. Bhuyan, B. K., *Cancer Res.* **30**, 2017 (1970).
1062. Rosenkranz, H. S., and Carr, H. S., *Cancer Res.* **30**, 112 (1970).
1063. Reusser, F., *J. Bacteriol.* **105**, 580 (1971).
1064. Hinz, M., Katsilambros, N., Maier, V., Schatz, H., and Pfeiffer, E. F., *FEBS Lett.* **30**, 225 (1973).
1065. Anderson, T., Schein, P. S., McHenamin, M. G., and Cooney, D. A., *J. Chem. Invest.* **54**, 672 (1974).
1066. Fiscor, G., Zuberi, R. I., Suami, T., and Machinami, T., *Chem.-Biol. Interact.* **8**, 395 (1974).
1067. Arison, R. N., and Peudale, E. L., *Nature (London)* **214**, 1254 (1967).
1068. Gatenbeck, S., and Sierakewicz, J., *Antimicrob. Agents Chemother.* p. 308 (1973).
1069. Meronuck, R. A., Steele, J. A., Mirocha, S. J., and Christensen, S. M., *Appl. Microbiol.* **23**, 613 (1972).
1070. Kaczka, E. A., Gitterman, C. O., Dulaney, E. L., Smith, M. C., Hendlin, D., Woodruff, H. B., and Folkers, K., *Biochem. Biophys. Res. Commun.* **14**, 54 (1964).
1071. Carrasco, L., and Vasquez, D., *Biochim. Biophys. Acta* **319**, 209 (1973).
1072. Takahashi, S., Nitta, K., Okami, Y., and Umezawa, H., *J. Antibiot., Ser. A* **14**, 107 (1961).
1073. Arima, K., Kohsaka, M., Tamura, G., Imanaka, H., and Sakai, H., *J. Antibiot.* **25**, 437 (1972).
1074. Nishioka, Y., Beppu, T., Kohsaka, M., and Arima, K., *J. Antibiot.* **25**, 660 (1972).
1075. Tavitian, A., Uretsky, S. C., and ACS, G., *Biochim. Biophys. Acta* **157**, 33 (1968).
1076. Tavitian, A., Uretsky, S. C., and ACS, G., *Biochim. Biophys. Acta* **179**, 50 (1969).
1077. Hamelin, R., Larsen, C. J., and Tavitian, A., *Eur. J. Biochem.* **35**, 350 (1973).
1078. Weiss, J. W., and Pitot, M. C., *Cancer Res.* **34**, 501 (1974).
1079. Peries, J., Canivet, M., Godard, C., Salle, H., and Tavitian, A., *J. Gen. Biol.* **23**, 1347 (1974).
1080. Uematsu, T., and Suhadolnik, R. J., *Arch. Biochem. Biophys.* **162**, 614 (1974).
1081. Miyamura, S., and Niwayama, S., *Antibiot. Chemother. (Washington, D.C.)* **9**, 497 (1959).
1082. Nakatani, S., *J. Antibiot., Ser. B* **14**, 167 (1961).

1083. Freeman, G. G., and Morrison, F. I., *Nature* (*London*) 162, 30 (1948).
1084. Perlman, D., Lummis, W. L., and Geiersbach, H. J., *J. Pharm. Sci.* 58, 633 (1969).
1085. Matsuura, S., *J. Antibiot.*, *Ser. A* 18, 43 (1965).
1086. U.S. Patent 3,336,289 (1967).
1087. Owen, S. P., and Smith, C. G., *Cancer Chemother. Rep.* 36, 19 (1964).
1088. Acs, G., Reich, E., and Mori, A., *Proc. Natl. Acad. Sci. U.S.A.* 52, 493 (1964).
1089. Bhuyan, B. K., *Cancer Res.* 31, 1923 (1971).
1090. Henderson, J. F., and Khoo, M. K. Y., *J. Biol. Chem.* 240, 3104 (1965).
1091. Bloch, A., Leonard, R. J., and Nichol, C. A., *Biochim. Biophys. Acta* 138, 10 (1967).
1092. DeBarros, C., Goncalves, de Lima, O., Delle-Monache, F., Ferrari, F., and Marrini-Betolo, G. B., *Farmaco, Ed. Sci.* 52, 721 (1970).
1093. Brockmann, H., Lenk, W., Schwantije, G., and Zeeck, A., *Chem. Ber.* 102, 126 (1969).
1094. Belgian Patent 645,730 (1964). *C.A.* 63, 14643 (1965).
1095. Rudaya, C. M., Singal, E. M., Ilinskaya, S. A., and Soloveva, N. K., *Antibiotiki* (*Moscow*) 16, 969 (1971).
1096. Zhdanovich, Y. V., Lokshin, G. B., Kuzovkov, A. D., and Rudaya, C. M., *Khim. Prir. Soedin.* p. 646 (1971).
1097. Rubtsova, G. V., Eselevich, M. M., Votrin, I. I., Debov, S. S., and Sazykin, J. O., *Biokhimiya* 37, 531 (1972).
1098. Eselevich, M. M., Torbokchina, L. I., Bobikov, E. V., and Sazykin, J. O., *Antibiotiki* (*Moscow*) 18, 15 (1973).
1099. Fuska, J., Nemec, P., and Kuhr, I., *J. Antibiot.* 25, 208 (1972).
1100. Boeckman, R. K., Jr., Fayos, J., and Clardy, J., *J. Am. Chem. Soc.* 96, 5954 (1974).
1101. Fuska, J., Ivanitskaya, L. P., Horáková, K., and Kuhr, I., *J. Antibiot.* 27, 931 (1974).
1102. Horáková, K., personal communication (1974).
1103. Härri, E., Löffler, W., Sigg, H. P., Stähelin, H., Stoll, C., Tamm, C., and Wiesinger, D., *Helv. Chim. Acta* 65, 839 (1962).
1104. Achimi, R., Mueller, B., and Tamm, C., *Helv. Chim. Acta* 57, 1442 (1974).
1105. Rüsch, M. E., and Stähelin, E., *Arzneim.-Forsch.* 15, 893 (1965).
1106. Carrasco, L., Barbacid, M., and Vasquez, D., *Biochim. Biophys. Acta* 312, 368 (1973).
1107. Hitoshi, M., Nakoto, M., and Katayama, T., *J. Chem. Soc., Perkin Trans. 1* p. 1819 (1973).
1108. Katagiri, K., Sato, K., Hayakawa, S., Matsushima, T., and Minato, H., *J. Antibiot.* 23, 420 (1970).
1109. Strauss, D., *Abstr. Pap. Antibiot. Congr. Prague* p. 145 (1964).
1110. Farkas-Himsley, H., *IRCS Libr. Compend.* 2, 1117 (1974).
1111. Fleck, W., Strauss, D., Koch, W., Kramer, P., and Prauser, H., German Patent 2,243,554 (1974).
1112. Fleck, W., Straus, D., Koch, W., and Prauser, H., *Z. Allg. Mikrobiol.* 14, 551 (1974).
1113. Romanenko, E. A., Plotnikov, N. P., Bondareva, A. S., Sedakova, L. A., Bukhtina, M. A., Firsova, G A., and Maevski, M. M., *Antibiotiki* (*Moscow*) 14, 525 (1969).
1114. Ando, K., Suzuki, S., Takatsuki, A., Arima, K., and Tamura, G., *J. Antibiot.* 21, 582 (1968).

1115. Ando, K., Tamura, G., and Arima, K., *J. Antibiot.* **21**, 587 (1968)
1116. Takatsuki, A., Tamura, G., and Arima, K., *J. Antibiot.* **22**, 151 (1969).
1117. Urry, W. H., Wehrmeister, H. L., Hodge, E. B., and Hidy, P. H., *Tetrahedron Lett.* p. 3109 (1966).
1118. Fuska, J., and Fusková, A., in preparation (1975).
1119. Wiley, P. F., Herr, R. R., and McKellar, F. A., and Argoudelis, S. A., *J. Org. Chem.* **30**, 2330 (1965).
1120. Argoudelis, A. D., Bergy, M. E., and Pyke, T. R., *J. Antibiot.* **24**, 543 (1971).
1121. Reusser, F., *J. Bacteriol.* **108**, 30 (1971).
1122. Kreuger, W. C., Pschigoda, L. M., and Reusser, F., *J. Antibiot.* **26**, 424 (1973).
1123. Sezikawa, Y., Inoue, E., and Kagino, K., *J. Antibiot., Ser. A* **15**, 236 (1962).
1124. Farmer, P. B., and Suhadolnik, R. J., *Biochemistry* **11**, 911 (1972).
1125. Shannon, W. M., Westbrook, L., and Schabel, F. M., Jr., *Proc. Soc. Exp. Biol. Med.* **145**, 542 (1974).
1126. Brink, J. J., and LePage, G. A., *Cancer Res.* **24**, 1042 (1964).
1127. Moore, E. C., and Cohen, S. S., *J. Biol. Chem.* **242**, 2116 (1967).
1128. Nichols, W. W., *Cancer Res.* **24**, 1502 (1964).
1129. Kihlman, B. A., and Odmark, G., *Hereditas* **55**, 71 (1966).
1130. Ševčik, V., Podojil, M., Kyselova, M., and Vrtišková, A., *Česk. Mikrobiol.* **1**, 263 (1956).
1131. Japanese Patent 61/11,297 (1961); *C.A.* **56**, 3927 (1962).
1132. Johnson, I. S., Baker, L. A., and Wright, H. F., *Ann. N.Y. Acad. Sci.* **76**, 861 (1958).
1133. Levenberg, B., Melnick, I., and Buchanan, J. M., *J. Biol. Chem.* **225**, 163 (1957).
1134. Chu, S. Y., and Henderson, J. F., *Biochem. Pharmacol.* **21**, 401 (1972).
1135. Hartman, S. C., and McGrath, T. F., *J. Biol. Chem.* **248**, 8506 (1973).
1136. Handschumacher, R., Bates, C. J., Chang, P. K., Andersen, A. T., and Fischer, G. A., *Science* **161**, 92 (1968).
1137. Bedair, B. J., and Fuerst, R., *Antimicrob. Agents Chemother.* p. 260 (1961).
1138. Rao, K. V., *Antibiot. Chemother.* (*Washington, D.C.*) **12**, 123 (1962).
1139. Young, C. W., Robinson, P. F., and Sackton, B., *Biochem. Pharmacol.* **12**, 855 (1963).
1140. Nomura, S., Yamamoto, H., Umezawa, I., Matsumae, A., and Hata, T., *J. Antibiot.* **20**, 55 (1967).
1141. Hata, T., Nomura,, S., and Umezawa, I., *Antimicrob. Agents Chemother.* p. 543 (1966).
1142. Thirumalachar, M. J., and Rahalkar, P. W., *Prog. Antimicrob. Anticancer Chemother., Proc. Int. Congr. Chemother., 6th, 1969* Part I, p. 70 Tokyo (1970).
1143. Sato, K., and LePage, G. A., *Cancer Res.* **25**, 477 (1965).
1144. U.S. Patent 3,117,916 (1964), *C.A.* **60**, 8597 (1964).
1145. U.S. Patent 3,318,783 (1967)., *C.A.* **67**, 42884 (1967).
1146. Chakrabarty, S. L., and Nandi, P., *Experientia* **27**, 595 (1971).
1147. Brazhnikova, M. G., and Kudinova, M. K., *Antibiotiki* (*Moscow*) **15**, 675 (1970).
1148. Schmitz, H., Heineman, B., Lein, J., and Hooper, I. R., *Antibiot. Chemother.* (*Washington, D.C.*) **10**, 740 (1960).
1149. Sato, K., Komiyama, K., Kietao, C., Iwai, Y., Atsumi, K., Oiwa, R., Katagiri, M., Umezava I., Omura, S., and Hata, T., *J. Antibiot.* **27**, 620 (1974).

1150. Arcamone, F., DiMarco, A., Gaetani, M., and Scotti, M., *G. Microbiol.* **9,** 83 (1961).
1151. Fuska, J., Kuhr, I., Nemec, P., and Fusková, A., *J. Antibiot.* **27,** 123 (1974).
1152. Fuska, J., Horáková, K., Veselý, P., Nemec, P., and Ujházy, V., *Prog. Chemother. Proc. Int. Congr. Chemother., 8th, 1973* Vol. III, p. 835 (1974).
1153. Soong, P., Au, A. A., and Bin, H., *Hua Hsueh* **3,** 127 (1964); *C.A.* **62,** 11626 (1964).
1154. Shomura, T., Tsuruoka, T., Watanabe, H., Inoue, S., and Niida, T., Japan Kokai 73/72,394 (1973); *C.A.,* **80,** 35821 (1974).
1155. French Patent 1,503,233 (1967); *C.A.* **69,** 85438 (1968).
1156. French Patent 1,503,235 (1967); C.A. **69,** 95102 (1968).
1157. German Patent 2,011,582 (1970); *C.A.* **73,** 129583 (1970).
1158. German Patent 2,011,982 (1970). *C.A.* **73,** 129582 (1970).
1159. Oki, T., Matsuzawa, Y., Numata, K., and Takamatsue, A., *J. Aitibiot.* **26,** 701 (1973).
1160. Kosmachev, A. E., *Mikrobiologiya* **31,** 66 (1962).
1161. Ishida, N., Okada, T., and Kamata, A., German Patent 2,392,404 (1974).
1162. Haňka, L. J., and Dietz, A., *Antimicrob. Agents Chemother.* p. 425 (1973).
1163. Haňka, L. J., Martin, D. G., and Neil, G. L., *Cancer Chemother. Rep.* **57,** 141 (1973).
1164. Martin, D. G., Duchamp, D. J., and Chidester, G. G., *Tetrahedron Lett.* p. 2549 (1973)
1165. Jayaram, H. N., Cooeny, D. A., Ryan, J. A., Neil, G. L., Dion, R. L., and Bono, V. N., *Cancer Chemother. Rep.* Part 1, **59,** 481 (1975).
1166. Martin, D. G., Haňka, L. J., and Neil, G. L., *Cancer Chemother. Rep.* **58,** 935 (1974).
1167. Szyba, K., and Mordarski, M., *Acta Microbiol. Pol., Ser. A* **5,** 81 (1973).
1168. Furumai, T., Takeda, K., Tani, K., Matsuzawa, N., Shimizu, Y., Ohashi, Y., and Okuda, T., *J. Antibiot.* **26,** 70 (1973).
1169. Ito, Y., Ohashi, Y., Kawabe, S., Sakurrawa, M., Ogawa, T., Egawa, Y., and Okuda, T., *J. Antibiot.* **26,** 77 (1973).
1170. Tsai, J. S., Su, T. T., Pao, C. C., Liang, S. P., and Kurylowicz, W., *Med. Dosw. Mikrobiol.* **10,** 105 (1958).
1171. Evans, J. S., and Mengel, G. D., *Antimicrob. Agents Chemother.* p. 238 (1968).
1172. Japanese patent 73/28 079 (1973). *C.A.* **80,** 46524 (1974).
1173. German Patent 1,939,045 (1970); *C.A.* **73,** 42918 (1970).
1174. Kruglyak, E. B., Ukholina, R. S., Sveshnikova, M. A., Proshlyakova, V. V., and Kovsharova, I. N., *Antibiotiki (Moscow)* **7,** 588 (1962).
1175. Arison, B. H., and Beck, J. L., *Tetrahedron Lett.* **29,** 2743 (1973).
1176. Rossolimo, O. K., Ivanitskaya, L. P., Kovsharova, I. M., Konstantinova, M. V., Lavrova, M. F., Manafova, N. A., Sveshnikova, M. S., *Antibiotiki (Moscow)* **14,** 849 (1969).
1177. Rossolimo, O. K., Maksimova, T. S., Sveshnikova, M. A., Olkhovatova, O. L., Kovsharova, I. N., and Proshlyakova, V. V., *Antibiotiki (Moscow)* **16,** 320 (1971).
1178. Preobrazhenskaya, T. P., Ukholina, R. S., Nekhaeva, N. R., Filikheva, V. A., Gavrilova, G. V., Kudinova, M. K., Borisova, V. N., Petukhova, N. M., Kovsharova, I. N., Proshlyakova, V. V., and Rossolimo, O. K., *Antibiotiki (Moscow)* **18,** 963 (1973).
1179. Kudinova, M. K., Babenko, G. A., Ukholina, R. S., Maksimova, T. S.,

Nechaeva, N. P., Terekhova, L. P., and Rossolimo, O. K., *Antibiotiki* (*Moscow*) **13**, 201 (1968).

1180. Gauze, G. G., Dudnik, Yu. V., Loshkareva, N. B., and Zbarsky, I. V., *Antibiotiki* (*Moscow*) **11**, 423 (1966).

1181. Khuskivadze, B. K., *Antibiotiki* (*Moscow*) **13**, 340 (1968).

1182. Laiko, A. V., *Antibioteiki* (*Moscow*) **9**, 711 (1964).

1183. Shorin, V. A., Rossolimo, O. K., Lyashenko, V. A., and Shapovalova, S. P., *Antibiotiki* (*Moscow*) **6**, 979 (1961).

SUBJECT INDEX

A

Amino acids
  resolution of racemic mixtures
    by immobilized enzymes, 226–231
    by immobilized microbial cells, 226
  synthesis of
    by immobilized cells
      glutamic acid, 238–239
    by immobilized enzymes
      alanine, 239
      aspartic acid, 231–232
      citrulline, 237–238
      ornithine, 237–238
      tryptophan, 232–236
      tyrosine, 236–237
Antitumor compounds
  antibiotics, antitumor, 260–274
  microbial products, antitumor, 276–334
  tumor types
    chicken, 275
    man, 275
    mouse, 274–275
    rat, 275

C

Cephalosporin acylation
  by immobilized cells, 219–224
  by immobilized enzymes, 219–223
Cephalosporin deacylation
  by immobilized cells, 225
  by immobilized enzymes, 218–219, 226
Cephalosporins
  acetate as precursor, 177–178
  amino acids as precursors, 172, 177,
    179, 180, 181, 184
  cephalosporin C
    biosythesis, 190–197
    metabolism
      acetylhydrolase, 196
      arylamidase, 196
    dihydrothiazine ring formation, 189,
      190
  fermentation studies

cephalosporin C by *C. acremonium*,
    172–176
  cephamycins, 176
  genetic studies with producing orga-
    nisms, 169–172
  occurrence of analogs by mutants of
    producers, 162–169
  organisms producing, 160–165
  peptides as intermediates, 186–189
  sulfur metabolism and biogenesis, 183–
    185
  tripeptide intermediates in biogenesis,
    176–177
Chenodeoxycholic acid formation using
  immobilized enzymes, 251
Coenzyme A formation by immobilized
  cells, 248–249

G

Gramicidin S production by immobilized
  enzymes, 251

I

Immobilized microbial cells
  preparation 203–206

M

Malic acid formation
  by immobilized cells, 247, 248
  by immobilized enzymes, 248

N

N-oxidation by immobilized enzymes,
  242

P

Pertussis vaccine
  activity as immunizing agent, 2
  adenylate cyclase and, 2
  adverse reactions, 2, 6, 7
  agglutinins, 72–73
  assays, 47, 48, 49, 50, 65, 66

bacteriocins and, 31
β adrenergic blockade, 12, 13, 14
benefits vs. side effects, risk of, 3, 7,
    45, 46, 73
biologically active components, 9, 45
clinical potency related to mouse test,
    3, 6, 46, 47, 48, 49
clinical testing, 4, 6, 7, 52, 53
costs, 47
degradation of cultures, 30, 31, 58
genetics and production of by cultures,
    1, 2, 27–41, 58–64
growth inhibitors of producing orga-
    nisms, 35, 36
growth requirements of producing cul-
    tures, 31, 32, 33, 34, 36, 37, 38,
    40, 41
heat labile toxin and, 19, 20
hemagglutinin, 20
HSF (histamine sensitivity factor), 3,
    11, 12
interrelationships between biological
    activities, 21, 22, 23
lipopolysaccharide endotoxin and, 19
LPF (lymphocytosis promoting fac-
    tor), 3, 13, 14, 15, 16, 17
mechanism of infection and, 4
pathogenicity, 23
pathology of disease and, 3, 10, 23
possible improvements, 5, 6, 44, 48,
    49, 51, 53
problems in standardization, 1
safety of, 45, 46, 57, 73
sensitization to histamine and, 10, 11,
    12
toxicity, 73–77
use as adjuvant agent, 17, 18, 69
Penicillin acylation by immobilized
    enzymes, 213–216
Penicillin deacylation by immobilized
    enzymes, 206–212
Poly I:C formation by immobilized
    enzymes, 242
Protein hydrolysis by immobilized
    enzymes, 243

R

Rennets, microbial
    analytical methods, 142–144
    commercial products, 137

future prospects, 152
lipase and, 138
method of preparation, 137
microorganisms used for, 137, 146,
    147, 149
milk clotting enzymes in, 139, 142
producers, 137
proteolytic activity of, 138
search for new microbial sources, 146,
    149
    bacteria as sources, 148, 150
    fungi as sources, 150, 151
specificity of microbial rennets, 139–
    142
toxicity, 142
use in cheese preparation, 144–147

S

Sorbosone formation by immobilized
    cells, 246, 247
Steroid transformation
    by immobilized enzymes, 240, 241

U

Urocanic acid, formation by immobilized
    microbial cells, 243

V

Vinegar
    Babylonians, production by, 83–85
    Biblical references, 86
    biological oxidation of ethanol to
        acetic acid, 120–124
    composition of, 124–126
    cider as source, 103, 104
    European economic community, pro-
        duction of, 108, 109
    malt, 105
    microbiology of, 110–117
    medicinal use, 87, 88
    medieval, preparation of, 87
    nutrition and metabolism of produc-
        ing organisms, 117–120
    production processes, 90
        cavitator, 98
        circulating generator, 95, 96
        concentration, 99, 100

English process, 94, 95
Frings acetator, 96–98
Orlean process, 91, 92
quick process, 92–94
Tower fermentor, 98, 99
Robert E. Kellen Company market
    surveys, 105, 109, 110
Romans, preparation by, 85, 86

solvent use, 89
United States and British production,
    105–107
uses, 107
volume produced in United States, 82
white distilled vinegar production,
    100–103
wine and production of, 104, 105

CONTENTS OF PREVIOUS VOLUMES

Volume 1

Protected Fermentation
Miloš Herold and Jan Nečásek

The Mechanism of Penicillin Biosynthesis
Arnold L. Demain

Preservation of Foods and Drugs by Ionizing Radiations
W. Dexter Bellamy

The State of Antibiotics in Plant Disease Control
David Pramer

Microbial Synthesis of Cobamides
D. Perlman

Factors Affecting the Antimicrobial Activity of Phenols
E. O. Bennett

Germfree Animal Techniques and Their Applications
Arthur W. Phillips and James E. Smith

Insect Microbiology
S. R. Dutky

The Production of Amino Acids by Fermentation Processes
Shukuo Kinoshita

Continuous Industrial Fermentations
Philip Gerhardt and M. C. Bartlett

The Large-Scale Growth of Higher Fungi
Radcliffe F. Robinson and R. S. Davidson

AUTHOR INDEX–SUBJECT INDEX

Volume 2

Newer Aspects of Waste Treatment
Nandor Porges

Aerosol Samplers
Harold W. Batchelor

A Commentary on Microbiological Assaying
F. Kavanagh

Application of Membrane Filters
Richard Ehrlich

Microbial Control Methods in the Brewery
Gerhard J. Hass

Newer Development in Vinegar Manufactures
Rudolph J. Allgeier and Frank M. Hildebrandt

The Microbiological Transformation of Steroids
T. H. Stoudt

Biological Transformation of Solar Energy
William J. Oswald and Clarence G. Golueke

SYMPOSIUM ON ENGINEERING ADVANCES IN FERMENTATION PRACTICE

Rheological Properties of Fermentation Broths
Fred H. Deindoerfer and John M. West

Fluid Mixing in Fermentation Processes
J. Y. Oldshue

Scale-up of Submerged Fermentations
W. H. Bartholemew

Air Sterilization
Arthur E. Humphrey

Sterilization of Media for Biochemical Processes
Lloyd L. Kempe

Fermentation Kinetics and Model Processes
Fred H. Deindoerfer

Continuous Fermentation
  W. D. Maxon

Control Applications in Fermentation
  George J. Fuld

AUTHOR INDEX—SUBJECT INDEX

Volume 3

Preservation of Bacteria by Lyophilization
  Robert J. Heckly

Sphaerotilus, Its Nature and Economic
  Significance
  Norman C. Dondero

Large-Scale Use of Animal Cell Cultures
  Donald J. Merchant and C. Richard
  Eidam

Protection Against Infection in the Micro-
  biological Laboratory: Devices and
  Procedures
  Mark A. Chatigny

Oxidation of Aromatic Compounds by
  Bacteria
  Martin H. Rogoff

Screening for and Biological Character-
  izations of Antitumor Agents Using
  Microorganisms
  Frank M. Schabel, Jr., and Robert F.
  Pittillo

The Classification of Actinomycetes in
  Relation to Their Antibiotic Activity
  Elio Baldacci

The Metabolism of Cardiac Lactones by
  Microorganisms
  Elwood Titus

Intermediary Metabolism and Antibiotic
  Synthesis
  J. D. Bu'Lock

Methods for the Determination of Organic Acids
  A. C. Hulme

AUTHOR INDEX—SUBJECT INDEX

Volume 4

Induced Mutagenesis in the Selection of
  Microorganisms
  S. I. Alikhanian

The Importance of Bacterial Viruses in
  Industrial Processes, Especially in
  the Dairy Industry
  F. J. Babel

Applied Microbiology in Animal Nutrition
  Harlow H. Hall

Biological Aspects of Continuous Cultivation of Microorganisms
  T. Holme

Maintenance and Loss in Tissue Culture
  of Specific Cell Characteristics
  Charles C. Morris

Submerged Growth of Plant Cells
  L. G. Nickell

AUTHOR INDEX—SUBJECT INDEX

Volume 5

Correlations between Microbiological
  Morphology and the Chemistry of
  Biocides
  Adrian Albert

Generation of Electricity by Microbial
  Action
  J. B. Davis

Microorganisms and the Molecular Biology of Cancer
  G. F. Gause

Rapid Microbiological Determinations
  with Radioisotopes
  Gilbert V. Levin

The Present Status of the 2,3-Butylene
  Glycol Fermentation
  Sterling K. Long and Roger Patrick

Aeration in the Laboratory
  W. R. Lockhart and R. W. Squires

Stability and Degeneration of Microbial Cultures on Repeated Transfer
Fritz Reusser

Microbiology of Paint Films
Richard T. Ross

The Actinomycetes and Their Antibiotics
Selman A. Waksman

Fusel Oil
A. Dinsmoor Webb and John L. Ingraham

AUTHOR INDEX—SUBJECT INDEX

Volume 6

Global Impacts of Applied Microbiology: An Appraisal
Carl-Göran Hedén and Mortimer P. Starr

Microbial Processes for Preparation of Radioactive Compounds
D. Perlman, Aris P. Bayan, and Nancy A. Giuffre

Secondary Factors in Fermentation Processes
P. Margalith

Nonmedical Uses of Antibiotics
Herbert S. Goldberg

Microbial Aspects of Water Pollution Control
K. Wuhrmann

Microbial Formation and Degradation of Minerals
Melvin P. Silverman and Henry L. Ehrlich

Enzymes and Their Applications
Irwin W. Sizer

A Discussion of the Training of Applied Microbiologists
B. W. Koft and Wayne W. Umbreit

AUTHOR INDEX—SUBJECT INDEX

Volume 7

Microbial Carotenogenesis
Alex Ciegler

Biodegradation: Problems of Molecular Recalcitrance and Microbial Fallibility
M. Alexander

Cold Sterilization Techniques
John B. Opfell and Curtis E. Miller

Microbial Production of Metal–Organic Compounds and Complexes
D. Perlman

Development of Coding Schemes for Microbial Taxonomy
S. T. Cowan

Effects of Microbes on Germfree Animals
Thomas D. Luckey

Uses and Products of Yeasts and Yeast-like Fungi
Walter J. Nickerson and Robert G. Brown

Microbial Amylases
Walter W. Windish and Nagesh S. Mhatre

The Microbiology of Freeze-Dried Foods
Gerald J. Silverman and Samuel A. Goldblith

Low-Temperature Microbiology
Judith Farrell and A. H. Rose

AUTHOR INDEX—SUBJECT INDEX

Volume 8

Industrial Fermentations and Their Relations to Regulatory Mechanisms
Arnold L. Demain

Genetics in Applied Microbiology
S. G. Bradley

Microbial Ecology and Applied Microbiology
Thomas D. Brock

The Ecological Approach to the Study of Activated Sludge
Wesley O. Pipes

Control of Bacteria in Nondomestic Water Supplies
Cecil W. Chambers and Norman A. Clarke

The Presence of Human Enteric Viruses in Sewage and Their Removal by Conventional Sewage Treatment Methods
Stephen Alan Kollins

Oral Microbiology
Heiner Hoffman

Media and Methods for Isolation and Enumeration of the Enterococci
Paul A. Hartman, George W. Reinbold, and Devi S. Saraswat

Crystal-Forming Bacteria as Insect Pathogens
Martin H. Rogoff

Mycotoxins in Feeds and Foods
Emanuel Borker, Nino F. Insalata, Colette P. Levi, and John S. Witzeman

AUTHOR INDEX–SUBJECT INDEX

Volume 9

The Inclusion of Antimicrobial Agents in Pharmaceutical Products
A. D. Russell, June Jenkins, and I. H. Harrison

Antiserum Production in Experimental Animals
Richard M. Hyde

Microbial Models of Tumor Metabolism
G. F. Gause

Cellulose and Cellulolysis
Brigitta Norkrans

Microbiological Aspects of the Formation and Degradation of Cellulose Fibers
L. Jurášek, J. Ross Colvin, and D. R. Whitaker

The Biotransformation of Lignin to Humus—Facts and Postulates
R. T. Oglesby, R. F. Christman, and C. H. Driver

Bulking of Activated Sludge
Wesley O. Pipes

Malo-lactic Fermentation
Ralph E. Kunkee

AUTHOR INDEX–SUBJECT INDEX

Volume 10

Detection of Life in Soil on Earth and Other Planets. Introductory Remarks
Robert L. Starkey

For What Shall We Search?
Allan H. Brown

Relevance of Soil Microbiology to Search for Life on Other Planets
G. Stotzky

Experiments and Instrumentation for Extraterrestrial Life Detection
Gilbert V. Levin

Halophilic Bacteria
D. J. Kushner

Applied Significance of Polyvalent Bacteriophages
S. G. Bradley

Proteins and Enzymes as Taxonomic Tools
Edward D. Garber and John W. Rippon

Mycotoxins
Alex Ciegler and Eivind B. Lillehoj

Transformation of Organic Compounds by Fungal Spores
Claude Vézina, S. N. Sehgal, and Kartar Singh

Microbial Interactions in Continuous Culture
Henry R. Bungay, III and Mary Lou Bungay

Chemical Sterilizers (Chemosterilizers)
Paul M. Borick

Antibiotics in the Control of Plant Pathogens
M. J. Thirumalachar

AUTHOR INDEX–SUBJECT INDEX

CUMULATIVE AUTHOR INDEX–CUMULATIVE TITLE INDEX

Volume 11

Successes and Failures in the Search for Antibiotics
Selman A. Waksman

Structure–Activity Relationships of Semisynthetic Penicillins
K. E. Price

Resistance to Antimicrobial Agents
J. S. Kiser, G. O. Gale, and G. A. Kemp

Micromonospora Taxonomy
George Luedemann

Dental Caries and Periodontal Disease Considered as Infectious Diseases
William Gold

The Recovery and Purification of Biochemicals
Victor H. Edwards

Ergot Alkaloid Fermentations
William J. Kelleher

The Microbiology of the Hen's Egg
R. G. Board

Training for the Biochemical Industries
I. L. Hepner

AUTHOR INDEX–SUBJECT INDEX

Volume 12

History of the Development of a School of Biochemistry in the Faculty of Technology, University of Manchester
Thomas Kennedy Walker

Fermentation Processes Employed in Vitamin C Synthesis
Miloš Kulhánek

Flavor and Microorganisms
P. Margalith and Y. Schwartz

Mechanisms of Thermal Injury in Nonsporulating Bacteria
M. C. Allwood and A. D. Russell

Collection of Microbial Cells
Daniel I. C. Wang and Anthony J. Sinskey

Fermentor Design
R. Steel and T. L. Miller

The Occurrence, Chemistry, and Toxicology of the Microbial Peptide-Lactones
A. Taylor

Microbial Metabolites as Potentially Useful Pharmacologically Active Agents
D. Perlman and G. P. Peruzzotti

AUTHOR INDEX–SUBJECT INDEX

Volume 13

Chemotaxonomic Relationships Among the Basidiomycetes
Robert G. Benedict

Proton Magnetic Resonance Spectroscopy—An Aid in Identification and Chemotaxonomy of Yeasts
P. A. J. Gorin and J. F. T. Spencer

Large-Scale Cultivation of Mammalian Cells
R. C. Telling and P. J. Radlett

Large-Scale Bacteriophage Production
K. Sargent

Microorganisms as Potential Sources of Food
Jnanendra K. Bhattacharjee

Structure–Activity Relationships Among Semisynthetic Cephalosporins
M. L. Sassiver and Arthur Lewis

Structure–Activity Relationships in the Tetracycline Series
Robert K. Blackwood and Arthur R. English

Microbial Production of Phenazines
J. M. Ingram and A. C. Blackwood

The Gibberellin Fermentation
E. G. Jeffreys

Metabolism of Acylanilide Herbicides
Richard Bartha and David Pramer

Therapeutic Dentrifrices
J. K. Peterson

Some Contributions of the U.S. Department of Agriculture to the Fermentation Industry
George E. Ward

Microbiological Patents in International Litigation
John V. Whittenburg

Industrial Applications of Continuous Culture: Pharmaceutical Products and Other Products and Processes
R. C. Righelato and R. Elsworth

Mathematical Models for Fermentation Processes
A. G. Frederickson, R. D. Megee, III, and H. M. Tsuchija

AUTHOR INDEX–SUBJECT INDEX

Volume 14

Development of the Fermentation Industries in Great Britain
John J. H. Hastings

Chemical Composition as a Criterion in the Classification of Actinomycetes
H. A. Lechevalier, Mary P. Lechevalier, and Nancy N. Gerber

Prevalence and Distribution of Antibiotic-Producing Actinomycetes
John N. Porter

Biochemical Activities of *Nocardia*
R. L. Raymond and V. W. Jamison

Microbial Transformations of Antibiotics
Oldrich K. Sebek and D. Perlman

In Vivo Evaluation of Antibacterial Chemotherapeutic Substances
A. Kathrine Miller

Modification of Lincomycin
Barney J. Magerlein

Fermentation Equipment
G. L. Solomons

The Extracellular Accumulation of Metabolic Products by Hydrocarbon-Degrading Microorganisms
Bernard J. Abbott and William E. Gledhill

AUTHOR INDEX–SUBJECT INDEX

Volume 15

Medical Applications of Microbial Enzymes
Irwin W. Sizer

Immobilized Enzymes
K. L. Smiley and G. W. Strandberg

Microbial Rennets
Joseph L. Sardinas

Volatile Aroma Components of Wines and Other Fermented Beverages
A. Dinsmoor Webb and Carlos J. Muller

Correlative Microbiological Assays
Ladislav J. Haňka

Insect Tissue Culture
W. F. Hink

Metabolites from Animal and Plant Cell Culture
Irving S. Johnson and George B. Boder

Structure–Activity Relationships in Coumermycins
John C. Godfrey and Kenneth E. Price

Chloramphenicol
Vedpal S. Malik

Microbial Utilization of Methanol
Charles L. Cooney and David W. Levine

Modeling of Growth Processes with Two Liquid Phases: A Review of Drop Phenomena, Mixing, and Growth
P. S. Shah, L. T. Fan, I. C. Kao, and L. E. Erickson

Microbiology and Fermentations in the Prairie Regional Laboratory of the National Research Council of Canada 1946–1971
R. H. Haskins

AUTHOR INDEX–SUBJECT INDEX

Volume 16

Public Health Significance of Feeding Low Levels of Antibiotics to Animals
Thomas H. Jukes

Intestinal Microbial Flora of the Pig
R. Kenworthy

Antimycin A, a Piscicidal Antibiotic
Robert E. Lennon and Claude Vézina

Ochratoxins
Kenneth L. Applegate and John R. Chipley

Cultivation of Animal Cells in Chemically Defined Media, A Review
Kiyoshi Higuchi

Genetic and Phenetic Classification of Bacteria
R. R. Colwell

Mutation and the Production of Secondary Metabolites
Arnold L. Demain

Structure–Activity Relationships in the Actinomycins
Johannes Meienhofer and Eric Atherton

Development of Applied Microbiology at the University of Wisconsin
William B. Sarles

AUTHOR INDEX–SUBJECT INDEX

Volume 17

Education and Training in Applied Microbiology
Wayne W. Umbreit

Antimetabolites from Microorganisms
David L. Pruess and James P. Scannell

Lipid Composition as a Guide to the Classification of Bacteria
Norman Shaw

Fungal Sterols and the Mode of Action of the Polyene Antibiotics
J. M. T. Hamilton-Miller

Methods of Numerical Taxonomy for Various Genera of Yeasts
I. Campbell

Microbiology and Biochemistry of Soy Sauce Fermentation
F. M. Young and B. J. B. Wood

Contemporary Thoughts on Aspects of Applied Microbiology
P. S. S. Dawson and K. L. Phillips

Some Thoughts on the Microbiological Aspects of Brewing and Other Industries Utilizing Yeast
G. G. Stewart

Linear Alkylbenzene Sulfonate: Biodegradation and Aquatic Interactions
William E. Gledhill

The Story of the American Type Culture Collection—Its History and Development (1899–1973)
William A. Clark and Dorothy H. Geary

Microbial Penicillin Acylases
E. J. Vandamme and J. P. Voets

SUBJECT INDEX

Volume 18

Microbial Formation of Environmental Pollutants
Martin Alexander

Microbial Transformation of Pesticides
Jean-Marc Bollag

Taxonomic Criteria for Mycobacteria and Nocardiae
S. G. Bradley and J. S. Bond

Effect of Structural Modifications on the Biological Properties of Aminoglycoside Antibiotics Containing 2-Deoxystreptamine
Kenneth E. Price, John C. Godfrey, and Hiroshi Kawaguchi

Recent Developments of Antibiotic Research and Classification of Antibiotics According to Chemical Structure
János Bérdy

SUBJECT INDEX

Volume 19

Culture Collections and Patent Depositions
T. G. Pridham and C. W. Hesseltine

Production of the Same Antibiotics by Members of Different Genera of Microorganisms
Hubert A. Lechevalier

Antibiotic-Producing Fungi: Current Status of Nomenclature
C. W. Hesseltine and J. J. Ellis

Significance of Nucleic Acid Hybridization to Systematics of Actinomycetes
S. G. Bradley

Current Status of Nomenclature of Antibiotic-Producing Bacteria
Erwin F. Lessel

Microorganisms in Patent Disclosures
Irving Marcus

Microbiological Control of Plant Pathogens
Y. Henis and I. Chet

Microbiology of Municipal Solid Waste Composting
Melvin S. Finstein and Merry L. Morris

Nitrification and Denitrification Processes Related to Waste Water Treatment
D. D. Focht and A. C. Chang

The Fermentation Pilot Plant and Its Aims
D. J. D. Hockenhull

The Microbial Production of Nucleic Acid-Related Compounds
Koichi Ogata

Synthesis of L-Tyrosine-Related Amino Acids by β-Tyrosinase
Hideaki Yamada and Hidehiko Kumagai

Effect of Toxicants on the Morphology and Fine Structure of Fungi
Donald V. Richmond

SUBJECT INDEX

A 6
B 7
C 8
D 9
E 0
F 1
G 2
H 3
I 4
J 5